MATHEMATICAL RESEARCH MATHEMATISCHE FORSCHUNG

Inverse and Improperly Posed Problems in Differential Equations

edited by G. Anger

Band 1

Akademie-Verlag · Berlin

In this series original contributions of mathematical research in all fields are contained, such as

— research monographs
— collections of papers to a single topic
— reports on congresses of exceptional interest for mathematical research.

This series is aimed at promoting quick information and communication between mathematician of the various special branches.

In diese Reihe werden Originalbeiträge zu allen Gebieten der mathematischen Forschung aufgenommen wie

— Forschungsmonographien
— Sammlungen von Arbeiten zu einem speziellen Thema
— Berichte von Tagungen, die für die mathematische Forschung besonders aktuell sind.

Die Reihe soll die schnelle Information und gute Kommunikation zwischen den Mathematikern der verschiedenen Fachgebiete fördern.

Manuscripts
in English and German comprising at least 100 pages and not more than 500 pages can be admitted to this series. With respect to a quick publication the manuscripts are reproduced photomechanically. Authors who are interested in this series please turn directly to the 'Akademie-Verlag'. Here you will get more detailed information about the form of the manuscripts and the modalities of publication.

Manuskripte
in englischer und deutscher Sprache, die mindestens 100 Seiten und nicht mehr als 500 Seiten umfassen, können in diese Reihe aufgenommen werden. Im Interesse einer schnellen Publikation werden die Manuskripte auf fotomechanischem Weg reproduziert. Autoren, die an der Veröffentlichung entsprechender Arbeiten in dieser Reihe interessiert sind, wenden sich bitte direkt an den Akademie-Verlag. Sie erhalten dort genauere Informationen über die Gestaltung der Manuskripte und die Modalitäten der Veröffentlichung.

Inverse and Improperly Posed Problems in Differential Equations

Mathematical Research **Mathematische Forschung**

Wissenschaftliche Beiträge
herausgegeben von der
Akademie der Wissenschaften der DDR
Zentralinstitut für Mathematik und Mechanik

Band 1

Inverse and Improperly Posed
Problems in Differential Equations

Inverse and Improperly Posed Problems in Differential Equations

Proceedings of the Conference on Mathematical and Numerical Methods held in Halle/Saale (GDR) from May 29 to June 2, 1979

---

edited by Prof. Dr. Gottfried Anger
Martin-Luther-Universität Halle–Wittenberg

Akademie-Verlag · Berlin 1979

Erschienen im Akademie-Verlag, DDR-108 Berlin, Leipziger Str. 3—4
Lektor: Dr. Reinhard Höppner

Lizenznummer: 202 · 100/426/79
Umschlag: Rolf Kunze
Gesamtherstellung: VEB Druckerei „Thomas Müntzer", 582 Bad Langensalza
Bestellnummer: 762 744 1 (2182/1) · LSV 1065
Printed in GDR
DDR 37,— M

P R E F A C E

The Department of Mathematics of the Martin Luther University Halle-Wittenberg was the sponsor of a Conference on Mathematical and Numerical Methods for Inverse and Improperly Posed Problems in Differential Equations, which was held in Halle (Saale) from May 29 to June 2, 1979. The aim of this conference was to bring together both pure and applied mathematicians from diverse areas in science and technology. The invited lectures and some short communications are published in these Proceedings.

It is well known that in the past some famous mathematicians were of the opinion that improperly posed problems were not of interest in mathematics and its applications. During the last twenty years this situation has changed completely. Most problems in applications, for instance in physics, chemistry, biology, engineering and so on, are inverse problems. Previously these problems were studied as direct problems. Often the solution an inverse problem does not depend continuously on the measured data and is therefore improperly posed. Each physical system must be stable, i.e. it must depend continuously on the exterieur data accessible to measurements. This is one reason why improperly posed problems have been the subject to such extensive study recently. The practical importance of inverse and improperly posed problems is such that they may be considered among the pressing problems of current mathematical research.

Many problems remain unsolved in the study of inverse problems. Before attacking a specific inverse problem we should know the solution of the corresponding direct problem. Therefore new theories for the solution of inverse problems are a further development of the hitherto existing mathematical theories for direct problems.

The organizer takes this opportunity to thank all mathematicians and physisists who took part in the work of the conference for their contributions. The organizer would like to express his warmest thanks to the Martin Luther University for its constant encouragement and for sponsoring this conference. He is indebted to many members of the University for freely offering their help and advice, especially to Prof. Dr. W. Walsch, Dr. H.-H. Buchsteiner, R. Aderhold, A. Juhl, E. Pfrepper and B. Thaler. The organizer also thanks the Akademie-Verlag Berlin, especially Mrs. R. Helle and Dr. R. Höppner, for its help in preparing these Proceedings.

Halle (Saale), May 1979 — Gottfried Anger

C O N T E N T S

SOLUTION OF AN INVERSE PROBLEM FOR THE HEAT EQUATION BY METHODS OF MODERN POTENTIAL THEORY

G. ANGER and R. CZERNER *)

We study an inverse problem for the heat equation. The right-hand side (sources) is to be determined from the values u on the boundary $\partial\Omega$ of a domain $\Omega \subset R^n$, where $x_n = t$ is the time coordinate. The sources are positive measures on $\overline{\Omega}$; the functions considered are potentials of a positive measure. Because this problem cannot be solved uniquely, we study the set of all positive measures on $\overline{\Omega}$ which create outside $\overline{\Omega}$ the same potential. This set is convex and weakly compact. Therefore we can apply the theorem of Choquet - Krein - Milman. In this theorem the extremal points are important. We characterize classes of extremal points (measures) in the case $\overline{\Omega} = \{x \in R^n, 0 \leq x_n \leq T\}$ by an approximation property and by geometrical properties. If the support of an extremal point (measure) is known the solution of the inverse problem is uniquely determined. The approximation property by adjoint potentials must have a physical meaning which we do not as yet know (Theorem 1 and Theorem 5).

1. One of the most important problems in physics is the determination of sources (right-hand side of a differential equation) which give rise to a given field. If we know the field in the whole space then the sources are uniquely determined [1], [7], [16]. In general the field is only known outside a domain $\Omega \subset R^n$, which contains the sources. Without further information the sources cannot be determined uniquely [5], [7]. Therefore we consider all sources which create the same potential (field) outside $\overline{\Omega}$ as a first step of a theory. If A is the (continuous) solution of the direct boundary (initial) value problem with boundary values f, then Af(x) can be represented by the adjoint mapping (balayage principle) A^*. The inverse $(A^*)^{-1}$ (inverse balayage principle) of A^* describes our first step. In [2], [4], [5], [7] we study this problem for the Laplace equation (see also [16]) with applications in geophysics. Distillation processes can be described by the same method (inverse balayage principle) [14]. In [3], [4], [8] we obtained special results for the heat equation. In [12] one can find some results for the wave equation. There are several inverse problems in continuum mechanics and solid state

---

*) Prof. Dr. habil. Gottfried Anger, Sektion Mathematik der Martin - Luther - Universität Halle-Wittenberg, Universitätsplatz 6, DDR - 401 Halle(Saale).
Dipl.-Math. Regina Czerner, Rechenzentrum der Martin - Luther - Universität Halle-Wittenberg, Weinbergweg, DDR - 401 Halle(Saale).

physics, in which the right-hand side of the polyharmonic equation is to be determined [13]. The determination of dipole sources is important in brain research, electrocardiography, design of antennas [6] and so on. The paper [6] is an overview on inverse problems. In this paper 106 references are listed.

In the present paper we study certain classes of extremal measures of an inverse problem for the heat (diffusion) equation

(1) $$\frac{\partial u}{\partial x_n} - \Delta_{n-1} u = f,$$

where $x_n = t$ is the time coordinate and

(2) $$\Delta_n u = \sum_{j=1}^{n} \frac{\partial^2 u}{\partial x_j^2}$$

is the Laplace operator. The fundamental solution of the heat equation has the form

(3) $$G_W(x,y) = \frac{1}{(2\sqrt{\pi(x_n - y_n)}\,)^{n-1}} \exp\left(-\frac{\sum_{j=1}^{n-1}(x_j - y_j)^2}{4(x_n - y_n)}\right) \quad \text{if } x_n > y_n,$$

$$G_W(x,y) = 0 \quad \text{if } x_n \leqq y_n,$$

and for the Laplace equation $-\Delta_n u = f$

(4) $$G_L(x,y) = \frac{1}{2\pi} \log \frac{1}{|x - y|}, \quad x \neq y,\ n = 2,$$

(5) $$G_L(x,y) = \frac{1}{(n-2)\,\sigma_n} \frac{1}{|x - y|^{n-2}}, \quad x \neq y,\ n \geqq 3,$$

where σ_n is the surface area of the sphere of radius $r = 1$.

2. Let $G : R^n \times R^n \longmapsto R$ be a locally integrable fundamental solution of a linear differential equation $P(x,D)u = f$, i.e. G satisfies the equation

(6) $$\int G(x,y) P^*(x,D)\, \varphi(x)\,dx = \varphi(y)$$

for all $\varphi \in C_0^\infty(R^n)$ and all $y \in R^n$, where $P^*(x,D)$ is the adjoint differential operator. In the case of the Laplace equation the Newtonian kernel G_L is symmetric. We define the potential of a measure μ at a point $x \in R^n$ by [1], [7], [16]

(7) $$G\mu(x) := \int G(x,y)\, d\mu(y)$$

and the adjoint potential $G^*\mu$ by

(8) $G^*\mu(y) := \int G(x,y)\, d\mu(x)$.

Let $\mu \geq 0$ be a measure satisfying supp $\mu \subset \overline{\Omega}$. We consider all positive measures ν satisfying supp $\nu \subset \overline{\Omega}$ and

(9) $G^*\nu(x) = G^*\mu(x)$ for all $x \notin \overline{\Omega}$.

We denote by $D(\overline{\Omega})$ the space of all $u \in C(\overline{\Omega})$ which are solutions of the adjoint differential equation $P^*(x,D)u = 0$. The relation (9) is equivalent to [1], [7], [16],

(10) $\int u\, d\nu = \int u\, d\mu$ for all $u \in D(\overline{\Omega})$.

The set of all positive measures ν having the property (9) or (10) is denoted by $\mathcal{L}(\mu)$. For the Laplace equation and the heat equation (see also the following relations (12), (13)) it follows from $u(y) = 1$ on $\partial\Omega$ that $u(x) = 1$ on $\overline{\Omega}$ and from this we obtain

$$\mathcal{L}(\mu) \subset B := \{\nu : \|\nu\| \leq \|\mu\|\} \subset C^*(\overline{\Omega}).$$

Therefore $\mathcal{L}(\mu)$ is, as a closed subset of a weakly compact set B, also a weakly compact set. It is not difficult to prove $\mathcal{L}(\mu)$ is convex. Following the theorem of Choquet - Krein - Milman [9] the set ex $\mathcal{L}(\mu)$ of all extremal measures (points) of $\mathcal{L}(\mu)$ is of special interest. Let

(11) $H(\overline{\Omega}) := \{ h \in C(\overline{\Omega}) : h(x) = \sum_{k=1}^{N} c_k\, G(x^k,x),\ x^k \notin \overline{\Omega} \}$.

The relation $H(\overline{\Omega}) \subset D(\overline{\Omega})$ holds. The following theorem was proved in [11]; see also [16].

<u>Theorem 1:</u> A measure $\nu \in \mathcal{L}(\mu)$ is extremal if and only if $H(\overline{\Omega})$ is dense in $L^1(\nu)$, i.e. $\overline{H(\overline{\Omega})} = L^1(\nu)$ in the norm of $L^1(\nu)$.

These considerations are valuable for the Laplace equation and the heat equation. The definition of $\mathcal{L}(\mu)$ for other differential equations can be found in [13], [14] [16]. In [5] the extremal points of $\mathcal{L}(\delta_{x^0})$ relative to the Laplace equation are characterized in R^2.

3. In this paper we consider an inverse problem for the heat equation and the special domain $\Omega_T := \{x \in R^n,\ 0 < x_n < T\}$. Let $\partial\Omega_T := \{x \in R^n,\ x_n = T\}$. If supp ν is compact the potential $G_W\nu$ tends to zero for $x \to \omega$ [1], where ω is the point at infinity. Therefore we introduce the space $C_\omega(\overline{\Omega}_T)$ of all continuous functions $f : \overline{\Omega}_T \longmapsto R$ vanishing at infinity. With respect to uniform con-

vergence this space is a Banach space. The solution of the initial value problem for the adjoint heat equation with respect to the hyperplane $x_n = T$ is given by

$$(12)\qquad u(x) = \int u(z) G_w(z,x)\, dz,$$

where $z = (z_1, \ldots, z_{n-1}, T)$, $dz = dz_1 \cdots dz_{n-1}$ and $|u| \leq M$ on $\partial\Omega_T$. The relation $G_w\nu(x) = G_w\mu(x)$ for $x_n > T$ can be written in the form

$$(13)\qquad \int u(y)\, d\nu(y) = \int u(y)\, d\mu(y) ,$$

where $u(y) = G_w(x,y)$, $x_n > T$. The set $H(\partial\Omega_T)$ of restrictions of functions $h \in H(\overline{\Omega}_T)$ on $\partial\Omega_T$ is dense in $C_\omega(\partial\Omega_T)$ [3] . If $H(\partial\Omega_T) \neq C_\omega(\partial\Omega_T)$ then there exists a measure $\mu \neq 0$ on $\overline{\Omega}_T$, such that $\mu(G_w(x,\cdot)) = G_w\mu(x) = 0$ for all $x_n > T$ (consequence of the Hahn - Banach Theorem). From this fact it follows that $\mu = 0$. Using the maximum - minimum principle for the solutions of the adjoint heat equation the equation (13) holds for all $u \in C_\omega(\overline{\Omega}_T)$, which are solutions of the adjoint heat equation in Ω_T . Let

$$J(\partial\Omega_T) := \{ f \in C_\omega(\overline{\Omega}_T) : f(y) = 0 \text{ on } \partial\Omega_T \},$$

$$D(\overline{\Omega}_T) := \{ u \in C_\omega(\overline{\Omega}_T) : -\frac{\partial u}{\partial x_n} - \Delta_{n-1} u = 0 \},$$

and

$$(14)\qquad \begin{aligned} Af(x) &:= \int f(z) G_w(z,x)\, dz \quad \text{if } x_n < T, \\ Af(x) &:= f(x) \quad \text{if } x_n = T, \end{aligned}$$

where $x \in \overline{\Omega}_T$. Then $A : C_\omega(\overline{\Omega}_T) \longrightarrow C_\omega(\overline{\Omega}_T)$ is a projection, i.e.

$$(15)\qquad \begin{aligned} Af_1 &= f_1 \quad \text{if } f_1 \in D(\overline{\Omega}_T) , \\ Af_2 &= 0 \quad \text{if } f_2 \in J(\partial\Omega_T). \end{aligned}$$

Therefore $C_\omega(\overline{\Omega}_T) = D(\overline{\Omega}_T) \oplus J(\partial\Omega_T)$. Each $\mu \in C^*_\omega(\Omega_T)$ can be represented by a measure $\tilde{\mu} \in C^*_\omega(R^n)$ satisfying $\|\tilde{\mu}\| < \infty$, whose support is on $\overline{\Omega}_T$. The adjoint mapping A^* is defined by

$$(16)\qquad (Af, \mu) = (f, A^*\mu) \text{ for all } f \in C_\omega(\overline{\Omega}_T) \text{ and all } \mu \in C^*(\overline{\Omega}_T),$$

where $(f, \mu) = \int f\, d\mu$. If $f = u \in D(\overline{\Omega}_T)$, then (16) has the form

$$(17)\qquad \int u\, d\mu = \int u\, d(A^*\mu) = \int u\, d\nu .$$

The mapping $\mu \longmapsto A^*\mu = \nu$ is called the balayage (sweeping-out) process relative to the initial value problem. The measure ν is called the balayaged (swept-out) measure. If $\mu = \delta_{x^0}$ we get

$$(18)\qquad u(x^0) = \int u(y)\, d\mu^T_{x^0}(y) = \int u(z) G_w(z,x^0)\, dz,$$

where ${}^{T}\mu_{x^0} := A^{*}\delta_{x^0}$. The swept-out measure $\nu = A^{*}\mu$ can be written in the form

(19) $\nu(f) = \int {}^{T}\mu_x(f)\, d\mu(x) = \int(\int f(y)\, d^{T}\mu_x(y))\, d\mu(x) = \int(\int G_w(z,x)\, d\mu(x)) f(z)\, dz.$

The same relations hold for the solution of the initial value problem of the heat equation relative to the hyperplane $x_n = \tau$. The solution is given for $x_n > \tau$ by

$$u(x) = \int u(z) G_w(x,z)\, dz =: \int u(y)\, d\mu_x^{\tau}(y) .$$

4. We shall now study some extremal measures of the convex set

(20) $\mathcal{L}(\delta_{x^0}) := \{ \nu \in C_\omega(\overline{\Omega}_T) : \nu \geqq 0,\ \int u\, d\delta_{x^0} = \int u\, d\nu ,\ u \in D(\overline{\Omega}_T) \}$,

where $x^0 = 0$ and $\delta_{x^0}(f) = f(0)$. There exists a solution $u \in D(\overline{\Omega}_T)$ satisfying $u(x) = 1$ on $\overline{\Omega}_T$. Therefore $\int d\nu = 1$ for all $\nu \in \mathcal{L}(\delta_{x^0})$. In [3] it was proved that

$$\mu_{x^0}^{\tau} \in \mathrm{ex}\, \mathcal{L}(\delta_{x^0}) \text{ for all } \tau \text{ satisfying } 0 < \tau \leqq T$$

and $\delta_{x^0} \in \mathrm{ex}\, \mathcal{L}(\delta_{x^0})$. We now consider the following measures

(21) $\nu = \nu_K + \gamma \in \mathcal{L}(\delta_{x^0})$,

where $\mathrm{supp}\, \nu_K \subset K \subset \Omega_T$, K compact and $\mathrm{supp}\, \gamma \subset \partial\Omega_T$. We want to determine the structure of γ. Let $u \in D(\overline{\Omega}_T)$ and $h = \mathrm{rest}_{\partial\Omega_T} u$. We get, because the relation

$$\int u\, d\delta_{x^0} = \int f\, d\mu_{x^0}^{T} = \int u\, d\nu_K + \int u\, d\gamma = \int h\, d(A^{*}\nu_K) + \int h\, d\gamma$$

holds,

(22) $\int h\, d\gamma = \int u\, d\delta_{x^0} - \int u\, d\nu_K = \int h\, d\mu_{x^0}^{T} - \int h\, d(A^{*}\nu_K)$.

The measure γ is uniquely determined, because the set of all $h = \mathrm{rest}_{\partial\Omega_T} u$, $u \in D(\overline{\Omega}_T)$, is equal to $C_\omega(\partial\Omega_T)$. Let

(23) $g(z) := G_w(z,0) - \int G_w(z,x)\, d\nu_K(x)$,

where $z = (z_1, \ldots, z_{n-1}, T)$. Applying (19) to (22) we get for the measure (21)

(24) $\nu(f) = \nu_K(f) + \gamma(f) = \int f\, d\nu_K + \int g(z) f(z)\, dz$,

where $f \in C_\omega(\overline{\Omega}_T)$. If $\nu \geqq 0$ then $\nu_K \geqq 0$ and $\gamma \geqq 0$. From $\gamma \geqq 0$ it follows that $g \geqq 0$. A consequence of the last inequality is that $\nu_K \geqq 0$ has to satisfy the inequality

(25) $G_w(z,0) \geqq \int G_w(z,x)\, d\nu_K(x)$ for $z_n = T$ and $1 - \int d\nu_K \geqq 0$.

At first we consider the heat equation in R^2. Let

$$\Omega_T = \{x \in R^2,\ 0 < x_2 < T\},\ \partial\Omega_T = \{x \in R^2,\ x_2 = T\} .$$

We study extremal measures of the form

(26) $\nu_a := a\delta_{\tilde z} + \gamma \in \mathcal{B}(\delta_{x^o}),\ a > 0,\ \tilde z = (b,\tau),$

and

(27) $\nu_{a'} = a'\delta_{y^1} + a'\delta_{y^2} + \gamma \in \mathcal{B}(\delta_{x^o}),\ y^1 = (-b,\tau),\ y^2 = (b,\tau),\ b, a' > 0.$

The measure (26) is an extremal element of $\mathcal{B}(\delta_{x^o})$ if $g_{a_o}(z_1^o) = 0$, where

(28) $z_1^o = bT/\tau\ ,\ a_o = (1 - \tau/T)^{1/2}\exp(-\frac{b^2}{4\tau})\ .$

Then

(29) $g_{a_o}(z_1) = \frac{1}{2\sqrt{\pi T}}\exp(-\frac{z_1^2}{4T})\ (1 - \exp(-\frac{(z_1\tau - bT)^2}{4T\ (T-\tau)}))\ .$

In the case $0 < a < a_o$ we obtain the important inequality

(30) $g_a(z_1) = g_{a_o}(z_1) + b\ G_w(z,\tilde z) \geqq d\ G_w(z,\tilde z)\ ,\ z = (z_1,T),\ b = a_o - a.$

Inequality (30) is important in paragraph 5. The measure (27) is extremal if

(31)
$$b < (2(1 - \tau/T))^{1/2}$$
$$a_o = \frac{1}{2}(1 - \tau/T)^{1/2}\exp(\frac{b^2}{4(T-\tau)})\ .$$

One can find a proof for (31) in [3]. The study of extremal measures of the form

(32) $\nu_K = a_1\delta_{y^1} + a_2\delta_{y^2}\ ,\ y^1 = (y_1^{(1)}, t_1),\ y^2 = (y_1^{(2)}, t_2),$

$K = \{y^1\}\cup\{y^2\}$ is more difficult. From $g(z_1) = g'(z_1) = 0$ it follows that

(33) $a_1 = (1 - t_1/T)^{1/2}\ \exp(-\frac{z_1^2}{4T})\ \exp(\frac{(z_1 - y_1^{(1)})^2}{4(T - t_1)})\ A_1\ ,$

(34) $a_2 = (1 - t_2/T)^{1/2}\ \exp(-\frac{z_1^2}{4T})\ \exp(\frac{(z_1 - y_1^{(2)})^2}{4(T - t_2)})\ A_2\ ,$

where

$$A_1 = (\frac{z_1 - y_1^{(2)}}{T - t_2} - \frac{z_1}{T})\ /\ (\frac{z_1 - y_1^{(2)}}{T - t_2} - \frac{z_1 - y_1^{(1)}}{T - t_1})$$

$$A_2 = \left(\frac{z_1}{T} - \frac{z_1 - y_1^{(1)}}{T - t_1} \right) / \left(\frac{z_1 - y_1^{(2)}}{T - t_2} - \frac{z_1 - y_1^{(1)}}{T - t_1} \right).$$

From $a_1 > 0$, $a_2 > 0$ it follows that the root z_1 must be an element of the interval

$$\left(\frac{y_1^{(2)}T}{t_2}, \frac{y_1^{(1)}T}{t_1}\right) \text{ respectively } \left(\frac{y_1^{(1)}T}{t_1}, \frac{y_1^{(2)}T}{t_2}\right).$$

The following condition (35) is sufficient for $g(z_1) = g'(z_1) = 0$, $g''(z_1) > 0$

$$(35) \quad \frac{T(y_1^{(1)}t_2 - y_1^{(2)}t_1)^2}{8t_1t_2(T-t_1)(T-t_2)} < \frac{2t_1t_2(y_1^{(1)}t_2 - y_1^{(2)}t_1)}{(t_2-t_1)(y_1^{(1)}t_2+y_1^{(2)}t_1)+2t_1t_2((y_1^{(1)}-y_1^{(2)})T+y_1^{(2)}t_1-y_1^{(1)}t_2)}$$

Remark 1: If the conditions (33), (34) and (35) are satisfied the measure (32) is extremal. If $t_1 = t_2 = \tau$, $y_1^{(1)} = b$, $y_1^{(2)} = -b$ we obtain the inequality of (31).

One can find a proof for the relations (33), (34), (35) and Remark 1 in [10].

We now consider an example of the heat equation in R^3, where $x_3 = t$ is the time coordinate. In R^2 the set of the roots of g has dimension zero. We shall show that in R^3 the set of the roots of g can have dimension 1. Let

$$K := \left\{(z_1,z_2,t) \in R^3 : z_1 = r\cos\alpha,\ z_2 = r\sin\alpha,\ t = \text{const.}\right\}$$

be a circle having the centre at $(0,0,t)$, $0 < t < T$. We shall construct a measure ν_K on K such that the corresponding function $g \geqq 0$ on

$$\partial\Omega_T := \left\{(x_1,x_2,x_3) \in R^3,\ x_3 = T\right\}$$

is zero on a circle whose centre is at $(0,0,T)$. In this case the function g is of the form

$$(36) \quad g(z_1,z_2) = \frac{1}{4\pi}\left(\frac{1}{T}\exp\left(-\frac{z_1^2+z_2^2}{T}\right) - \int_0^{2\pi}\frac{a(\alpha)}{T-t}\exp\left(-\frac{(z_1-r\cos\alpha)^2+(z_2+r\sin\alpha)^2}{4(T-t)}\right)r\,d\alpha\right.$$

$$= \frac{1}{4\pi}\left(\frac{1}{T}\exp\left(-\frac{z_1^2+z_2^2}{T}\right) - \frac{r}{(T-t)}\exp\left(-\frac{z_1^2+z_2^2+r^2}{4(T-t)}\right)A(z_1,z_2)\right).$$

Let $z_1 = \rho\cos\varphi$, $z_2 = \rho\sin\varphi$, $a(\alpha) = a = \text{const.}$ Then

$$(37)\qquad g(z_1,z_2) = \frac{1}{4\pi}\Big(\exp(-\frac{\varrho^2}{4T}) - \frac{r}{T-t}\exp(-\frac{\varrho^2+r^2}{4(T-t)})\, A_o(\varrho,r)\Big),$$

where

$$A_o(\varrho,r) = a\int_0^{2\pi} \exp(\frac{\varrho r\cos\psi}{2(T-t)})\, d\psi = 2\pi a\, I_o(\frac{\varrho r}{2(T-t)}) .$$

I_o is the Bessel function of order zero. The function g is constant for ϱ= const. From $g'(\varrho) = g(\varrho) = 0$ it follows that

$$(38)\qquad I_1(\frac{\varrho r}{2(T-t)}) = \frac{\varrho t}{rT}\, I_o(\frac{\varrho r}{2(T-t)})$$

and

$$(39)\qquad a = \frac{T-t}{2\pi Tr\, I_o(\varrho r/2(T-t))}\exp(-\frac{\varrho^2}{4} + \frac{\varrho^2+r^2}{4(T-t)}) .$$

The second derivative $g''(\varrho)$ is positive if

$$(40)\qquad \varrho^2 > \frac{T^2}{t^2}(r^2 - \frac{4t(T-t)}{T}) .$$

Now we have to study the solutions of (38), i.e. of $g'(\varrho) = 0$. The number $\varrho = 0$ is a solution of this equation. The function $g''(\varrho)$ attains its minimum if

$$(41)\qquad r^2 < 4\pi(1 - t/T) .$$

The equation

$$(42)\qquad I_1(x) = kx\, I_o(x),\quad x = \frac{\varrho r}{2(T-t)},\quad k = \frac{2t(T-t)}{r^2T},$$

is equivalent to

$$(43)\qquad I_2(x) = 2(\tfrac{1}{2} - k)\, I_o(x).$$

We have to use the relation $I_o(x) - \frac{2}{x} I_1(x) = I_2(x)$. From $I_1(x) < I_o(x)$ for all $x \in R$ we obtain $\varrho < rT/t$. If $k \geq 1/2$ equation (42) has no solution $x \neq 0$. Let $0 < k < 1/2$. Then (42) has a uniquely determined solution $x_k \neq 0$. From $k_1 < k_2$ it follows that $x_{k_2} < x_{k_1}$. In the case $x_k \neq 0$ we obtain from (40)

$$r^2 > \frac{4t(T-t)}{T} .$$

Therefore a solution $\varrho_o \neq 0$ of $g(\varrho) = g'(\varrho) = 0$, $g''(\varrho) > 0$, has to satisfy (38) and

$$(44)\qquad 0 < \frac{T^2}{t}(r^2 - \frac{4t(T-t)}{T}) < \varrho_o^2 < \frac{r^2T^2}{t^2} .$$

It is interesting to compare these results with the results (31) for R^2, to see the structural similarity.

5. Our aim is to prove an inequality in R^n which is similar to inequality (30). This inequality holds for measures $\nu = \nu_K + \gamma \in \mathcal{L}(\delta_x o)$. By Theorem 1 the measure $\nu \in \mathrm{ex}\,\mathcal{L}(\delta_x o)$ is extremal if and only if $H(\bar{\Omega}_T)$ is dense in $L^1(\nu)$, i.e. for each $f \in L^1(\nu)$ and each $\varepsilon > 0$ there exists a $u_m \in H(\bar{\Omega}_T)$ such that

$$(45) \qquad \|f - u_m\|_{L^1(\nu)} = \int |f - u_m| d\nu_K + \int |f - u_m| d\gamma < \varepsilon .$$

In the case of the Laplace equation we proved the following theorem [5] : If $\Omega \subset R^n$ is a bounded domain and $g > 0$ on $\partial\Omega$, then from $\int |u_m| d\gamma \longrightarrow 0$ (i.e. $f = 0$ on $\partial\Omega$) it follows that $u_m(x)$ tends to zero uniformly on each compact set $K \subset \Omega$. If g is positive the boundary $\partial\Omega$ is a "barrier" for the function $f \in H(\bar{\Omega})$. If $g(y^o) = 0$, $y^o \in \partial\Omega$, we can penetrate into Ω using functions $f \in H(\bar{\Omega})$. In the case of the heat equation this fact can be important, for instance in diffusion through a membrane (see Theorem 4 and Theorem 5).

The boundary $\partial\Omega_T \subset R^n$ is not compact. The infimum of the function $g > 0$ on $\partial\Omega_T$ is zero. Therefore we have to give another proof, different from that for the case of the Laplace equation. For the heat equation a compact boundary was considered in [15] . Let

$$(46) \qquad g_a(z) := G_w(z,0) - a\int G_w(z,x)\, d\nu_K(x) = G_w(z,0)(1 - a\, h(z)),\ a > 0,$$

where

$$h(z) := \int (G_w(z,x)/G_w(z,0))\, d\nu_K(x)$$

If $a < 1$ then $g_a > g \geq 0$ on $\partial\Omega_T$. From

$$G_w(z,x)/G_w(z,0) = C(T,\tau)\exp\Big(-\sum_{j=1}^{n-1} z_j^2 + 2T\sum_{j=1}^{n-1} z_j x_j - T\sum_{j=1}^{n-1} x_j^2\Big)\frac{1}{4(T-\tau)T}$$

it follows that the function h tends to zero for $|z| \longrightarrow \infty$. For $a = 1$ we obtain

$$g_1(z) = g(z) = G_w(z,0)(1 - h(z))$$

and from $g > 0$ we obtain the inequality $1 > h(z) > 0$. There exists at least one point $z^o \in \partial\Omega_T$ such that

$$\sup\{h(z),\ z \in \partial\Omega_T\} = h(z^o).$$

The function h is analytic for $\operatorname{supp}\nu_K \subset \Omega_T$. In the case $\Omega_T \subset R^2$ the function h is an analytic function of the real variable z_1 vanishing at infinity. Therefore the relation $h(z) = h(z^0)$ holds for finite many points $z^1, \dots, z^k$.

<u>Theorem 2:</u> Let $g : \partial\Omega_T \longmapsto R$ be positive. Then there exists a constant $b > 0$ such that the inequality

$$(47) \qquad g(z) \geqq b \int G_w(z,x)\, d\nu_K(x)$$

holds on $\partial\Omega_T$.

<u>Proof:</u> We consider the function

$$g_a(z) = G_w(z,0)(1 - a\, h(z)),\ a > 0,$$

defined in (46). If $a_0 = 1/h(z^0)$ then $g_{a_0}(z^0) = 0$. From $g = g_1 > 0$ it follows that $a_0 = 1 + b$, $b > 0$, and $g_{a_0} \geqq 0$. We obtain

$$g(z) = g_1(z) = G_w(z,0)(1 - a_0 h(z) + b\, h(z)) = g_{a_0}(z) + b \int G_w(z,x)\, d\nu_K(x)$$
$$\geqq b \int G_w(z,x)\, d\nu_K(x)\ .$$

<u>Theorem 3:</u> Let $g : \partial\Omega_T \longmapsto R$ be positive and $u_m \in H(\bar{\Omega}_T)$ satisfying

$$\int |u_m(z)|\, d\gamma(z) = \int g(z)\, |u_m(z)|\, dz \longrightarrow 0 \text{ for } m \longrightarrow \infty$$

(i.e. $f(z) = 0$ on $\partial\Omega_T$ in formula (45)). Then

$$\int |u_m(x)|\, d\nu_K(x) \longrightarrow 0 \text{ and } u_m(0) \longrightarrow 0 \text{ for } m \longrightarrow \infty.$$

<u>Proof:</u> The following inequality holds

$$(48) \qquad \int |u_m(x)|\, d\nu_K(x) = \int \left| \int G_w(z,x) u_m(z)\, dz \right| d\nu_K(x) \leqq \int \left(\int G_w(z,x)\, |u_m(z)|\, dz \right) d\nu_K(x)$$
$$\leqq \int \left(\int G_w(z,x)\, d\nu_K(x) \right) |u_m(z)|\, dz \leqq \frac{1}{b} \int g(z)\, |u_m(z)|\, dz\ .$$

Therefore from

$$\int g(z)\, |u_m(z)|\, dz \longrightarrow 0 \text{ it follows that } \int |u_m(x)|\, d\nu_K(x) \longrightarrow 0$$

for $m \longrightarrow \infty$. We now consider the relation

$$g(z) + \int G_w(z,x)\, d\nu_K(x) = G_w(z,0)\ .$$

Then

$$u_m(0) = \int G_w(z,0) u_m(z)\, dz = \int \left(g(z) + \int G_w(z,x)\, d\nu_K(x) \right) u_m(z)\, dz$$
$$= \int g(z) u_m(z)\, dz + \int u_m(x)\, d\nu_K(x) \leqq \int g(z)\, |u_m(z)|\, dz + \int |u_m(x)|\, d\nu_K(x)\ .$$

From this inequality we obtain $u_m(0) \longrightarrow 0$.

6. For extremal measures the following theorems hold in R^n.

Theorem 4: Let $g \geqq 0$ and $\nu = \nu_K + \gamma \in \mathrm{ex}\,\mathcal{L}(\delta_{xo})$. If $g > 0$ then $\nu_K = 0$; if $\nu_K \neq 0$ then g must have at least one root.

Proof: The measure ν is an element of ex $\mathcal{L}(\delta_{xo})$ if and only if $H(\bar{\Omega}_T)$ is dense in $L^1(\nu)$. We consider the following special function $f_o \in L^1(\nu)$:

$$f_o(x) = 0 \text{ on } \partial\Omega_T \text{ and } f_o(x) = 1 \text{ on supp } \nu_K .$$

Then there exists a sequence of functions $u_m \in H(\bar{\Omega}_T)$ such that

$$\int |0 - u_m(z)|g(z)\,dz \longrightarrow 0, \quad \int |1 - u_m(x)|d\nu_K(x) \longrightarrow 0 .$$

From (48) it follows that $\int |u_m(x)|\,d\nu_K(x) \longrightarrow 0$, which is a contradiction to $\int u_m(x)\,d\nu_K(x) \longrightarrow 1$. If $\nu_K \neq 0$ then g must have at least one root.

Theorem 5: Let $\Omega_T \subset R^2$ and $g : \partial\Omega_T \mapsto R$ be the function defined in (23). If g has a root in y^o then there exists a sequence of functions $v_m \in H(\bar{\Omega}_T)$ such that

(49) $$\int g(z)v_m(z)\,dz \longrightarrow 0$$

and

(50) $$v_m(x) \longrightarrow d_1\, G_w(x,y^o) + d_2 \frac{\partial G_w(x,y^o)}{\partial y_1}$$

Proof: We consider the special functions

(51)
$$h_m(z_1) = \frac{m}{2}\, d_1 \quad \text{if} \quad z_1^o - \frac{1}{m} < z_1 < z_1^o + \frac{1}{m}$$
$$h_m(z_1) = 0 \quad \text{otherwise}$$

(52)
$$s_m(z_1) = -m^2\, d_2 \quad \text{if} \quad z_1^o - \frac{1}{m} < z_1 < z_1^o$$
$$s_m(z_1) = m^2\, d_2 \quad \text{if} \quad z_1^o \leqq z_1^o + \frac{1}{m}$$
$$s_m(z_1) = 0 \quad \text{otherwise}$$

These functions satisfy the following conditions

(53) $$\int h_m(z_1)\,dz_1 \longrightarrow d_1, \int (z_1 - z_1^o)h_m(z_1)dz_1 \longrightarrow 0, \int (z_1 - z_1^o)^2\, h_m(z_1)dz_1 \longrightarrow 0,$$

(54) $\int s_m(z_1)\,dz_1 \longrightarrow 0, \int (z_1 - z_1^o) s_m(z_1) dz_1 \longrightarrow d_2, \int (z_1 - z_1^o)^2 s_m(z_1)\,dz_1 \longrightarrow 0.$

Because $\overline{H(\partial\Omega_T)} = C_\omega(\partial\Omega_T)$ we can construct two sequences of functions u_m , $\tilde{u}_m \in$ $\in H(\bar{\Omega}_T)$ having the same properties. From $v_m := u_m + \tilde{u}_m$ and

(55) $$v_m(x) = \int G_w(z,x) v_m(z)\,dz_1 = G_w(z^o,x)\int v_m(z)\,dz_1 + $$

$$+ \frac{\partial G_w(z^o,x)}{\partial z_1}\int (z_1 - z_1^o) v_m(z)\,dz_1 + \int (z_1 - z_1^o)^2 R_{G_w}(z^*,x) v_m(z)\,dz_1$$

it follows that

$$v_m(x) \longrightarrow d_1\, G_w(z^o,x) + d_2 \frac{\partial G_w(z^o,x)}{\partial z_1} \quad \text{in } \Omega_T .$$

<u>Remark 2:</u> In the case of k roots of $g \geqq 0$ in $z^1, \ldots , z^k$ we can construct a sequence of functions $v_m = u_{m1} + \ldots + u_{mk} + \tilde{u}_{m1} + \ldots + \tilde{u}_{mk}$ such that

(56) $$v_m(x) \longrightarrow d_{11}\, G_w(z^1,x) + \ldots + d_{1k}\, G_w(z^k,x) +$$

$$+ d_{21} \frac{\partial G_w(z^1,x)}{\partial z_1} + \ldots + d_{2k} \frac{\partial G_w(z^k,x)}{\partial z_1} .$$

There are 2k independent parameters $d_{11}, \ldots , d_{1k} , d_{21}, \ldots, d_{2k}$ in formula (56). Let K be a set of 2k points. Under certain conditions on K it is possible to approximate each value $\int f\, d\nu_K$ by functions $\int v_m\, d\nu_K$.

7. We now want to determine the structure of the extremal measures $\nu = \nu_K + \gamma \in$ $\in \mathcal{L}(\delta_{xo})$ in R^2. In this case $g \geqq 0$ on $\partial\Omega_T$. Using Taylor's formula relative G_w we obtain

$$u(x) = \int G_w(z,x) u(z)\,dz_1 = G_w(z^o,x)\int u(z)\,dz_1 + \frac{\partial G_w(z^o,x)}{\partial z_1}\int (z_1 - z_1^o) u(z)\,dz_1$$

$$+ \int (z_1 - z_1^o)^2 R_{G_w}(\tilde{z},x) dz_1 ,$$

where $u \in H(\bar{\Omega}_T)$ and

$$R_{G_w}(\tilde{z},x) = \frac{1}{2}\,\partial^2\, G_w(\tilde{z},x) \,/\, \partial z_1^2 .$$

Let $g(z_1^o) = 0$, $g(z_1) > 0$ for $z_1 \neq z_1^o$. Then we get, if

$$I_1 := \{ z_1 \in R,\ z_1^o - \varepsilon \leqq z_1 \leqq z_1^o + \varepsilon \},\ I_2 := R \setminus I_1,$$

(57) $$u(x) = \int_{I_2} G_w(z,x)u(z)\,dz_1 + G_w(z^o,x)\int_{I_1} u(z)\,dz_1 + \frac{\partial G_w(z^o,x)}{\partial z_1}\int_{I_1}(z_1-z_1^o)u(z)\,dz_1$$

$$+ \int_{I_1}(z_1 - z_1^o)^2\, R_{G_w}(\tilde{z},x)u(z)\,dz_1 \;.$$

Let $\nu_i = \text{rest}_{K_i}\nu_K$ be the restriction of ν_K on $K_i \subset K$. Then

$$\int G_w(z,x)\,d\nu_K(x) \geqq \int G_w(z,x)\,d\nu_i(x) \text{ on } \partial\Omega_T.$$

On I_2 the inequality

$$g(z_1) = G_w(z,0)(1 - h(z_1)) > 0$$

holds. Let $h(z_1^*) < 1$ be the maximum of h on I_2 and $1/h(z_1^*) = 1 + b^*$, $b^* > 0$. Then the inequality (47) holds on I_2. Therefore we obtain

(58) $$\int_{I_2} g(z)\,|u(z)|\,dz_1 \geqq \frac{1}{b^*}\int_{I_2}\Big(\int G_w(z,x)\,d\nu_K(x)\Big)|u(z)|\,dz_1 =$$

$$\geqq \frac{1}{b^*}\int_{I_2}\Big(\int G_w(z,x)\,d\nu_i(x)\Big)|u(z)|\,dz_1 \;.$$

On the compact interval I_1 the following inequalities hold

$$g(z) = (z_1 - z_1^o)^2\, g''(z) \leqq M_1\,(z_1 - z_1^o)^2,$$

$$\frac{(z_1 - z_1^o)^2}{2}\frac{\partial^2 G_w(z,x)}{\partial z_1^2} \leqq M_2\,(z_1 - z_1^o)^2 \leqq M\, g(z), \quad x \in \text{supp}\,\nu_K \;.$$

Then

(59) $$\Big|\int(z_1 - z_1^o)^2\, R_{G_w}(z,x)\,u(z)\,dz_1\Big| \leqq M^* \int g(z)\,|u(z)|\,dz_1 \;.$$

We obtain for $u = u_m$

(60) $$\int u_m(x)\,d\nu_i(x) = d_{i1}\int_{I_1} u_m(z)\,dz_1 + d_{i2}\int_{I_1}(z_1 - z_1^o)u_m(z)\,dz_1 + c_m^i \;,$$

where

$$d_{i1} = \int G_w(z^o,x)\,d\nu_i(x)\;,\quad d_{i2} = \int \frac{\partial G_w(z^o,x)}{\partial z_1}\,d\nu_i(x)\;,$$

$$c_m^i = \int_{I_2}\Big(\int G_w(z,x)\,d\nu_i(x)\Big)u_m(z)\,dz_1 + \int_{I_1}(z_1 - z_1^o)^2\Big(\int R_{G_w}(z,x)\,d\nu_i(x)\Big)u_m(z)\,dz_1$$

__Theorem 6:__ Let $g : \partial\Omega_T \longmapsto R$ be the function defined in (23) having the property $g(z_1^o) = 0$, $g(z_1) > 0$ for $z_1 \neq z_1^o$. Then the support of ν_K consists at most of two points. In the case of k roots the support ν_K consists at most of 2k points.

Proof: This Theorem 6 was proved for the Laplace equation in [5] . Let $g(z_1^o) = 0$, $g(z_1) > 0$ for $z_1 \neq z_1^o$ and $K_i = \{x^i\}$, i = 1, 2, 3. Then

$$d_{i1} = c_i\, G_w(z^o,x^i), \quad d_{i2} = c_i \frac{\partial G_w(z^o,x^i)}{\partial z_1} .$$

The determinant of the system (60), i = 1,2 , is equal to

$$\Delta^* = \frac{c_1c_2}{8\pi\sqrt{(T-t_1)(T-t_2)}} \exp\left(-\frac{(z_1-x_1^{(1)})^2}{4(T-t_1)}\right) \exp\left(-\frac{(z_1-x_1^{(2)})^2}{4(T-t_2)}\right) Q(z,x^1,x^2) ,$$

where

$$Q(z,x^1,x^2) = -\frac{z_1 - x_1^{(2)}}{T - t_2} + \frac{z_1 - x_1^{(1)}}{T - t_1} .$$

From (33) and (34) it follows that $Q(z,x^1,x^2) \neq 0$, because $0 \leq a_1, a_2 < \infty$. Therefore

$$\int u_m(z)\, dz_1 \text{ and } \int (z_1 - z_1^o) u_m(z)\, dz_1$$

can be calculated from the system (60) for i = 1,2 . Let $\|f - u_m\|_{L^1(\nu)} \longrightarrow 0$. Then

$$\int |f - u_m| d\gamma = \int |f - u_m| g\, dz_1 \longrightarrow 0 .$$

From (58) and (59) it follows that the sequence (c_m^i) , i = 1, 2 , converges. Also $\int u_m\, d\nu_i$ converges to $\int f\, d\nu_i$. Therefore

(61) $\int u_m(z)\, dz_1$ and $\int (z_1 - z_1^o) u_m(z)\, dz_1$

converge. Let $K_3 = \{x^3\}$. The value $\int f\, d\nu_3$ can be arbitrary, because $\nu \in \mathrm{ex}\, \mathcal{L}(\delta_{x^o})$
From (60) it follows that this is not possible.

Remark 3: The whole proof of Theorem 6 can be found in [10]. Many problems remain unsolved, for instance in R^3, $n \geq 3$. In future much work must be done to study completely our inverse problem for the heat equation.

REFERENCES

[1] ANGER, G.: Funktionalanalytische Betrachtungen bei Differentialgleichungen unter Verwendung von Methoden der Potentialtheorie. Berlin 1967

[2] ANGER, G.: Direct and inverse problems in potential theory. In Colloquium on Non-Linear Evolution Equations and Potential Theory. Prague 1975, pp. 11 - 44

[3] ANGER, G.: Ein inverses Problem für die Wärmeleitungsgleichung, I. In Proceedings of the Third Finnish-Romanien Seminar on Complex Analysis, Bucuresti 1976

[4] ANGER, G.: Convex sets in inverse problems. In Trudy Vsesojuznoj Konferencii po Uravnenijam s Castymi Proizvodnymi, Moscow 1976 (Russian)

[5] ANGER, G.: Uniquely determined mass distributions in inverse problems. In Veröffentlichungen des Zentralinstitutes für Physik der Erde, Potsdam, 52, 2 (1977), pp. 633 - 656

[6] ANGER, G.: Some remarks on inverse problems. In Colloquium on Elliptic Differential Equations, Rostock 1977, to be published in the Rostocker Mathematisches Kolloquium

[7] ANGER, G.: Lectures on potential theory and inverse problems. Summer School held at the Sektion Geowissenschaften der Bergakademie Freiberg (Sachsen), June 6 - 16, 1978; to be published in the series Nationalkomitee für Geodäsie und Geophysik bei der Akademie der Wissenschaften der DDR, Geodätische und Geophysikalische Reihe, Berlin 1979

[8] ANGER, G. and CZERNER, R.: The extreme measures in the inverse problem of the heat equation. The same Proceedings as [6]

[9] CHOQUET, G.: Lectures on Analysis I, II, III. New York - Amsterdam 1969

[10] CZERNER, R.: Charakterisierung von konvexen Mengen beim inversen Problem der Wärmeleitungsgleichung. Dissertation, Martin-Luther-Universität Halle - Wittenberg, Halle (Saale) 1979

[11] DOUGLAS, R. G.: On the extremal measures and subspace density. Michi. math. J. 11 (1964), pp. 243 - 246

[12] JUHL, A.: An inverse problem for the wave equation. The same Proceedings as [6]

[13] MICHEL, B.: About inverse problems in continuum mechanics and solid state physics. These Proceedings on Inverse and Improperly Posed Problems in Differential Equations

[14] KARR, A. F. and PITTENGER, A. O.: An inverse balayage problem for Brownian motion. Department of Mathematical Sciences, The John Hopkins University, Technical Report No. 277, Baltimore, October 1977

[15] PFREPPER, E.: Das inverse Problem für die Wärmeleitungsgleichung. Diplomarbeit, Sektion Mathematik der Martin-Luther-Universität Halle - Wittenberg, Halle (Saale) 1978

[16] SCHULZE, B.-W und WILDENHAIN, G.: Methoden der Potentialtheorie für elliptische Differentialgleichungen beliebiger Ordnung. Akademie-Verlag Berlin and Birkhäuser-Verlag Basel-Stuttgart 1977

ON AN INVERSE EIGENVALUE PROBLEM WITH TRUNCATED SPECTRAL DATA

VICTOR BARCILON[1] AND G. TURCHETTI[2]

Summary.

The inverse problem for a vibrating string whose first N eigenvalues are given is considered. It is shown that among all the strings which share this truncated spectrum, the density $\hat{\rho}(x)$ of that string which minimizes a linear functional $F[\rho]$ cannot be strictly positive. For the case N=1, we can actually prove that $\hat{\rho}(x)$ is in fact a delta function.

1. Introduction.

For most inverse problems arising in Geophysics, the available data set is much smaller than that required for the construction of a unique solution. Typically, this data set consists only of a finite number of measurements.

In view of the nature of the data set, the inverse problem becomes one of characterization of a class of compatible solutions rather than one of reconstruction of a single solution. Such a characterization can be accomplished by means of extremal solutions which provide global bounds for all the solutions in that class. This approach was used by Parker (1974, 1975) in his investigation of the inverse problem for gravity anomalies which led him to the development of the theory of ideal bodies.

The same approach can be adopted for inverse eigenvalue problems. Using the vibrating string as a prototype, the typical problem can be stated as follows: Given the first N natural frequencies of vibration $\{\bar{\omega}_n\}_{n=1}^N$, characterize the class of strings (of unit length, taut by a unit tension) which have this truncated spectrum in common. For future reference, the normal mode problem for these strings which are labelled by their densities $\rho(x)$, is:

$$\left.\begin{aligned} &u'' + \omega^2\rho u = 0 , \\ &u(0)\cos\alpha - u'(0)\sin\alpha = 0 , \\ &u(1)\cos\gamma + u'(1)\sin\gamma = 0 . \end{aligned}\right\} \quad (1)$$

[1]Dept. of Geophysical Sciences, The University of Chicago, Chicago, IL. 60637
[2]Istituto di Fisica "A. Righi", Universita degli Studi di Bologna, Via Irnerio 46 40126 Bologna, Italy.

We shall denote by $\omega_n[\rho]$ the computation of the nth eigenvalue for a specific string $\rho(x)$.

Following the spirit of Parker's ideal body theory, we can make the task of characterizing the class of solutions more specific by introducing the notion of an ideal string $\hat{\rho}(x)$ (Barcilon 1979). This ideal string $\hat{\rho}(x)$, minimizes the generalized moment

$$F[\rho] = \int_0^1 f(x)\rho(x)dx \ , \tag{2}$$

subject to the natural frequencies constraints

$$\omega_n[\rho] = \bar{\omega}_n \qquad n=1,2,\ldots,N \tag{3}$$

and the inequality constraint

$$\rho(x) \geq 0 \ . \tag{4}$$

In the definition of $F[\rho]$, the function $f(x)$ is a given non-negative function.

This problem was first considered by one of the authors (Barcilon, 1979). The results of that paper are however incorrect and we propose to reexamine the problem anew. We shall prove that $\hat{\rho}(x)$ cannot be strictly positive throughout the interval [0,1]. In fact, we suspect that

$$\hat{\rho}(x) = \sum_{n=1}^{N} m_n \delta(x-x_n) \ , \tag{5}$$

where

$$m_n > 0 \tag{6a}$$

and

$$0 \leq x_1 < x_2 < \ldots < x_n \leq 1 \ . \tag{6b}$$

In other words, we conjecture that the ideal string is made up of N point masses $\{m_n\}_{n=1}^{N}$. For the case N=1, we shall be able to prove that this conjecture is true.

2. Structure of the ideal string.

Let us denote by R the class of strings whose first N eigenvalues coincide with $\{\bar{\omega}_n\}_{n=1}^{N}$, i.e.

$$R = \{\rho(x): \ \rho(x) \geq 0, \ \omega_n[\rho] = \bar{\omega}_n, \ n=1, 2, \ldots, N\} \tag{7}$$

The ideal string $\hat{\rho}(x)$ is obviously a member of that class. We shall consider nearly ideal strings with densities

$$\sigma(x;\varepsilon) = (\sqrt{\hat{\rho}(x)} + \varepsilon\eta(x))^2 , \tag{8}$$

where ε is a small parameter and $\eta(x)$ is a yet unspecified function. For a given $\eta(x)$, the nth eigenvalue associated with one such nearly ideal string is:

$$\Omega_n(\varepsilon) = \omega_n[\sigma(x;\varepsilon)]. \tag{9}$$

It is easy to show that

$$\left.\frac{d\Omega_n}{d\varepsilon}\right|_{\varepsilon=0} = -\bar{\omega}_n \int_0^1 \sqrt{\hat{\rho}(x)}\,\eta(x)\hat{u}_n^2(x)dx \tag{10}$$

where $\hat{u}_n(x)$ is the nth normalized eigenfunction of the ideal string. For these nearly ideal strings to be in R, the above derivative must vanish: we shall impose these conditions and require that $\eta(x)$ be such that

$$\int_0^1 \eta(x)\sqrt{\hat{\rho}(x)}\;\hat{u}_n^2(x)dx = 0, \quad n=1, 2, \ldots N. \tag{11}$$

The function $\eta(x)$ is therefore an arbitrary function lying in the subspace orthogonal to that spanned by $\{\sqrt{\hat{\rho}}\,\hat{u}_n^2\}_{n=1}^N$.

On the other hand, since $F[\sigma(x;\varepsilon)]$ is a minimum for $\varepsilon=0$, we must also have

$$\left.\frac{dF}{d\varepsilon}\right|_{\varepsilon=0} = 2\int_0^1 \eta(x)\sqrt{\hat{\rho}(x)}f(x)dx = 0. \tag{12}$$

Consequently, $\sqrt{\hat{\rho}}f$ must be orthogonal to $\eta(x)$ hence must lie in the space spanned by $\{\sqrt{\hat{\rho}}\hat{u}_n^2\}_1^N$, i.e.

$$\sqrt{\hat{\rho}(x)}\;(f(x) - \sum_{n=1}^{N} f_n\,\hat{u}_n^2(x)) = 0 . \tag{13}$$

Now, if $\hat{\rho}(x)$ was positive throughout the interval [0,1], (13) would imply that f(x) can be represented by a finite sum of $\hat{u}_n^2(x)$. In general such a representation is not possible since $\{\hat{u}_n^2\}_{n=1}^N$ does not form a complete set. Therefore, the density of the ideal string must vanish somewhere in the interval.

In retrospect, this result is easy to understand. Obviously, in order to minimize $F[\rho]$, we must construct a string with a density as small as possible over as large a fraction of the interval [0,1] as possible, without violating the eigenvalue and inequality constraints. This suggests that the structure of the ideal string might be the following:

$$\hat{\rho}(x) = \sum_{n=1}^{N} m_n \delta(x-x_n) \quad . \tag{14}$$

3. <u>The case N=1</u>

For the case of a single eigenvalue constraint, we shall be able to prove that

$$\hat{\rho}(x) = m_1 \delta(x-x_1) \quad . \tag{15}$$

The proof is in two parts. First, we partition the interval (0,1) in I equal intervals and consider the problem (2)-(4) for the subset of strings with piecewise constant density over these intervals. We shall show that the desired string has a non-zero density over a single interval. Next, we consider the limit $I \to \infty$, and show that the mass of the string tends to a constant. This approach is closely related to that used by Sabatier (1977a,b) in his analysis of linear inverse problems with positivity constraints.

We shall represent a given string either by a piecewise constant function $\rho(x)$, i.e.

$$\rho(x) = \rho_i \qquad \frac{i-1}{I} < x < \frac{i}{I} \, , \tag{16}$$

or by a point in R^I with coordinates $(\rho_1, \rho_2, \ldots, \rho_I)$. The subset of R corresponding to these piecewise constant strings is represented geometrically by the first sheet of that portion of the hypersurface in R^I with the equation

$$\omega_1[\vec{\rho}] - \bar{\omega}_1 = 0 \, , \tag{17}$$

for which $\rho_i \geq 0$, $i = 1, 2, \ldots, I$. Finally, the optimization criterion can be written thus:

$$F = \vec{f} \cdot \vec{\rho} \tag{18}$$

where the components of the vector $\vec{f}$ are given by

$$f_i = \int_{\frac{i-1}{I}}^{\frac{i}{I}} f(x)dx \quad . \tag{19}$$

Consequently, our first task is to solve the following nonlinear programming problem: Find the vector $\hat{\vec{\rho}}$ which minimizes

$$F = \vec{f} \cdot \hat{\vec{\rho}} \tag{20a}$$

subject to the constraints

$$\omega_1[\hat{\vec{\rho}}] - \bar{\omega}_1 = 0 , \tag{20b}$$

and

$$\hat{\vec{\rho}} \geq 0 . \tag{20c}$$

The fact that (20) is a nonlinear programming problem complicates matters a great deal. However, for this single eigenvalue constraint case, we shall be able to show that the surface $\omega_1[\vec{\rho}] - \bar{\omega}_1 = 0$ is concave. With this property, the solution can be found in a standard manner.

In order to study the curvature of the surface $\omega_1[\vec{\rho}] - \omega_1 = 0$, we consider two nearby points $\vec{\rho}^{(0)}$ and $\vec{\rho}$ on this surface. The difference between these two vectors can be written in terms of the local tangent and normal vectors, namely

$$\vec{\rho} - \vec{\rho}^{(0)} = \epsilon\vec{t} + \epsilon^2\vec{n} \tag{21}$$

where

$$\vec{t} \cdot \nabla\omega_1 = 0 \tag{22}$$

and

$$\vec{n} = C\ \nabla\omega_1\ ; \tag{23}$$

C is a constant we shall determine and ϵ a small ordering parameter. Since $\omega_1[\vec{\rho}]$ is an analytic function of $\vec{\rho}$, we can write

$$\omega_1[\vec{\rho}^{(0)} + \epsilon\vec{t} + \epsilon^2\vec{n}] = \omega_1[\vec{\rho}^{(0)}] + \epsilon t_i \frac{\partial\omega_1}{\partial\rho_i}$$

$$+ \frac{\epsilon^2}{2}\left[t_j t_k \frac{\partial^2\omega_1}{\partial\rho_j\partial\rho_k} + 2n_j \frac{\partial\omega_1}{\partial\rho_j}\right] + O(\epsilon^3),$$

where the summation convention has been used. Since $\vec{\rho}$ and $\vec{\rho}^{(0)}$ lie on the surface $\omega_1 - \bar{\omega}_1 = 0$ and since $\vec{t}$ is tangent to this surface, we deduce that

$$\vec{n} = -\frac{1}{2}\frac{\kappa}{|\nabla\omega_1|^2}\nabla\omega_1 \tag{23'}$$

where

$$\kappa = t_j t_k \frac{\partial^2 \omega_1}{\partial \rho_j \partial \rho_k} + O(\varepsilon) . \tag{24}$$

Abandoning temporarily the geometric language, we look upon $\vec{\rho}$ as a string with density

$$\rho(x) = \rho_i^{(0)} + \varepsilon t_i - \frac{\varepsilon^2}{2} \frac{\kappa}{|\nabla\omega_1|^2} \omega_{1,i} + O(\varepsilon^3) \qquad \frac{i-1}{I} < x < \frac{i}{I} \tag{25}$$

The eigenfunction corresponding to the first eigenvalue can also be written as a power series in ε, namely

$$u_1(x) = u_1^{(0)}(x) + \varepsilon v_1(x) + \varepsilon^2 w_1(x) + \ldots \tag{26}$$

The first term is the eigenfunction associated with $\rho^{(0)}(x)$, viz.

$$\left.\begin{aligned} & u^{(0)''} + \bar{\omega}^2 \rho^{(0)} u = 0 , \\ & u^{(0)}(0) \cos\alpha - u^{(0)'}(0) \sin\alpha = 0 , \\ & u^{(0)}(1) \cos\gamma + u^{(0)'}(1) \sin\gamma = 0 , \end{aligned}\right\} \tag{27}$$

where the subscript 1 has been omitted. On the other hand, v and w satisfy:

$$v'' + \bar{\omega}^2 \rho^{(0)} v + \bar{\omega}^2 t u^{(0)} = 0 \tag{28}$$

and

$$w'' + \bar{\omega}^2 \rho^{(0)} w + \bar{\omega}^2 t v + \bar{\omega}^2 n u^{(0)} = 0 . \tag{29}$$

The boundary conditions for $v(x)$ and $w(x)$ are identical to those for $u^{(0)}(x)$. Multiplying (28) by $u^{(0)}$ and integrating, yields

$$\int_0^1 t(x)\, u^{(0)^2}(x) dx = 0 . \tag{30}$$

Reverting to geometric pictures in R^I, we can write (30) thus:

$$\vec{t} \cdot \vec{U} = 0 \tag{30'}$$

where $\vec{U}$ is a vector with components

$$U_i = \int_{\frac{i-1}{I}}^{\frac{i}{I}} u^{(0)^2}(x) dx . \tag{31}$$

Since $\vec{t}$ is an arbitrary tangent vector it follows that $\vec{U}$ is aligned with $\nabla\omega_1$. This implies that

$$\nabla\omega_1 > 0 . \tag{32}$$

Similarly, multiplying (29) by $u^{(0)}$ and integrating, we obtain

$$\int_0^1 t(x)u^{(0)}(x)v(x)dx + \int_0^1 n(x)u^{(0)^2}(x)dx = 0 . \tag{33}$$

A more useful form of this equation can be obtained by deriving an expression for the first integral in terms of v solely. Omitting some straightforward calculations, we rewrite (33) as follows:

$$\frac{\bar{\omega}^2}{2}\frac{|\vec{U}|}{|\nabla\omega_1|}\kappa = \int_0^1 v'^2(x)dx - \bar{\omega}^2\int_0^1 \rho(x)v^{(0)^2}(x)dx . \tag{34}$$

Classical results associated with Rayleigh's variational characterization of the first eigenvalue imply that the right hand side of (34) is positive. Therefore

$$t_i t_j \frac{\partial^2\omega_1}{\partial\rho_i\partial\rho_j} > 0 \tag{35}$$

and consequently the surface $\omega_1[\vec{\rho}] - \bar{\omega}_1 = 0$ is concave.

The programming problem (20) is therefore a concave programming problem: hence the points at which the surface $\omega_1[\vec{\rho}] - \bar{\omega}_1 = 0$ intersects the various axes will play an important role in the construction of the solution. Let us denote the intersection of the surface with the ith axis by ρ_i^*. This point stands for a string with zero density everywhere except on the ith interval where it is equal to ρ_i^*.

The equation of the hyperplane passing through all the points $\vec{\rho}_i^*$ is:

$$\vec{\rho}_p = \sum_{i=1}^{I} \chi_i \vec{\rho}_i^* \tag{36}$$

where $\chi_i \geq 0$ and

$$\sum_{i=1}^{I} \chi_i = 1 .$$

Since the surface $\omega_1[\vec{\rho}] = \bar{\omega}_1$ lies above this plane, we can represent any point on this plane thus:

$$\vec{\rho} = \vec{\rho}_p + \hat{\nu}h \tag{37}$$

where $\vec{\rho}_p$ is a point on the plane, $\hat{\nu}$ the unit normal to the plane pointing toward the surface and h the distance between the plane and the surface. Note that h is therefore a function of $\chi_1, \ldots, \chi_I$.

Making use of this representation, the linear form to be minimized becomes

$$F = \sum_{i=1}^{I} \chi_i \; (\vec{f} \cdot \vec{\rho}_i^{\,*}) + (\hat{\nu} \cdot \vec{f})h \tag{38}$$

If

$$\vec{f} \cdot \vec{\rho}_j^{\,*} = \min_{i=1,\ \ldots,\ I} (\vec{f} \cdot \vec{\rho}_i^{\,*}) \ , \tag{39}$$

it is an easy matter to prove that the minimum of F is reached by setting

$$\chi_i = \delta_{ij} \ ,$$

and that this minimum is equal to $\vec{f} \cdot \vec{\rho}_j^{\,*}$ since

$$h(0, \ldots, 1, \ldots 0) = 0 \ .$$

Thus, for the restricted subclass of piecewise constant strings, the minimum of F is achieved for a string with a zero density everywhere except on a single interval. As the partition is made finer and finer. the subset of piecewise constant string becomes larger and larger and in the limit coincides with R. Thus, provided that we can show that $\rho_j^*/I \to m$ as $I \to \infty$, we would have proved that the ideal string has a delta function structure.

To that effect, let us consider a generic string ρ_i^*. The eigenfunction $u_1^*(x)$ corresponding to this string can be written as follows:

$$u_1^*(x) = \begin{cases} r_i I(A_i \sin r_i + B_i \cos r_i)\,(x + \tan\alpha), & x < \frac{i-1}{I} \\ A_i \cos[r_i(Ix-i)] + B_i \sin[r_i(Ix-i)], & \frac{i-1}{I} < x < \frac{i}{I} \\ r_i I\, B_i\, [(x-1)\tan\gamma], & x > \frac{i}{I}\ . \end{cases} \tag{40}$$

The coefficients A_i and B_i are chosen in such a way as to make $u_1^*(x)$ continuous. From those homogeneous equations for A_i and B_i we deduce the desired equations for ρ_i^*:

$$\tan r_i = r_i \frac{I(1+\tan\alpha + \tan\gamma) - 1}{r_i^2 (I+I\tan\gamma - i)\,(I\tan\alpha + i - 1) - 1} \tag{41}$$

where

$$r_i = \frac{\bar{\omega}_1 \sqrt{\rho_i^*}}{I} \tag{42}$$

We are only interested in the first root of (41). For $I \gg 1$, this root is small and we can replace the tangent in (41) by its argument. Therefore

$$1 \approx \frac{1 + \tan\alpha + \tan\gamma}{I r_i^2\, (1+\tan\gamma - \frac{i}{I})\,(\tan\alpha + \frac{i-1}{I})}\ . \tag{41'}$$

Clearly

$$r_i = O(\frac{1}{\sqrt{I}})\ ,$$

and consequently

$$\frac{\rho_i^*}{I} = O(1). \tag{43}$$

Acknowledgments: One of the authors (V.B.) has greatly benefitted from discussions with Professor P.C. Sabatier during the 1978 Recontre R.C.P. "Problèmes Inverses" held in Montpellier. This research was initiated during a stay by V.B. at the University of Bologna made possible partly by the Consiglio Nazionale delle Ricerche. Finally, V.B. would like to acknowledge the support of ONR under contract N00014-76-C-0034.

References:

Barcilon, V., 1979. Ideal solution of an inverse normal mode problem with finite spectral data.. Geophys. J. Roy. Astr. Soc., to appear.

Parker, R.L., 1974. Best bounds on density and depth from gravity data, Geophysics, 39, 644-649.

Parker, R.L., 1975. The theory of ideal bodies for gravity interpretation, Geophys. J. R. Astr. Soc., **42**, 315-334.

Sabatier, P.C., 1977a. Positivity constraints in linear inverse problems - I. General theory, Geophys. J.R. Astr. Soc., **48**, 415-442.

Sabatier, P.C., 1977b. Positivity constraints in linear inverse problems - II. Applications, Geophys. J.R. Astr. Soc., **48**, 443-460.

Quasilinear Parabolic Systems
with Non-linear Boundary Conditions

John R. Cannon
The University of Texas at Austin
Austin, Texas 78712

and

Richard E. Ewing
The Ohio State University
Columbus, Ohio 43210

Abstract: Parabolic systems of partial differential equations, which are quasilinear in divergence form with high order coefficient matrices which are neither symmetric nor sparse, are considered. For an initial-boundary value problem which involves a non-linear boundary condition, the existence of a weak solution can be obtained via the method of Galerkin. Uniqueness and continuous dependence upon the data require additional regularity of the weak solution.

0. Introduction. In [1], Cannon, Ford and Lair described a complicated flow (diffusion) process involving several immiscible compressible fluid chemical species in a horizontal linear medium that could be described by the equation

$$\frac{\partial U}{\partial t} = \frac{\partial}{\partial x}[C(x,t,U)\frac{\partial U}{\partial x} + B(x,t,U)] + F(x,t,U) \ , \tag{0.1}$$

where U, B, and F are vectors, and C is a matrix. It was noted that if several space variables became involved that the generalization of C would involve four indices rather than two. The Neumann problem for the general case of several space variables with a quasi-linear system of equations analogous to (0.1) and boundary conditions which involved the natural flux at the boundary set equal to known data was analyzed. Weak existence, uniqueness and stability were established via methods similar to Ladyzenskaja, Solonnikov and Uralceva [3]. An application to problems in petroleum engineering was discussed.

The purpose of this paper is to report a generalization of the results of [1] to the case of non-linear boundary conditions which removes some restrictive assumptions made in the existence proof given in [1].

1. <u>Mathematical Problem</u>. We shall consider an initial-boundary value problem for the differential system

$$\frac{\partial U^i(x,t)}{\partial t} = \sum_{\ell=1}^{n} \frac{\partial}{\partial x_\ell}\{\sum_{j=1}^{k}\sum_{m=1}^{n} C_{ij}^{\ell m}[x,t,U(x,t)]\frac{\partial U^j(x,t)}{\partial x_m} + B_i^\ell[x,t,U(x,t)]\} + F_i[x,t,U(x,t)],\quad i=1,\ldots,k \tag{1.1}$$

where $x=(x_1,x_2,\ldots,x_n)$ lies in a bounded domain $\Omega \subset E_n$, $0<t<T$ and a solution $U=(U^1,U^2,\ldots,U^k)$ maps $Q_T \equiv \Omega \times (0,T)$ into E_k subject to the initial condition

$$U^i(x,0) = U_0^i(x), \qquad x \in \Omega, \qquad i=1,\ldots,k. \tag{1.2}$$

The boundary S of Ω and its inner-directed unit normal $\nu=(\nu_1,\nu_2,\ldots,\nu_n)$ are assumed as smooth as may be needed, and (1.1) is considered subject to the boundary conditions

$$\sum_{\ell=1}^{n}\{\sum_{j=1}^{k}\sum_{m=1}^{n} C_{ij}^{\ell m}[x,t,U(x,t)]\frac{\partial U^j(x,t)}{\partial x_m} + B_i^\ell[x,t,U(x,t)]\} \times \nu_\ell(x,t) = g_i[x,t,U(x,t)],\quad i=1,\ldots,k, \tag{1.3}$$

where $g=(g_1,g_2,\ldots,g_k)$ is a given map of $S_T \times E_k$ $(S_T \equiv S \times [0,T])$ into E_k.

Formulations will be abbreviated through the combined use of summation convention on repeated indices, $\dot{u}$ for $\partial u/\partial t$, $u_{,m}$ for $\partial u/\partial x_m$ and suppression of arguments. Thus, the differential system (1.1) becomes

$$\dot{U}^i = (C_{ij}^{\ell m} U^j_{,m} + B_i^\ell)_{,\ell} + F_i, \qquad Q_T, \tag{1.4}$$

and the boundary system (1.3) becomes

$$(C_{ij}^{\ell m} U^j_{,m} + B_i^\ell)\nu_\ell = g_i, \qquad S_T. \tag{1.5}$$

The Euclidean norm of appropriate dimension will be denoted by $|\cdot|$. Thus, $|x|^2 = x_1^2 + x_2^2 + \ldots + x_n^2$ while $|U|^2 = (U^1)^2 + \ldots + (U^k)^2$. Summation convention allows these norms to be written as $|x|^2 = (x_m)^2 = x^2 = x_m^2$ and $|U|^2 = (U^i)^2 = U^2$, respectively. If $q \geq 1$ and $r \geq 1$, let $L_{q,r}(Q_T)$ be the Banach space of measurable functions from Q_T to E^k with finite norm

$$(1.6) \qquad \|U\|^r_{q,r,Q_T} \equiv \int_0^T \|U(\cdot,t)\|^r_{q,\Omega}\, dt$$

where

$$(1.7) \qquad \|U(\cdot,t)\|^q_{q,\Omega} \equiv \int_\Omega |U(x,t)|^q\, dx$$

shows $\|U(\cdot,t)\|_{q,\Omega}$ as the $L_q(\Omega)$ norm of the E^k norm of $U(\cdot,t)$ for each $t \in [0,T]$. If $q = r = 2$, $(\cdot,\cdot)_{2,Q_T}$ will denote the natural inner product induced by (1.6), and we let $W_2^{1,0}(Q_T)$ be the Hilbert space implied by

$$(1.8) \qquad (U,V)_{W_2^{1,0}(Q_T)} \equiv (U,V)_{2,Q_T} + (U_{,m}, V_{,m})_{2,Q_T} ,$$

while $W_2^{1,1}(Q_T)$ is the Hilbert space implied by

$$(1.9) \qquad (U,V)_{W_2^{1,1}(Q_T)} \equiv (U,V)_{W_2^{1,0}(Q_T)} + (\dot{U},\dot{V})_{2,Q_T} .$$

Finally, $V_2(Q_T)$ is the Banach space consisting of elements of $W_2^{1,0}(Q_T)$ having a finite norm

$$(1.10) \qquad |U|^2_{Q_T} \equiv \|U_x\|^2_{2,Q_T} + \operatorname*{ess\,sup}_{t\in[0,T]} \|U(\cdot,t)\|^2_{2,\Omega}$$

where

$$(1.11) \qquad \|U_x\|^2_{2,Q_T} \equiv (U_{,m}, U_{,m})_{2,Q_T} .$$

In general, our notation consists of rather natural generalizations of that of Ladyzenskaja, Solonnikov and Uralceva [3]. For example, $V_2^{1,0}(Q_T)$ is the completion of $W_2^{1,1}(Q_T)$ in the norm of $V_2(Q_T)$.

Solution of (1.4) will be developed and considered in a weak sense. If $V \in W_2^{1,1}(Q_T)$ and $V(x,T) \equiv 0$, multiply (1.4) by V^i (sum by convention), integrate by parts and obtain

$$\int_{Q_T} [V^i_{,\ell}(C^{\ell m}_{ij}\, U^j_{,m} + B^\ell_i) - \dot{V}^i U^i]\,dxdt \\ = \int_\Omega V^i_0 U^i_0 dx + \int_{Q_T} V^i F_i dxdt - \int_{S_T} V^i g_i dSdt \tag{1.12}$$

as a motivation for a definition of weak solutions of (1.1) through (1.3).

Definition 1.1. We shall say that $U \in V_2(Q_T)$ is a weak solution of (1.1) through (1.3) if (1.12) is valid whenever $V \in W_2^{1,1}(Q_T)$ and $V(x,T) \equiv 0$.

2. Preliminaries. Certain auxiliary results will be needed as we proceed. If $(x,t) \in \Omega$ and $U(x,t) \in E_k$, a linear mapping is defined in E_{kn} by $\{C^{\ell m}_{ij}\}$. It has a matrix representation with rows obtained by $m = 1,2,\ldots,n$ and $j = 1,2,\ldots,k$ for each fixed pair (ℓ, i) of indices. It is important to note that the resulting matrix is often not symmetric.

Assumption 2.1. $C^{\ell m}_{ij} \in C(\bar{Q}_T \times E_k)$ such that

$$\lambda |y|^2 \le y^i_\ell\, C^{\ell m}_{ij}\, y^j_m \le \mu |y|^2 , \qquad y \in E_{kn} , \tag{2.1}$$

where λ and μ are positive constants. Moreover, M_c is a positive constant such that

$$|C^{\ell m}_{ij}\, y^j_m|^2 \le M_c^2 |y|^2 , \qquad y \in E_{kn} . \tag{2.2}$$

Also, $g_i \in C(S_T \times E_k)$ and M_g and K_g are positive constants such that

$$|g| \le M_g + K_g|U| \ . \tag{2.3}$$

Finally, $F_i \in C(\bar{Q}_T \times E_k)$ and $B_i^\ell \in C(\bar{Q}_T \times E_k)$ such that

$$|F| \le M_F + K_F|U| \ , \quad U \in E_k \ , \tag{2.4}$$

and

$$|B| \le M_B + K_B|U| \ , \quad U \in E_k \ , \tag{2.5}$$

for certain constants M_F, K_F, M_B and K_B .

Lemma 2.1. Suppose Assumption 2.1 is satisfied. If F and B are evaluated at $(x,t,U) \in Q_T \times E_k$ while g is evaluated at $(x,t,U) \in S_T \times E_k$, then

$$2|W^i F_i| \le \varepsilon^{-1} M_F^2 + (\varepsilon + K_F)|W|^2 + K_F|U|^2 \ , \tag{2.6}$$

$$2|W^i g_i| \le M_g^2 + (K_g + 1)|W|^2 + K_g|U|^2 \ , \quad \text{and} \tag{2.7}$$

$$2|W^i_{,\ell} B_i^\ell| \le \varepsilon^{-1}(M_B^2 + K_B|U|^2) + \varepsilon(1 + K_B)|W_x|^2 \tag{2.8}$$

where ε is an arbitrary positive number and $|W_x|^2 \equiv (W^i_{,\ell})^2$.

Proof. The proof which is elementary can be found in [1].

Lemma 2.2. Let $U^i \in V_2(Q_T)$, $i = 1,2,\ldots,k$. Then, for each ε. $0 < \varepsilon < 1$, there exists a positive constant C which depends upon ε such that

$$\|U\|^2_{2,S_t} \le \varepsilon\|U_x\|^2_{2,Q_t} + C\|U\|^2_{2,Q_t} \ . \tag{2.9}$$

Proof. The result (2.9) follows as a corollary of the argument of the trace theorem for Sobolev spaces [4].

For uniqueness and stability we shall need the following additional information with respect to the coefficients.

Assumption 2.2. Suppose Assumption 2.1 is satisfied, and there exist constants K_1, K_2, K_3 and K_4 such that

$$|F_i(V) - F_i(W)|^2 \le K_1 |Z|^2 , \tag{2.10}$$

$$|B_i^{\ell}(V) - B_i^{\ell}(W)|^2 \le K_2 |Z|^2 , \tag{2.11}$$

$$|C_{ij}^{\ell m}(V) - C_{ij}^{\ell m}(W)|^2 \le K_3 |Z|^2 \quad \text{and} \tag{2.12}$$

$$|g_i(V) - g_i(W)|^2 \le K_4 |Z|^2 \tag{2.13}$$

for each V and W in $V_2(Q_T)$ with $Z \equiv V - W$.

3. Galerkin Process. Let $\{v^{\alpha}: \alpha = 1,2,\ldots\} \subset W_2^1(\Omega)$ be a complete orthonormal basis for $L_2(\Omega)$. Although $\{v^{\alpha}\}$ is an infinite set, we consider only finite sums $(\alpha = 1,2,\ldots,\gamma)$ for the present in

$$W^i \equiv w_{\alpha}^i \, v^{\alpha} \tag{3.1}$$

where $\{w_{\alpha}^i: 1 \le i \le k,\ 1 \le \alpha \le \gamma\} \subset C^1[0,T]$ is to be determined. By requiring that (1.12) hold on the finite dimensional subspace suggested by (3.1), we obtain

$$\dot{w}_{\alpha}^i = -\left(\int_{\Omega} v^{\alpha}_{,\ell}\, C_{ij}^{\ell m}\, v^{\beta}_{,m} dx\right) w_{\beta}^j - \int_{\Omega} v^{\alpha}_{,\ell}\, B_i^{\ell} dx - \int_S v^{\alpha} g_i dS + \int_{\Omega} v^{\alpha}\, F_i dx : \ \{U \equiv W\}, \tag{3.2}$$

with the initial condition

$$w_{\alpha}^i(0) = \int_{\Omega} v^{\alpha}\, U_0^i dx , \tag{3.3}$$

where we adopt the convention that $\{U \equiv W\}$ means that W always replaces U in B, C, g and F until a further appearance of $\{U \equiv \cdot\}$ occurs.

Employing analysis similar to that in [1] we obtain the following result.

Lemma 3.1 Under Assumption 2.1 with γ a fixed positive integer $(\alpha = 1,2,\ldots,\gamma)$, the quasi-linear system (3.2) has a solution in $V_2(Q_T)$ satisfying the initial condition (3.3).

4. Existence. By using the Galerkin approximations of Lemma 3.1 and the results of Lemma 2.1 and Lemma 2.2, the analysis of [1] can be modified to obtain the existence of weak solutions to (1.1) through (1.3).

Theorem 4.1. Under Assumption 2.1, there is a weak solution $U \in V_2(Q_T)$ satisfying Definition 1.1.

Uniqueness and stability require a regularity result.

Theorem 4.2. Under Assumption 2.1, each weak solution of (1.1) through (1.3) belongs to $V_2^{1,0}(Q_T)$.

Proof. The proof involves direct modification of known analysis [3, pp. 156-159].

5. Uniqueness and stability. If another solution of (1.1) through (1.3) exists with a certain smoothness, then it is the same (a.e.) as the solution U in $V_2(Q_T)$ already obtained.

Lemma 5.1. Suppose Z^i, A_i, M_i^ℓ, P_i, Z_0^i are elements of $V_2^{1,0}(Q_T)$, $L_2(Q_T)$, $L_2(Q_T)$, $L_2(S_T)$, $L_2(\Omega)$, respectively, and satisfy the identity

$$(5.1) \qquad \int_{Q_T} Z^i \dot{V}^i \, dxdt = \int_{Q_T} [A_i V^i + M_i^\ell V^i_{,\ell}] dxdt + \int_{S_T} P_i V^i + \int_\Omega Z_0^i V_0^i \, dx$$

for every V in $W_2^{1,1}(Q_T)$ with $V(x,T) = 0$. Then

$$(5.2) \qquad \int_\Omega |Z(x,t)|^2 dx = -2 \int_{Q_t} [A_i Z^i + M_i^\ell Z^i_{,\ell}] dxdt - 2 \int_{S_t} P_i Z^i dSdt - \int_\Omega |Z_0|^2 dx$$

for almost all $t \in [0,T]$.

Proof. The proof is similar to [3, p. 141-143] and [1, p. 456-459].

Theorem 5.1. Suppose Assumption 2.2 is satisfied and W is a weak solution of (1.1) through (1.3) for which $|W_x|^2 \in L_{Q',R'}(Q_T)$ and

$$1/R' + n/Q' = n/2 + 1 \tag{5.3}$$

with

$$\begin{cases} Q' \in [n,\infty], & R' \in [1,\infty] \quad \text{for} \quad n > 2 \ , \\ Q' \in (1,\infty], & R' \in [1,\infty] \quad \text{for} \quad n = 2 \quad \text{and} \\ Q' \in [1,\infty], & R' \in [1,2] \quad \text{for} \quad n = 1 \ . \end{cases} \tag{5.4}$$

Then, any other weak solution U of (1.1) through (1.3) must equal W a.e. on Q_T provided that $R' < \infty$. If $R' = \infty$ and if $\|W_x\|_{2Q',2R'Q_T}$ is sufficiently small, then $U = W$ a.e. on Q_T.

Proof. The proof is similar to that presented in [1].

A smoothness condition as in uniqueness is needed in order to show that a solution of (1.1) through (1.3) is stable in the norm of the space $V_2(Q_T)$ with respect to variations of the auxiliary functions C, B, F, g and U_0.

Theorem 5.2. Suppose U is a solution (1.1) through (1.3) with $|U_x| \in L_{2Q',2R'}(Q_T)$ where $R' < \infty$ and Q' and R' satisfy (5.3) and (5.4). If $\tilde{U}$ is a solution of (1.1) through (1.3) with auxiliary functions $\tilde{C}, \tilde{B}, \tilde{F}, \tilde{g}$ and $\tilde{U}_0$ satisfying Assumption 2.2 with all constants same as for C, B, F, g and U_0 then there exists a positive constant K_0, depending only upon the constants of Assumptions 2.1 and 2.2, k, T, Ω and $|U_x|$, such that

$$\begin{aligned} |U-\tilde{U}|_{Q_T} \le K_0[\|U_0-\tilde{U}_0\|_{2,\Omega} &+ \|g(U)-\tilde{g}(U)\|_{2,S_T} + \|F(U)-\tilde{F}(U)\|_{2,Q_T} \\ &+ \|B(U)-\tilde{B}(U)\|_{2,Q_T} + \|C(U)-\tilde{C}(U)\|_{\infty,Q_T}] \ . \end{aligned} \tag{5.5}$$

Proof. The proof is similar to that presented in [1].

References

1. J.R. Cannon, W.T. Ford and A.V. Lair, Quasilinear parabolic systems, J. of Differential Equations, Vol. 20(1976), 441-472.

2. A. Friedman, Partial Differential Equations of Parabolic Type, Prentice-Hall, Englewood Cliffs, N.J., 1964.

3. O.A. Ladyzenskaja, V.A. Solonnikov and N.N. Uralceva, Linear and Quasi-linear Equations of Parabolic Type, A.M.S. Transl. Math. Mono., 23, Providence, R.I., 1968.

4. J.L. Lions and E. Magenes, Non-Homogeneous Boundary Value Problems and Applications, I, Springer-Verlag, N.Y.-Heidelberg-Berlin, 1972.

5. H.L. Royden, Real Analysis, 2nd Ed., Macmillan Company, Toronto, Ontario, Canada, 1968.

ABOUT THE STABILITY OF THE OPTIMAL

CONTROL SOLUTION OF INVSERSE PROBLEMS

Guy CHAVENT*

SUMMARY :

A definition of parameter identifiability (OLSI) well fitted to the commonly used output least square method is given. This definition requires the injectivity of the mapping parameter → observation (often taken as identifiability definition) but takes also in account model and measurement errors. A sufficient condition of identifiability is given and applied to an actual inverse problem in seismics.

*) IRIA-LABORIA - BP 105 - 78150 LE CHESNAY - FRANCE
Université PARIS-IX - 75116 PARIS Cedex 16 - FRANCE

I - INTRODUCTION

Motivated by important and diverse possible applications, the study of the inverse problem in partial differential equations has been rather intensively studied during the last years. Then were a-priori two possible points of view for attacking the problem :

- the "theoretical" point of view, which could be defined by saying that the concerned people are interested in gaining the more possible insight into the inverse problem using mainly a pencil, paper and their brain. This point of view is predominent in the area of theoretical physics [1], [2], [3]. A central point of a general theory of parameter estimations in P.D.E. would be a good theory of identifiability ; due to the difficulty of the problem, only very lacuneous answer have been given : some people have hinted on the importance of state observability to get identifiability, as for example in [4], [5], [6], [7], other people define the identifiability as the unicity of the parameter achieving a given observation (in the no model-error case), as for example in [6], [8], [9], [10].

- the "numerical" point of view, which is the point of view of the people beeing concerned mainly by the problem of getting on a computer satisfying estimation of an unknown parameter in an approximate model and from noisy data . Among those people, one of the most widely used method is the so-called "output least square method", as for instance in [10] to [23]. This method is, however, rather critically considered by the "theoretical" people, because it is very difficult to prove any theoretical result on it, except the existence of at least one solution of the minimization problem and the derivability of the criterion ([22], [23], [24]). Moreover it is possible to prove that in very simple cases the method works very poorly [21].

So the task of this paper is both ambitious and modest :

- ambitious because we shall try to construct a (tiny) bridge between the two above mentionned point of view : we shall dare to give an identifiability definition, well fitted to the output least square method and usable in presence of both model and measurement errors.

- modest because we shall be able to use this definition in only one (but at least one...) actual identification problem [19], [20], and because we are conscious that the sufficient given in § III are too restrictive for many practical applications. However we think that this kind of definition could be used, if a less restrictive sufficient condition were found, in cases where some "numerical identifiability" has been observed without any theoretical result beeing yet available [22].

II - THE OUTPUT LEAST SQUARE IDENTIFIABILITY (OLSI)

We recall first on figure 1 the "output least square method"

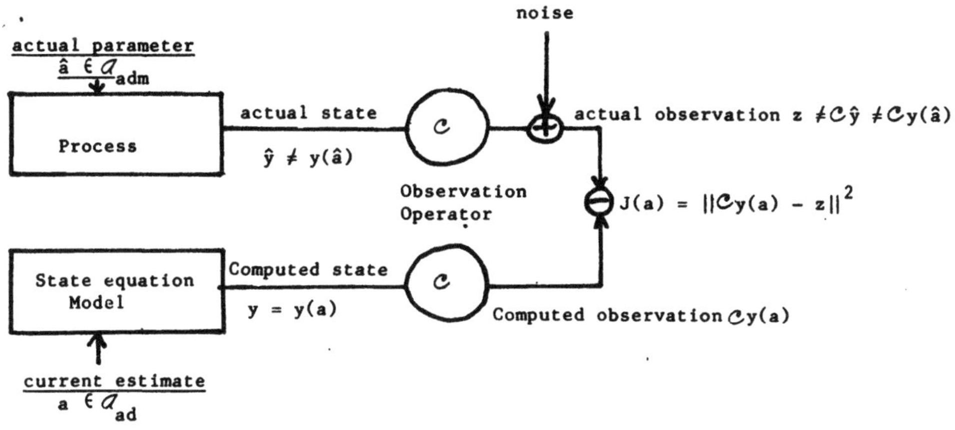

Figure 1 : The output least square method

Where $\mathcal{A}_{ad}$ is the set of admissible parameters, y(a) the state computed using the state equations with parameter a, $\mathcal{C}$ the observation operator, z the recorded observation and J(a) the output least square criterion to be minimized over $\mathcal{A}_{ad}$.

So the "output least square estimate" $\tilde{a}$ of $\hat{a}$ is given by :

(2.1) Find $\tilde{a} \in \mathcal{A}_{ad}$ such that $J(\tilde{a}) \leq J(a) \qquad \forall a \in \mathcal{A}_{ad}$

If we define the set $\mathcal{Z}$, as

(2.2) $\mathcal{Z}$ = image of $\mathcal{A}_{ad}$ by the mapping $a \to \mathcal{C}y(a)$

we can represent the output least square method using sets and mappings as in figure 2. As it appears clearly on that figure, all the discrepancy between simulation and reality, coming as well from the modelization error (the state equation represent only approximately the process) as from the measurement errors, result simply into the fact that the observation z does not belong to the set $\mathcal{Z}$. Then the output least square method can be conceptually divided into two consecutive steps :

(2.3)
- i) Project z in $\tilde{z}$ on the set $\mathcal{Z}$
- ii) find in $\mathcal{A}_{ad}$ an antecedent $\tilde{a}$ of $\tilde{z}$ for the mapping $a \to \mathcal{C}y(a)$

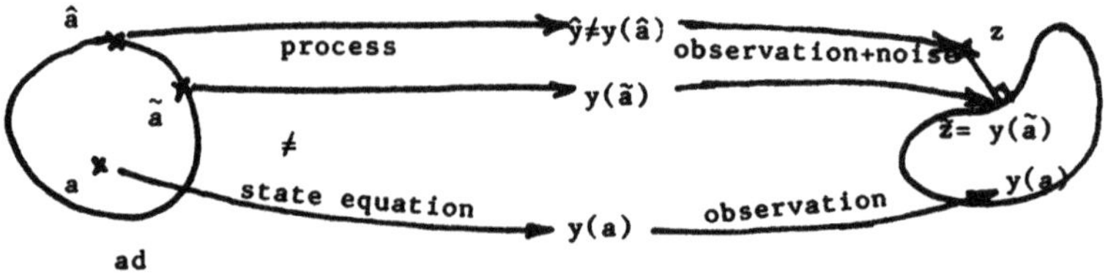

Figure 2 - The output least square method as a projection

As $\mathcal{Z}$ is always non-convex and the mapping $a \to \mathcal{C}y(a)$ non necessarily injective, each of the above step can lead to multiple determinations of the output least square estimate. So two necessary conditions for our output least square estimate to be unique are :

i) z has a unique projection $\tilde{z}$ on $\mathcal{Z}$

ii) the mapping $a \to \mathcal{C}y(a)$ is injective on $\mathcal{A}_{ad}$

So we shall suppose in the following that the mapping $a \to \mathcal{C}y(a)$ is injective. If not, we can either take a smaller set $\mathcal{A}_{ad}$ on which it is injective, or if this is not possible or do not make sense, replace $\mathcal{A}_{ad}$ by its quotient space by the equivalence relation $a \; R \; a' \Leftrightarrow \mathcal{C}y(a) = \mathcal{C}y(a')$. In that latter case, we can only hope to estimate the equivalence class of â, which can be done in some way by solving the optimization problem (2.1) by a gradient method starting with different initializations.

Concerning the unique projection property of z, we shall see in the next paragraph thay it may be true, if $\mathcal{A}_{ad}$ is convex and $a \to \mathcal{C}y(a)$ regular enough, as soon as z is not too far from $\mathcal{Z}$; this requests, as we have already seen, that the discrepencies beetwen reality and simulation are not too large, which is a reasonnable hypothesis to do !

We are now so far to give our identifiability definition :

Definition : With the notations of figure 1 and 2, we shall say that the parameter a is "output least square identifiable" (OLSI) in $\mathcal{A}_{ad}$ for the observation operator $\mathcal{C}$ iff there exists a neighbourhood $\mathcal{V}$ of $\mathcal{Z}$ such that, for every $z \in \mathcal{Z}$ the optimization problem (2.1) has a unique solution $\tilde{a}$ depending "continuously" on z.

As we have seen above, this definition clearly requires the mapping $a \to \mathcal{C}y(a)$ to be injective. An important feature of the above definition is that the estimated $\tilde{a}$ of â involved in it is computationnaly obtainable. An other feature resulting in a straight forward manner from the OLSI definition is the :

Proposition 1 : If the modelization and observation errors tend to zero, i.e. if $z \to z(\hat{a})$, and if a is OLSI in $\mathcal{A}_{adm}$ for $\mathcal{C}$, then the corresponding estimates $\tilde{a}$ tend to the true parameter $\hat{a}$.

It remains now to show that this definition is nonvoid ; we shall for that give in next paragraph a sufficient condition for OLSI, which will be then applied to an inverse problem in one 1-D wave equation.

III - A SUFFICIENT CONDITION FOR OLSI

Let $\mathcal{A}$ and $\mathcal{H}$ be the vector spaces where the parameter a and the observation belong. Let

(3.1) $\begin{cases} \mathcal{A}, \mathcal{H} \text{ be Hilbert spaces, with norm } \|\ \|_i \text{ and scalar product } (,)_i, \text{ where} \\ i = \mathcal{A} \text{ or } \mathcal{H}, \text{ or without subscript when no confusion is possible.} \end{cases}$

(3.2) $\begin{cases} \mathcal{A}_{ad} \subset \mathcal{A} \text{ be compact convex} \\ \mathcal{O} \text{ and open neighbourhood of } \mathcal{A}_{ad} \text{ in } \mathcal{A} \end{cases}$

We denote by T the mapping $a \to \mathcal{C}y(a)$, defined through the state equation and the observation operator, so that the subset Z of $\mathcal{H}$ defined in (3.2) is $T(\mathcal{A}_{ad})$. Let :

(3.3) $\begin{cases} T : \mathcal{O} \to \mathcal{H} \text{ be a } \mathcal{C}^2 \text{ mapping which is injective on } \mathcal{O} \text{ and such that } T'(a) \\ \text{is invertible for every } a \in \mathcal{A}_{ad} \end{cases}$

The OLSI sufficient condition shall result immediatly from the two following lemmas :

Lemma 1 : Under hypothesis (3.1) to (3.3), we can replace $\mathcal{O}$ by a smaller neighbourhood of $\mathcal{A}_{ad}$, which we shall still denote by $\mathcal{O}$, such that :

(3.4) $\mathcal{W} = T(\mathcal{O})$ is an open neighbourhood of $Z = T(\mathcal{A}_{ad})$

(3.5) T is a C^1 diffeomorphism from $\mathcal{O}$ on to $\mathcal{W}$

Proof : The result is a straight forward consequence of the compactness of $\mathcal{A}_{ad}$, of the implicit function theorem and of the hypothesis (3.3) on the mapping T.

Lemma 2 : Under hypothesis (3.1) to (3.3) there exists, for every $\lambda > 1$, a neighbourhood $\mathcal{V}$ of Z in $\mathcal{X}$ and $\alpha > 0$ such that every $z \in \mathcal{V}$ has a unique projection $P(z) \in Z$ and such that $\|P(z_1) - P(z_2)\| \leq \lambda \| z_1 - z_2\|$ whenever $z_1, z_2 \in \mathcal{V}$ and satisfy $\| z_1 - z_2\| \leq \alpha$.

Proof : Suppose that $\mathcal{O}$ has been reduced, according to lemma 1, so that (2.4) and (2.5) hold. Define then

$$(3.6) \qquad M = \sup_{a \in \mathcal{A}_{ad}} \|T''(a)\| \ , \ 0 < m < \inf_{a \in \mathcal{A}_{ad}} \| T'(a)^{-1}\|^{-1} = \inf_{z \in Z} \| T^{-1}(z)'\|^{-1}$$

and

$$(3.7) \qquad F = \{z \in \mathbb{D} \mid \| T^{-1}(z)'\|^{-1} \leq m\}$$

Due to (3.5), F is a closed subset of $\mathbb{D}$, and hence the set $F \cup \complement_{\mathcal{X}} \mathbb{D}$ is closed in $\mathcal{X}$, and does not intersect Z due to the definition (3.6) of m. Hence the continuous function $z \to d(z, F \cup \complement_{\mathcal{X}} \mathbb{D})$ is strictly positive on the compact set Z, and so is bounded below by some $\varepsilon > 0$; Hence

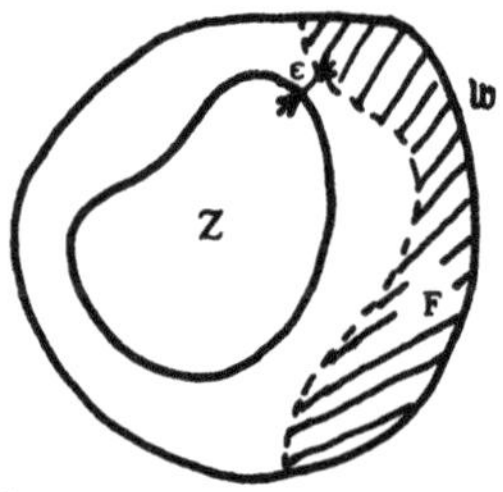

$$(3.8) \qquad \| T^{-1}(z)'\|^{-1} > m \quad \forall z \in \mathcal{V}_\varepsilon = \{z \in \mathcal{X} \mid d(z, Z) \leq \varepsilon\}$$

So we choose $\alpha \in \,]0, \varepsilon[$ and define, for any $\lambda > 1$:

$$(3.9) \qquad \mathcal{V} = \{z \in \mathcal{X} \mid d(z, Z)) \leq \mathrm{Min}(\frac{\varepsilon - \alpha}{2} \ , \ \frac{\lambda - 1}{\lambda} \frac{m^2}{M})\}$$

and we prove now that the above neighbourhood $\mathcal{V}$ of Z has the desired properties announced in Lemma 2.

So let $z_1, z_2 \in \mathcal{V}$ satisfying

$$(3.10) \qquad \| z_1 - z_2\| \leq \alpha$$

Each z_i admits at least one projection, say $\tilde{z}_i$, on Z $i = 1, 2$ and we get from (3.10), as $z_i \in \mathcal{V}$ $i = 1, 2$

$$(3.11) \qquad \| \tilde{z}_1 - \tilde{z}_2\| \leq \varepsilon$$

Let us first find a necessary condition for $\tilde{z}_i \in Z$ to be the projection of z_i onto Z.

Define $a_i \in \mathcal{A}_{ad}$ by $a_i = T^{-1}(\tilde{z}_i)$ $i = 1, 2$

$a \in \mathcal{A}_{ad}$ by $a = T^{-1}(z)$ for any $z \in Z$

If we choose :

(3.12) $$z \in Z \cap B(\tilde{z}_i, \varepsilon)$$

then the whole interval $[\tilde{z}_i, z]$ belongs to $\mathcal{V}_\varepsilon$ and we can use the finite increment theorem for T^{-1} between $\tilde{z}_i$ and z, which gives, using (3.8) :

(3.13) $$\| a - a_i \| \leqslant \frac{1}{m} \| z - \tilde{z}_i \|$$

Using the C^2 regularity of T on the convex $\mathcal{a}_{ad}$ gives

(3.14) $$z - \tilde{z}_i = T(a) - T(a_i) = T'(a_i)(a - a_i) + \mathcal{R} \text{ with } \|\mathcal{R}\| \leq \frac{M}{2} \| a - a_i \|^2$$

i.e. using (3.13) :

(3.15) $$\|\mathcal{R}\| \leq \frac{M}{2m^2} \| z - \tilde{z}_i \|^2$$

By definition of $\tilde{z}_i$, the function $a \to \| T(a) - z_i \|^2$ attains its minimum at a_i on the convex set $\mathcal{a}_{ad}$. Hence necessarily :

(2.16) $$(\tilde{z}_i - z_i, T'(a_i)(a - a_i)) \geq 0 \qquad \forall a \in \mathcal{a}_{ad}$$

which together with (3.14) gives

$$(\tilde{z}_i - z_i, z - \tilde{z}_i) \geq (\tilde{z}_i - z_i, \mathcal{R})$$

i.e. using (3.15) :

(3.17) $$(\tilde{z}_i - z_i, z - \tilde{z}_i) \geq -\frac{M}{2m^2} \| \tilde{z}_i - z_i \| \, \| z - \tilde{z}_i \|^2 \qquad \forall z \in Z \cap B(\tilde{z}_i, \varepsilon)$$

<u>which is the seeken necessary condition for $\tilde{z}_i$ to be the projection on Z of $\tilde{z}_i$.</u>

We can now prove the <u>local lipschitz continuity of the projection</u> operator P on Z :

$$\|\tilde{z}_1 - \tilde{z}_2\|^2 = (z_1 - z_2, \tilde{z}_1 - \tilde{z}_2) - (\tilde{z}_1 - z_1, \tilde{z}_2 - \tilde{z}_1) - (\tilde{z}_2 - z_2, \tilde{z}_1 - \tilde{z}_2)$$

Using (3.17) twice for the majoration of the two last terms gives

(3.18) $$\| \tilde{z}_1 - \tilde{z}_2 \|^2 \leq (z_1 - z_2, \tilde{z}_1 - \tilde{z}_2) + \frac{M}{2m^2} \|\tilde{z}_1 - \tilde{z}_2\|^2 \, (\| \tilde{z}_1 - z_1 \| + \| \tilde{z}_2 - z_2 \|)$$

As we have taken $z_1, z_2 \in \mathcal{V}$ we get from (3.9) :

$$\|\tilde{z}_1 - z_1\| + \|\tilde{z}_2 - z_2\| \leq \frac{\lambda-1}{\lambda} \cdot \frac{2m^2}{M}$$

which, together with (3.18) and the Schwarz inequality :

$$\|\tilde{z}_1 - \tilde{z}_2\|^2 \leq \|z_1 - z_2\| \, \|\tilde{z}_1 - \tilde{z}_2\| + \frac{\lambda-1}{\lambda} \|\tilde{z}_1 - \tilde{z}_2\|^2$$

i.e.

(3.19) $\quad \|\tilde{z}_1 - \tilde{z}_2\| \leq \lambda \|z_1 - z_2\|$

which ends the proof of lemma 2. ∎

Proposition 2 : Under the hypothesis (3.1) to (3.3), the parameter a is OLSI in $\mathcal{A}_{ad}$ for the observation operator $\mathcal{C}$.

Proof : It results immediatly from lemma 2 and from the definition of OLSI.

Remark : We are conscious that hypothesis (3.3) on the mapping $a \to \mathcal{C}y(a)$ is rather strong, and hence shall not be easy to verify in applications (though we give in next paragraph an actual case where the sufficient condition of prop. 2 applies). But we think that weaker sufficient conditions for OLSI could be given ,especially in the case where $\mathcal{V}$ is a "differential variety" of $\mathcal{H}$ of "smaller" dimension than $\mathcal{H}$. ∎

Remark : It is well know that the set of points z such that all the minimizing sequences for the distances $d(z, Z)$ are Cauchy (and hence having a unique projection on Z) is a $G - \delta$ dense in $\mathcal{H}$. Lemma 2 says that, if Z is the image of a convex set by a regular enough mapping T, this set contains at least a neighbourhood of Z. ∎

IV – A CASE OF OLSI IN REFLECTION SEISMIC

The propagation of plane waves in a plane layered medium is given by the 1-dimensional wave equation :

$$(4.1) \quad \begin{cases} a(x) \dfrac{\partial^2 y}{\partial t^2} - \dfrac{\partial}{\partial x}\left(a(x) \dfrac{\partial y}{\partial x}\right) = 0 & x \geq 0,\ t \geq 0 \\ - a(0) \dfrac{\partial y}{\partial x}(0, t) = g(t) & t > 0 \qquad \text{(boundary condition)} \\ y(x, 0) = \dfrac{\partial y}{\partial t}(x, 0) = 0 & \qquad \text{(initial condition)} \end{cases}$$

where :

(4.2) $\begin{cases} x \text{ is the travel time} \\ a \in L^{\infty}(R^{+}) \text{ is a strictly positive function,} \\ \text{piecewise constant on intervals of equal} \\ \text{travel time thickness } \Delta x, \\ (a \text{ is the acoustical impedance function}) \\ y(x,\ t) \text{ is the displacement at depth } x \text{ and} \\ \text{time } t, \\ g \text{ is the seismic source at the surface } x = 0, \\ \text{started at } t = 0 \end{cases}$

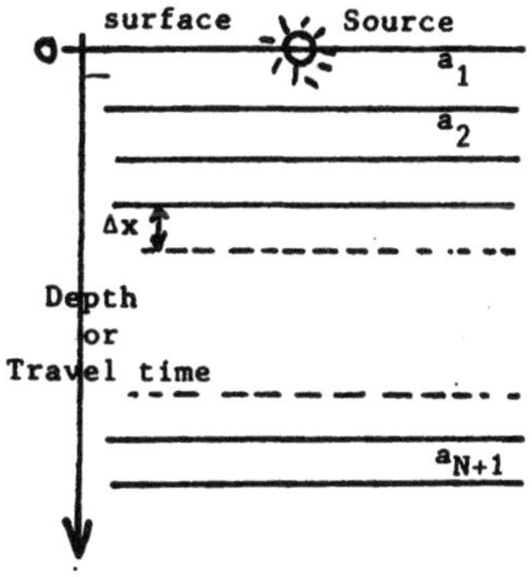

The displacement y is recorded at the surface x = 0 with a sampling interval $\Delta t = 2\Delta x$, so that the observation operator $\mathcal{C}$ and the observation space $\mathcal{H}$ are given by :

(4.3) $\begin{cases} \mathcal{H} = R^{N} \\ \mathcal{C} : y(x,\ t) \rightarrow \mathcal{C}y = (y(0,\ j\Delta t)\ ,\ j = 2,\ 3,\dots,N+1) \in \mathbb{R}^{N} \end{cases}$

The inverse problem in reflection seismics consists, supposing the source g(t) known, in recovering, from the above observations, the impedance distribution a. As the wave celerity in (4.1) is one, the observation (4.3) gives information only on the impedances lying above $(N+1)\Delta x$, i.e. $a_1, a_2, \dots, a_{N+1}$.

In order to get the same number of unknown a_j than of observation, we suppose (which is the case in marine seismics) :

(4.4) $a_1 > 0$ is known

and we take, as parameter space $\mathcal{A}$ and set $\mathcal{A}_{ad}$ of admissible impedances :

(4.5) $\begin{cases} \mathcal{A} = \mathbb{R}^{N} \\ \mathcal{A}_{ad} = \{a = (a_2,\dots,a_{N+1}) \in R^{N} \mid 0 < \underline{a} \leq a_i \leq \bar{a} \quad i = 2,\dots,N+1, \\ \qquad \sum_{i=1}^{N} |a_{i+1} - a_i| \leq v\} \end{cases}$

where $\underline{a}$, $\bar{a}$ are given lower and upper bound of the impedance, and v is given upper bound of the total variation of the seeken impedance distribution.

So let be given :

(4.6) $z \in \mathbb{R}^{N}$ = given observation (seismogram) then the inverse seismic problem may be formulated as :

(4.7) find $\tilde{a} \in \mathcal{A}_{ad}$ which minimizes $||\mathcal{C}y(a) - z||_N^2$ over $\mathcal{A}_{ad}$

for which one we prove :

Proposition 3 : With the hypothesis and notation (4.1) to (4.7) and if :

(4.8) $g(0) = 0, \; g(\Delta t) \neq 0$

then the impedance distribution a is OLSI and $\mathcal{A}_{ad}$ for the observation operator $\mathcal{C}$.

We refer to reference [17] for a complete proof of that proposition and a detailed theoretical and numerical study of the seismic inverse problem. We would like however to illustrate, to close this paragraph, how the numerical results for the above problem confirm the theoretically expected continuity of the mapping $z \to \tilde{a}$.

We have estimated[1)] a medium made of $N \simeq 400$ layers of travel-time thickness $\Delta x = 4$ ms from seismic recording z of duration 3 s. Eighty actual field recordings z are shown the figure 3. They have been recorded at regular time intervals on a ship moving with uniform speed, so that they are generated by 80 impedance distributions $\sigma \in R^{400}$ which should show some simularities, due to the approximative horizontal stratification of the earth in the vicinity of the recording points.

The inverse problem (4.7) was solved 80 times (using a gradient method) ; the 80 corresponding impedance distribution a are represented on figure 4, through the corresponding reflection coefficients sequence :

(4.9) $$r_i = \frac{a_i - a_{i+1}}{a_i + a_{i+1}} \qquad i = 1, 2, \ldots, N$$

It is clear on that figure 4 that the expected similarity between the 80 computed reflection coefficient sequences is present.

A more detailed analysis of this numerical example can be found in [20].

V - CONCLUSION

We have given a definition of parameter identifiability (OLSI) well fitted to the output least square method used in numerical computations. A sufficient condition for this OLSI to hold has been given, which we have applied to an actual parameter estimation problem in reflection seismics. It is hoped to apply this definition to other estimation problems.

1) This computation has been made in the frame of a research contract between IRIA and IFP (Institut Français du Pétrole) which we thank for permission of publishing them.

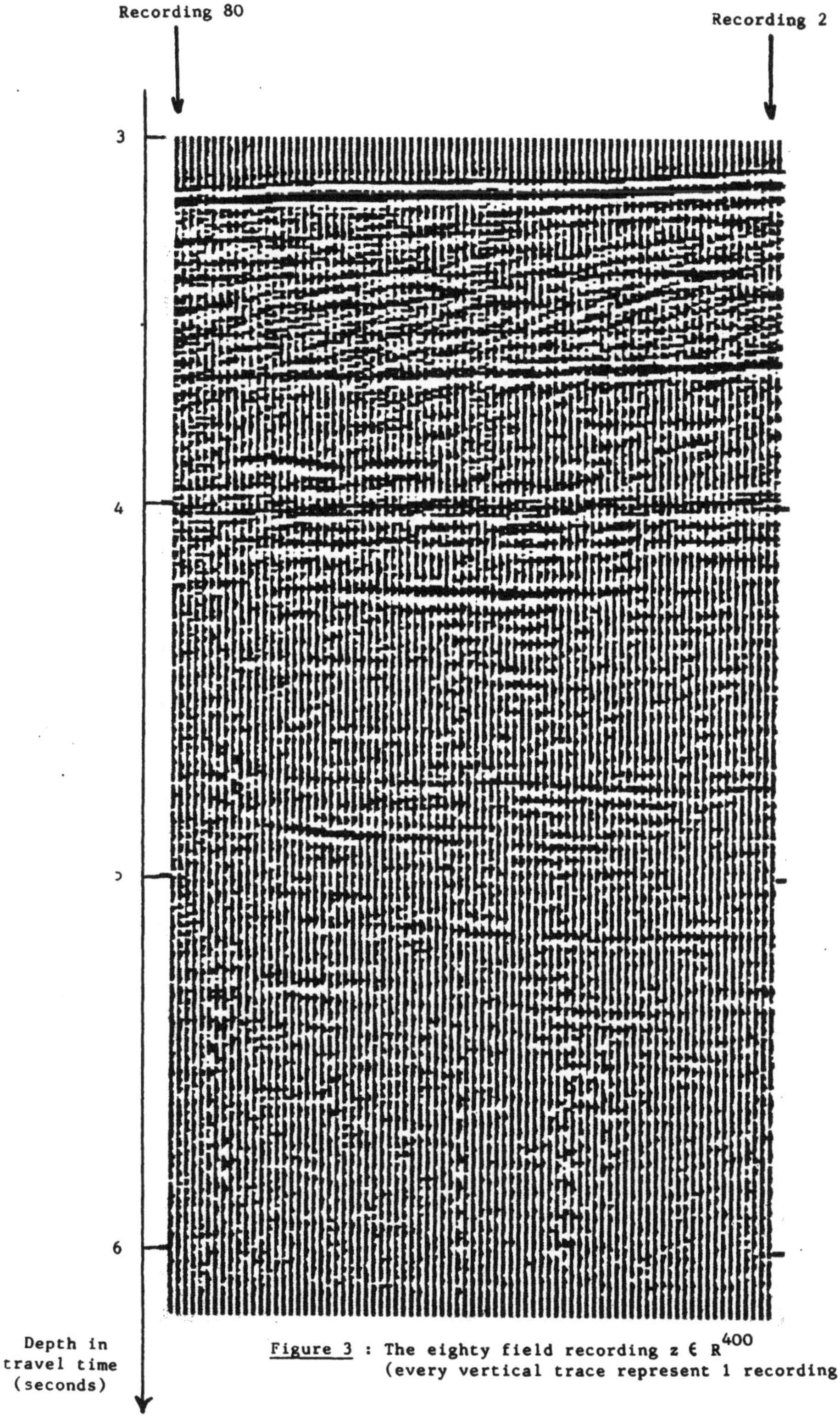

<u>Figure 3</u> : The eighty field recording $z \in R^{400}$
(every vertical trace represent 1 recording $z \in R^{400}$)

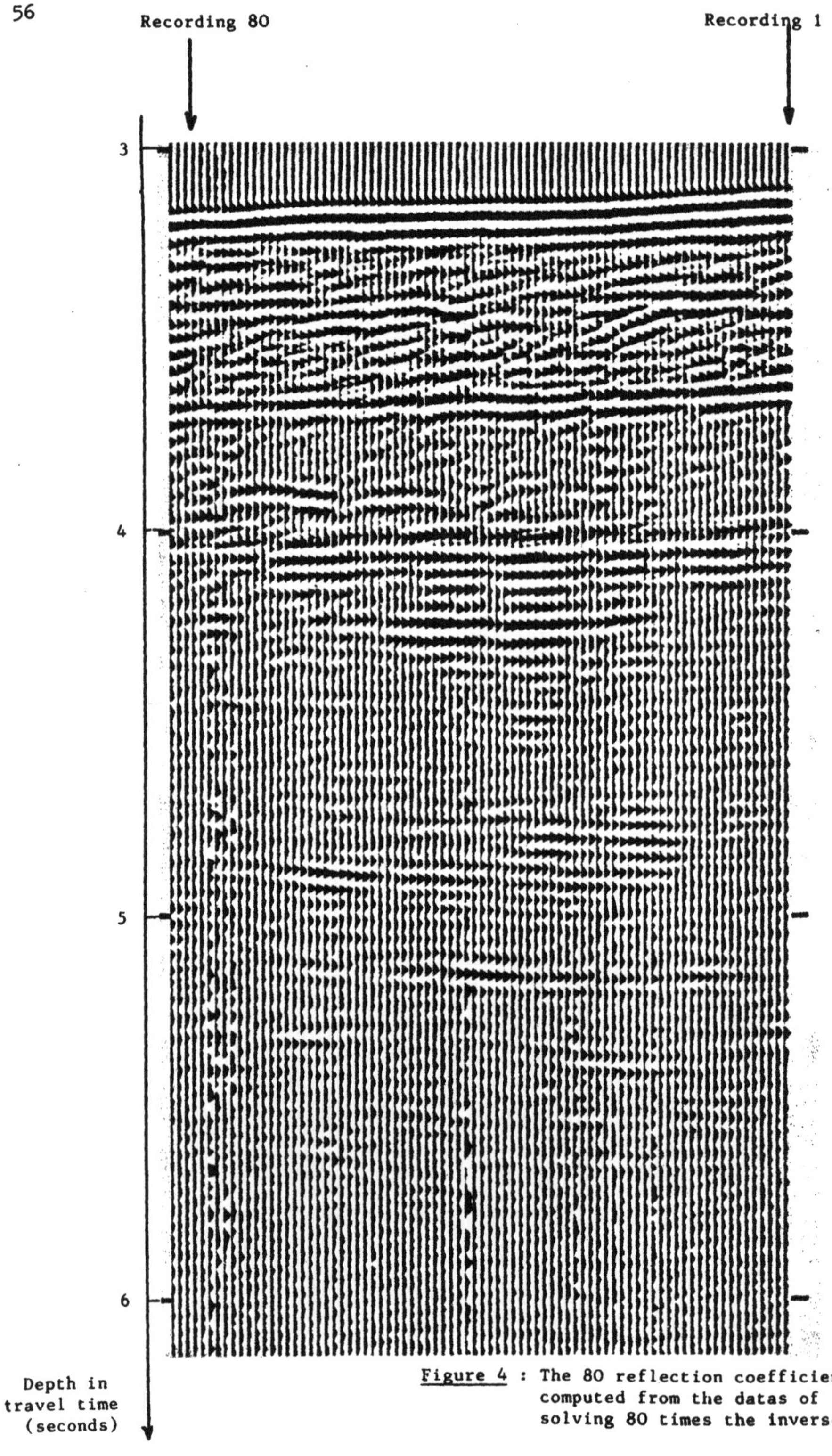

Figure 4 : The 80 reflection coefficients sequences computed from the datas of figure 3 by solving 80 times the inverse problem.

[1] K. CHADAN and P.C. SABATIER - "Inverse problems in quantum scattering theory" - Springer-Verlag - 1977.

[2] M.M. LAUTENTIEV - "Some improperly posed problems of mathematical physics" - Berlin - Heidelberg - New York - 1967 - (translated from Russian).

[3] G. ANGER - "Die Rolle der modernen Potential Theorie in der Theorie der inversen Aufgabenstellungen" - Gerlands Beitr. Geophysik - Lepzig 85 - 1976 - 1, S.1-20.

[4] P.K.C. WANG - "Control of distributed parameter systems" in "Advances in control systems" - Vol. 1 - C.T. LEONDES Editor - New York - 1963.

[5] R.E. GOODSON and R.E. KLEIN - "A definition and some results for distributed systems observability" - IEEE Trans. on Automatic Control - Vol. AC 15 - April 70.

[6] M.P. POLIS and R.E. GOODSON - "Parameter identification in distributed systems : a synthesizing overview" - Proc. of the IEEE - Vol. 64 - n° 1 - Jan. 1976.

[7] T.I. SEIDMAN - "The observability and prediction problem for diffusion processes" - in the 11th Annual Allerton Conf. on Circuit and System Theory - pp. 123-130 - Oct. 1973.

[8] K. MARTENSON - "Least square identifiability of dynamical systems" - Technical Report n° RB7344 - Dept. of Electrical Engineering - U.S.C. - Los Angeles - 1973.

[9] S. KITAMURA and S. NAGAKIRI - "Identifiability of spatially-varying and constant parameters in distributed systems of parabolic type" - in Siam Jl. on Control and Optimization - Vol. 15 - n° 5 - August 1977.

[10] G. CHAVENT - "Analyse fonctionnelle et identification de coefficients répartis dans les équations aux dérivées partielles" - Thèse - Paris 1971.

[11] G. CHAVENT - "Identification of functional parameters in partial differential equations" - in Identification of Parameters in distributed systems - Edited by R.E. Goodson and M. Polis - ASME - 345 East 47th Str. - New York - N.Y. 10017.

[12] G. CHAVENT, M. DUPUY - "History matching by use of optimal control theory" - Soc. of Petroleum Engineers Journal - Vol. 15 - N° 1 - Feb. 1975.

[13] W.H. CHEN and J.H. SEINFELD - "Estimation of spatially varying parameters in partial differential equations" - Int. Jl. of Control - Vol. 15 - 1972.

[14] C.M. BRAUNER and P. PENEL - "Identification of parameters in a non-linear self consistent system including a Burges equation" - Journal of Mathematics Analysis and Applications - Vol. 45 - n° 3 - March 1974.

[15] EL DOUGHERTY and D. KHAIRKHAH - "History matching of gas simulation models using optimal control theory" - Soc. Pet. Eng. - Preprint n° 5371 - 1975.

[16] M.L. WASSERMAN and A.S. EMANUEL - "History matching 3-D models using optimal control theory" - Journal of Canadian Petroleum Technology - Montreal - Oct.-Dec. 1976.

[17] G. CHAVENT and G. COHEN - "Numerical approximation and identification in a 1-D parabolic degenerated non-linear diffusion and transport equation" - in Optimization Techniques - Lecture Notes in Control and Information Sciences - Vol. 6 - Springer - 1978.

[18] M. BENAHMED - "Ajustement automatique de la perméabilité et/ou de la porosité dans un gisement monophasique régi par une équation parabolique quasi-linéaire" - Rapport Laboria n° 244 - Août 1977.

[19] A. BAMBERGER, G. CHAVENT, P. LAILLY - "About the stability of the inverse problem in the 1-D wave equation - Application to the interpretation of seismic profiles" - to appear in the Journal of Appl. Math. and Opt. - 1979.

[20] A. BAMBERGER, G. CHAVENT, CH. HEMON, P. LAILLY - "Inversion of normal incidence seismograms" - presented at the 48th Meeting of the Society of Exploration Geophysicists - San Francisco - Nov. 1978.

[21] T.I. SEIDMAN - "A non convergence for ill posed problems" -

[22] G. CHAVENT, P. LEMONNIER - "Identification de la non-linéarité d'une équation parabolique quasi-linéaire" - Journal of Applied Mathematics and Optimization - Vol. 1 - n° 2 - 1974.

[23] R. KLUGE and H. LANGMACH - "On some problem of determination of functional parameter in partial differential equations" - in "Distributed Parameter Systems : Modelling and Identification" - Edited by A. Ruberti - Lecture Notes in Control and Information Sciences - Vol. 1 - Springer - 1978.

[24] G. CHAVENT - "About the identification and modelling of miscible or immiscible displacements in porous media" - in "Distributed Parameter Systems : Modelling and Identification" - Edited by A. Ruberti - Lecture Notes in Control and Information Sciences - Vol. 1 - Springer - 1978.

[25] CARLOS S. KUBRUSLY - "Distributed parameter system identification, a survey" - Int. Journal Control - Vol. 26 - n° 4 - pp. 509-535 - 1977.

THE APPROXIMATION OF SOLUTIONS TO THE BACKWARDS HEAT EQUATION BY SOLUTIONS OF PSEUDOPARABOLIC EQUATIONS

by

DAVID COLTON*

Summary

It is well known that solutions of the backwards heat equation can be approximated by solutions of a pseudoparabolic equation depending on a small parameter ε. The emphasis in this paper is on the mathematical problems which arise in approximating solutions to initial-boundary value problems for this perturbed equation. The approximation procedure we propose is obtained through the development of a potential theory for pseudoparabolic equations, the asymptotic evaluation of certain contour integrals, and results based on a theorem of Levin in the theory of entire functions.

I. Introduction.

The problem of construccing solutions to the heat equation backwards in time is one of the classical improperly posed problems in partial differential equations. Mathematically the problem can be formulated (in R^3) in the following manner: For D a bounded domain in R^3 find $u(\underset{\sim}{x},t)$, $\underset{\sim}{x} \in R^3$, such that

$$\Delta_3 u = u_t \text{ in } D \times (0,t_0) \quad (1.1a)$$

$$u = 0 \text{ on } \partial D \times (0,t_0) \quad (1.1b)$$

$$u(\underset{\sim}{x},t_0) = \phi(\underset{\sim}{x}) \text{ in } D \quad (1.1c)$$

where $\phi(\underset{\sim}{x})$ is a prescribed function. As is well known ([15], [16]) in general no solution exists to (1.1a) - (1.1c) and if it does the solution does not depend continuously on the data $\phi(\underset{\sim}{x})$ in any reasonable norm. One approach to "solving" (1.1a) - (1.1c) is the method of quasi-reversibility as

*Professor of Mathematics, University of Delaware, Newark, Delaware, U.S.A.

initially developed by Lattes and Lions ([11]), and it is a version of this approach that we wish to discuss in this paper. The version we have in mind is to replace the problem (1.1a) - (1.1c) by the modified problem

$$\varepsilon \, \Delta_3 u_t - u_t + \Delta_3 u = 0 \text{ in } D \times (0,t_o) \tag{1.2a}$$

$$u = 0 \text{ on } \partial D \times (o,t_o) \tag{1.2b}$$

$$u(\underset{\sim}{x},t_o) = \phi(\underset{\sim}{x}) \text{ in } D \tag{1.2c}$$

where ε is a small positive parameter ([7], [9]). Three obvious questions immediately present themselves:

1) Does a unique solution to (1.2a) - (1.2c) exist for every $\varepsilon > 0$ and does it depend continuously on the data $\phi(\underset{\sim}{x})$?
2) As $\varepsilon \to 0$ does the solution of (1.2a) - (1.2c) approach the solution of (1.1a) - (1.1c) (if it exists!)?
3) What constructive methods are available for approximating the solution of (1.2a) - (1.2c) for $\varepsilon > 0$?

In most discussions of the method of quasi-reversibility attention is usually focused on questions 1) and 2), and the answer to 3) is normally "use finite difference approximations or partial eigenfunction expansions". However for multi-dimensional problems, large time intervals, and small values of the parameter ε, such methods are in general rather impractical. It is in fact with this difficulty in mind that we have chosen to use the third order perturbed equation (1.2a) instead of the fourth order equation

$$\varepsilon \, \Delta_3^2 u - u_t + \Delta_3 u = 0 \tag{1.3}$$

as proposed by Lattes and Lions ([11]). As will be seen, such a choice will enable us to use potential theoretic methods and asymptotic analysis to examine solutions of (1.2a) - (1.2c) for small ε, thus providing a more practical method for constructing approximations to solutions of the backwards heat equation in higher dimensions and for large time intervals.

Having tailored our model (1.2a) - (1.2c) to give a satisfactory answer to question 3) does not of course excuse us from dealing with the first two questions! However, the question of existence, uniqueness, and continuous dependence on the data is well known (c.f. [1]) and the second question has been answered by Ewing in the following theorem ([7]):

Theorem 1: Let $u(\underset{\sim}{x},t)$ be a solution of (1.1a), (1.1b) such that $||u(\underset{\sim}{x},t_o) - \phi(\underset{\sim}{x})||_{L^2} < \delta$, $||u(\underset{\sim}{x},0)||_{L^2} < M$, where δ, M are positive constants, and let $v(\underset{\sim}{x},t)$ be the solution of (1.2a) - (1.2c) for $\varepsilon = [\log {}^M/_\delta]^{-1}$. Then for every $t > 0$,

$$||u-v||_{L^2} \leq C(t)\varepsilon$$

where C(t) does not depend on ε.

An obvious drawback of Ewing's result is that the error bound depends logarithmically on δ instead of Hölder continuously as in the stabilized quasi-reversibility method of Millar ([15]). A similiar problem also arises if one uses quasi-reversibility methods in conjunction with (1.3), or, in a different direction, Tikhonov's regularization method for solving the backwards heat equation (c.f. [8], [18]). In our case this means that in order to achieve accurate results we must choose the parameter ε to be quite small and assume that the data at $t = t_o$ is measured with a high degree of accuracy.

Although our analysis in what follows is presented for the heat equation in R^3, analogous results can also be obtained in R^2 (see Section IV).

II. Potential Theory for the Pseudo-Heat Equation.

Equation (1.2a) is a particular example of an equation of pseudo-parabolic or Sobolev type and is usually referred to as the pseudo-heat equation. Equations of pseudoparabolic type appear in a variety of areas of application and have been the object of a considerable amount of attention in recent years. For information concerning this class of equations as well as an extensive bibliography we suggest consulting the recent book by Carroll and Showalter ([1]). In this section we shall develop a potential theory for the pseudo-heat equation ([3], [5]) with the aim of using these results in

Section III to construct approximations to the solution of (1.2a) - (1.2c). We shall develop this potential theory for the "forward" initial-boundary value problem

$$\varepsilon \Delta_3 u_t - u_t + \Delta_3 u = 0 \text{ in } D\times(0,t_o) \tag{2.1a}$$

$$u = f(\underset{\sim}{x},t) \text{ on } \partial D \times (0,t_o) \tag{2.1b}$$

$$u(\underset{\sim}{x},0) = 0 \text{ in } D \tag{2.1c}$$

where $f(\underset{\sim}{x},t)$ is assumed to be continuously differentiable and reserve until Section III the modifications and applications of these results to the study of (1.2a) - (1.2c).

We begin by defining the fundamental solution for (2.1a) by ([3], [5])

$$\Gamma(R,\ t-\tau) = -\frac{1}{\pi i R} \oint_{|\omega - \frac{1}{\sqrt{\varepsilon}}| = \delta} \exp[-\omega R + \frac{\omega^2(t-\tau)}{1-\varepsilon\omega^2}]\, d\omega \tag{2.2}$$

where $R = |\underset{\sim}{x} - \underset{\sim}{\xi}|$ for $\underset{\sim}{x}, \underset{\sim}{\xi} \in R^3$ and the path of integration is a circle of radius δ traversed counterclockwise about the point $\omega = \frac{1}{\sqrt{\varepsilon}}$. If we now assume that D is a bounded, simply connected domain with Lyapunov boundary ∂D and let $\rho(\underset{\sim}{\xi},\tau)$ denote a continuous density defined on $\partial D \times [0,t_o]$, we can define a pseudo-heat potential by

$$u(\underset{\sim}{x},t) = \frac{1}{2\pi}\int_0^t \int_{\partial D} \rho(\underset{\sim}{\xi},\tau)\, \frac{\partial^2}{\partial\nu\partial\tau}\Gamma(R,\ t-\tau)\, ds\, d\tau \tag{2.3}$$

where ν denotes the unit normal on ∂D pointing into D. In order for (2.3) to be a solution of (2.1a) - (2.1c) it is necessary to choose $\rho(\underset{\sim}{\xi},\tau)$ such that (2.1b) is satisfied. To this end we differentiate both sides of (2.3) with respect to t and let $\underset{\sim}{x} \to \partial D$. Then from the discontinuity properties of metaharmonic potentials we arrive at an integral equation for $\rho(\underset{\sim}{\xi},\tau)$ of the form

$$\varepsilon^{3/2}\frac{\partial f}{\partial t} = (\underset{\sim}{I} + \underset{\sim}{T} + \underset{\sim}{L}_1 + \underset{\sim}{L}_2)\rho \tag{2.4}$$

where $\underset{\sim}{T}$ is a Fredholm integral operator over ∂D, $\underset{\sim}{L}_1$ is a Volterra integral operator over $[0,t]$, and $\underset{\sim}{L}_2$ is a Volterra integral operator over $\partial D \times [0,t]$. The kernels of these integral operators depend on $\Gamma(R,t)$ and its derivatives (c.f.[3]). Due to the fact that $\varepsilon > 0$ we can deduce the following Theorem ([3]):

Theorem 2: Let $C(\partial D \times [0,t_o])$ be the Banach space of continuous functions defined over $\partial D \times [0,t_o]$ with respect to the maximum norm. Then $(\underset{\sim}{I} + \underset{\sim}{T} + \underset{\sim}{L}_1 + \underset{\sim}{L}_2)^{-1}$ exists as a bounded linear operator on $C(\partial D \times [0,t_o])$. Hence any solution of (2.1a) - (2.1c) can be represented in the form (2.3) for some continuous density $\rho(\underset{\sim}{\xi},\tau)$.

In order to actually construct $\rho(\underset{\sim}{\xi},\tau)$ from (2.4) it is necessary to determine accurate approximations to $\Gamma(R,t)$ and its derivatives for small values of ε and $R \geq 0$, $t \geq 0$. Clearly it suffices to consider the function

$$
\begin{aligned}
K(R,t) &= -\frac{1}{\pi i} \oint_{|\omega - \frac{1}{\sqrt{\varepsilon}}| = \delta} \exp\left[-\omega R + \frac{\omega^2 t}{1-\varepsilon\omega^2}\right] d\omega \\
&= -\frac{\exp(-t/\varepsilon)}{\sqrt{\varepsilon}\,\pi i} \oint_{|\mu-1| = \delta} \exp\left[-\mu \frac{R}{\sqrt{\varepsilon}} + \frac{t}{\varepsilon(1-\mu^2)}\right] d\mu .
\end{aligned}
\tag{2.5}
$$

The evaluation of $K(R,t)$ can be conveniently divided into three separate cases ([5]):

1) $R=0,\ t=O(1)$. In this case we expand $\exp\left(\frac{t}{\varepsilon(1-\mu^2)}\right)$ in its Taylor series in t and integrate termwise to arrive at

$$
K(0,t) = \frac{t \exp(-{}^{t}/_{\varepsilon})}{\varepsilon}\, \Phi({}^{1}/_{2}, 2; {}^{t}/_{\varepsilon}) \tag{2.6}
$$

where $\Phi(a,c;z)$ denotes the confluent hypergeometric function. From the asymptotic behaviour of $\Phi(a,c;z)$ for large z (c.f.[6]) we have from (2.6) that

$$K(0,t) = \frac{1}{\sqrt{\pi t}} \left[\sum_{n=0}^{N} \frac{(^1/_2)_n (^3/_2)_n}{n!} (\frac{\varepsilon}{t})^n + 0 (\frac{\varepsilon}{t})^{N+1} \right] \qquad (2.7)$$

$$\text{where } (\alpha)_n = \frac{\Gamma(\alpha+n)}{\Gamma(\alpha)} .$$

2) $\underline{R \geq 0,\ t = 0(\varepsilon)}$. We again expand $\exp\left(\frac{t}{\varepsilon(1-\mu^2)}\right)$ in powers of t and integrate termwise to obtain

$$K(R,t) = \frac{2 \exp(-^t/_\varepsilon)}{\sqrt{\varepsilon\pi}} \sum_{n=0}^{\infty} \frac{1}{(n+1)!n!} (\frac{t}{\varepsilon})^{n+1} (\frac{R}{2\sqrt{\varepsilon}})^{n+^1/_2} K_{n+^1/_2} (\frac{R}{\sqrt{\varepsilon}}) \qquad (2,8)$$

where $K_n(z)$ denotes the modified Bessel function. For $t = 0(\varepsilon)$, (2.8) can be readily approximated by truncating the series and applying the Clenshaw-Luke method of backward recursion ([13], Section 11.8). An estimate on the convergence ratio of (2.8) can be obtained by using the inequality

$$\left| (\frac{R}{2\sqrt{\varepsilon}})^{n+^1/_2} K_{n+^1/_2} (\frac{R}{\sqrt{\varepsilon}}) \right| \leq {}^1/_2 \ \Gamma(n+^1/_2) \qquad (2.9)$$

3) $\underline{R>0,\ t=0(1)}$. By deforming the circle $|\omega - \frac{1}{\sqrt{\varepsilon}}| = \delta$ in (2.5) onto the contour pictured below in Figure 1

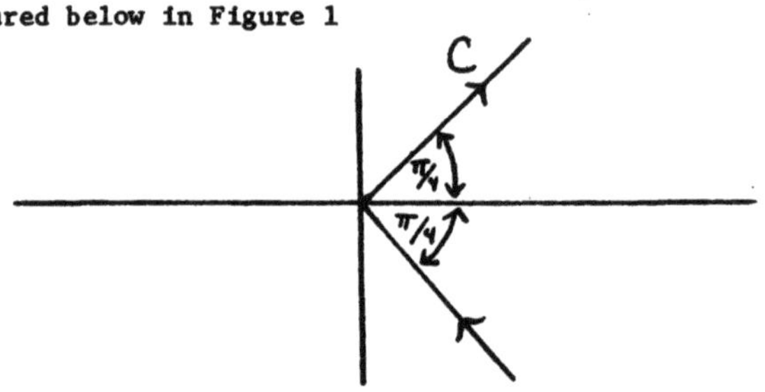

Figure 1

we can rewrite $K(R,t)$ in the form

$$K(R,t) = \frac{1}{\pi i \sqrt{t}} \int_C \exp[-\frac{R}{\sqrt{t}} z + z^2] \, g_\varepsilon(z)dz \qquad (2.10)$$

where

$$g_\varepsilon(z) = \exp\left[\frac{\varepsilon z^4}{t-\varepsilon z^2}\right]. \qquad (2.11)$$

By using ad hoc methods, it is now possible to obtain a complete asymptotic expansion for $K(R,t)$ as $\varepsilon \to 0$ in the form

$$K(R,t) = \frac{1}{\sqrt{\pi t}} \exp(-\frac{R^2}{4t})[1+d_1(\frac{\varepsilon}{t})+d_2(\frac{\varepsilon}{t})^2+ \dots + d_n(\frac{\varepsilon}{t})^n + 0((\frac{\varepsilon}{t})^{n+1})] \quad (2.12)$$

where the coefficients d_j are expressible in terms of Hermite polynomials ([5]). In particular

$$d_1 = \frac{1}{4^2} H_4(\frac{R}{2\sqrt{t}})$$

$$d_2 = \frac{1}{2\cdot 4^4} H_8 (\frac{R}{2\sqrt{t}}) + \frac{1}{4^3} H_6(\frac{R}{2\sqrt{t}}) \qquad (2.13)$$

with similar expressions holding for the higher order coefficients. In (2.13) $H_n(z)$ denotes Hermite's polynomial.

The expansions in all of the above cases may be differentiated termwise.

III. Approximation of Solutions to the "Backwards" Pseudo-Heat Equation.

We now discuss the problem of approximating solutions of initial-boundary value problems for the "backwards" pseudo-heat equation defined by (1.2a) - (1.2c). One approach proceeds as follows. By replacing t by t_o-t, and using the Fourier transform or Borel transform to construct a solution to the pure initial value problem, we can reduce problem (1.2a) - (1.2c) to an initial-boundary value problem of the form

$$\varepsilon\, \Delta_3 u_t - u_t - \Delta_3 u = 0 \text{ in } D \times (0,t_o) \qquad (3.1a)$$

$$u = f(\underset{\sim}{x},t) \text{ on } \partial D \times (0,t_o) \qquad (3.1b)$$

$$u(\underset{\sim}{x},0) = 0 \text{ in } D. \qquad (3.1c)$$

Note that (3.1a) - (3.1c) differs from (2.1a) - (2.1c) only in a sign change in the differential equation. Although this has no effect on the well-posedness of the problem, as we shall see it unfortunately has serious implications on the problem of constructing approximate solutions for small values of the (positive) parameter ε. This is, of course, not surprising for the limiting case as $\varepsilon \to 0$ is now an improperly posed problem, i.e. the backwards heat equation. If we follow the analysis of the previous section, it is seen that we can represent the solution of (3.1a) - (3.1c) in the form

$$u(\underset{\sim}{x},t) = \frac{1}{2\pi}\int_0^t \int_{\partial D} \rho(\underset{\sim}{\xi},\tau)\frac{\partial^2}{\partial\nu\partial\tau}\Gamma(R,\tau-t)\,ds\,d\tau \qquad (3.2)$$

where $\rho(\underset{\sim}{\xi},\tau)$ is determined as the solution of an integral equation of Fredholm-Volterra type and the notation is the same as in (2.3). We observe that the only difference between (3.2) and (3.3) is that the argument of the fundamental solution is $\tau-t$ instead of $t-\tau$. This change, however, unfortunately complicates the evaluation of $\Gamma(R,t)$ since we are now interested in the asymptotic behaviour of $\Gamma(R,t)$ for $t < 0$ instead of $t > 0$. Except for the expansion (2.8) the analysis of Section II breaks down and we are forced into adopting an ad hoc approach yielding expansions of $\Gamma(R,t)$ in terms of a series of generalized hypergeometric functions or a Bessel-Laguerre series (c.f.[4]). Such expansions are of limited value in the evaluation of $\Gamma(R,t)$ for $t = O(1)$ and hence we are led to look for other approximation procedures.

To present such an alternate approach we return to the formulation (1.2a) - (1.2c). We first note that if λj is a sequence

$$\lim_{j\to\infty} \frac{j}{\lambda_j^2} > 0$$

then in a sufficiently small complex neighbourhood of $[0,t_o]$ we can approximate any analytic function defined in this neighbourhood by a finite linear combination of functions taken from the set $\{e^{-\lambda_j^2 t}\}_{j=1}^{\infty}$. This result follows from a theorem of Levin in the theory of entire functions and the reader is referred to [12], p. 219, for details. From the results of [2] we can now conclude that any continuous function defined in $\bar{D}$ can be approximated in $\overline{D}$ by a linear combination of the functions

$$r^{-1/2} J_{n+1/2} (\lambda_j r) P_n^m(\cos \Theta) e^{im\phi}; \quad n=0,1,2,\ldots \quad -n\leq m\leq n \tag{3.4}$$

where $J_n(z)$ denotes Bessel's function and $P_n^m(z)$ an associated Legendre function. We observe that

$$u_{nmj} (\underset{\sim}{x},t) = r^{-1/2} J_{n+1/2}(\lambda_j r) P_n^m(\cos \Theta) \exp [im\phi - \frac{\lambda_j^2}{1 + \varepsilon\lambda_j^2} t] \tag{3.5}$$

is a solution of (1.2a). Although the set $\{u_{nmj}\}$ is a complete set of solutions when $\varepsilon=0([2])$, for $\varepsilon>0$ the order of the equation is increased and hence except for exceptional circumstances we would expect that extra functions would have to be introduced to supplement the above set. (An exceptional case occurs for example when D is a sphere and $\{\lambda_j^2\}$ are the eigenvalues of the Laplacian in D.) In particular from the Runge approximation property for pseudoparabolic equations ([17]) and the results of Section II (approximating the pseudo-heat potential defined on the cylinder $S \times [0,t_o] \supset D \times [0,t_o]$ by numerical quadrature) we have the following Theorem, where completeness is with respect to the L^2 norm over $D \times [0,t_o]$.

Theorem 3: Let $\{\lambda_j\}$ be as in (3.3) and let $\{\underset{\sim}{\xi}_j\}_{j=1}^{\infty}$ be a dense set of points on a sphere $S \supset D$. Then the functions

$$w_{nj}(\underset{\sim}{x},t) = \int_0^t \tau^n \frac{\partial^2}{\partial\nu\partial\tau} \Gamma(|\underset{\sim}{\xi}_j - \underset{\sim}{x}|, t-\tau)d\tau$$

$$u_{nmj}(\underset{\sim}{x},t) = r^{-1/2} J_{n+1/2} (\lambda_j r) P_n^m(\cos \Theta) \exp [im\phi - \frac{\lambda_j^2}{1+\varepsilon\lambda_j^2} t]$$

where $\nu = \underset{\sim}{\xi}_j / |\underset{\sim}{\xi}_j|$ form a complete set of solutions to (1.2a) defined in $D \times [0,t_o]$.

Now let $\{\Psi_n\}_{n=1}^{\infty}$ denote the complete set of functions defined in Theorem 3. In order to find an approximate solution to (1.2a) - (1.2c) we minimize the functional

$$\left\| \sum_{n=1}^{N} c_n \Psi_n(\underset{\sim}{x}, t_o) - \phi(\underset{\sim}{x}) \right\|_{L^2(D)} + \left\| \sum_{n=1}^{N} c_n \Psi_n(\underset{\sim}{x}, t) \right\|_{L^2(\partial D \times [0,t_o])} \tag{3.6}$$

and define the approximate solution to be $\sum_{n=1}^{N} c_n \Psi_n(\underset{\sim}{x}, t)$. The functions Ψ_n can be evaluated by using existing tables for the computation of Bessel and Legendre functions in combination with the results of Section II to approximate $\frac{\partial^2}{\partial\nu\partial\tau} \Gamma\ |\underset{\sim}{\xi} - x|, t-\tau)$. Due to the fact that (1.2a) - (1.2c) is well posed, the minimization procedure indicated above is stable. Following Gaponenko ([10]) we require that the minimization of (3.6) allow not only for the variation of the coefficients c_n but also for the basis functions Ψ_n themselves, i.e. by considering $\underset{\sim}{\xi}_j$, $j = 1, \ldots, N$, and λ_j, $j = 1, \ldots, N$, to be variables as well. Thus we are led to the problem of finding the minimum of a function of the 6 N variables $\mathrm{Re}\ c_n$, $\mathrm{Im}\ c_n$, $\underset{\sim}{\xi}_j = (\xi_j^{(1)}, \xi_j^{(2)}, \xi_j^{(3)})$, λ_j, where $1 \leq j \leq N$. A related minimization problem is considered in [10], together with its numerical implementation by the conjugate gradient method, and a similar approach for elliptic boundary value problems has also been discussed by Mathon ([14]).

IV. The Backwards Heat Equation in R^2.

The analysis of the previous sections can be extended to treat the backwards heat equation in R^2, and in this section we briefly outline the necessary modifications. Theorem 1 is valid in any number of dimensions. A potential theory for the two dimensional pseudo-heat equation can be developed exactly as in Section II where the fundamental solution is now defined by

$$\Gamma(R, t-\tau) = -\frac{1}{\pi i} \oint_{|\omega - \frac{1}{\sqrt{\varepsilon}}| = \delta} K_o(\omega R) \exp\left[\frac{\omega^2(t-\tau)}{1 - \varepsilon\omega^2}\right] d\omega \tag{4.1}$$

with $K_o(z)$ denoting a modified Bessel function of order zero. From [6] we have the representation

$$K_o(z) = \int_1^\infty e^{-zs} (s^2-1)^{-1/2} ds \tag{4.2}$$

and hence from (2.5) we can write

$$\Gamma(R,t) = \int_1^\infty K(Rs,t)(s^2 - 1)^{-1/2} ds. \tag{4.3}$$

Expansion formulae for $\Gamma(R,t)$ can now be obtained for $R>0$, $t=0(\varepsilon)$ from (2.8) and for $R>0$, $t=0(1)$ from (2.12) since these expansions can be integrated termwise with respect to R over the range $(1,\infty)$. In particular from (2.12) we have

$$\Gamma(R,t) = \frac{1}{\sqrt{\pi t}} \int_1^\infty \exp(-\frac{R^2 s^2}{4t}) (s^2 - 1)^{-1/2} ds + 0(\frac{\varepsilon}{t})$$

$$= \frac{1}{\sqrt{4\pi t}} \exp(-\frac{R^2}{8t}) \int_1^\infty \exp(-\frac{R^2 \mu}{8t}) (\mu^2 - 1)^{-1/2} d\mu + 0(\frac{\varepsilon}{t}) \tag{4.4}$$

$$= \frac{1}{\sqrt{4\pi t}} \exp(-\frac{R^2}{8t}) K_o(\frac{R^2}{8t}) + 0(\frac{\varepsilon}{t}).$$

The coefficients of the higher order terms can be expressed in terms of Whittaker functions. Finally, the analysis of Section III proceeds in the same manner except that the functions $r^{-1/2} J_{n+1/2} (\lambda_j r) P_n^m(\cos\theta)e^{im\phi}$ are now $J_n(\lambda_j r)e^{\pm in\theta}$.

Acknowledgement: This research was supported in part by N.S.F. Grant MCS78-02452 and AFOSR Grant 76-2879.

References

1. Carroll, R., and Showalter, R., Singular and Degenerate Cauchy Problems, Academic Press, New York, 1976.

2. Colton, D., Walsh's theorem for the heat equation, in Ordinary and Partial Differential Equations, W. N. Everitt and B. D. Sleeman, editors, Springer-Verlag Lecture Note Series Vol. 564, Springer-Verlag, Berlin, 1976, 54 - 60.

3. Colton, D., The exterior Dirichlet problem for $\Delta_3 u_t - u_t + \Delta_3 u = 0$, Applicable Analysis, 7(1978), 207 - 212.

4. Colton, D., and Wimp, J., The construction of solutions to the heat equation backward in time, Math. Methods in Applied Sciences, to appear.

5. Colton, D., and Wimp, J., Asymptotic behaviour of the fundamental solution to the equation of heat conduction in two temperatures, J. Math. Anal. Appl., to appear.

6. Erdélyi, A., et al., Higher Transcendental Functions, McGraw-Hill, New York, 1953.

7. Ewing, R., The approximation of certain parabolic equations backward in time by Sobolev equations, SIAM J. Math. Anal. 6 (1975), 283 - 294.

8. Franklin, J., On Tikhonov's method for ill-posed problems, Math. Comp. 28 (1974), 889-907.

9. Gajewski, H., and Zacharias, K., Zur Regularisierung einer Klasse nichtkorrekter Probleme bei Evolutionsgleichungen, J. Math. Anal. Appl. 38 (1972), 784 - 789.

10. Gaponenko, A., A numerical method of solving three-dimensional diffraction problems, USSR Comp. Math. and Math. Phys. 17 (1977), 270 - 277.

11. Lattes, R., and Lions, J. L., The Method of Quasi-reversibility, American Elsevier, New York, 1969.

12. Levin, B. J., Distribution of Zeros of Entire Functions, American Mathematical Society, Providence, 1964.

13. Luke, Y. L., Mathematical Functions and Their Approximations, Academic Press, New York, 1975.

14. Mathon, R., On the approximation of elliptic boundary value problems by fundamental solutions, Tech. Rep. 49, Dept. Computer Science, Univ. of Toronto, 1973.

15. Miller, K., Efficient numerical methods for backward solution of parabolic problems with variable coefficients, in Improperly Posed Boundary Value Problems, A. Carasso and A. P. Stone, editors, Pitman Research Notes in Mathematics Vol. 1, Pitman Publishing, London, 1975, 54-64.

16. Payne, L. E., Improperly Posed Problems in Partial Differential Equations, SIAM Publications, Philadelphia, 1975.

17. Rundell, W., and Stecher, W., A Runge approximation and unique continuation theorem for pseudoparabolic equations, SIAM J. Math. Anal., to appear.

18. Tikhonov, A. N. and Arsenin, V. Y., Solutions of Ill-Posed Problems, John Wiley, New York, 1977.

REGULARIZATION OF THE BACKWARD SOLUTION OF PARABOLIC PROBLEMS

LARS ELDÉN

Summary

The regularization method applied to the backward solution of parabolic problems is studied from an algorithmic point of view. It is shown that the discretized problem is equivalent to a sequence of sparse linear systems, and an efficient algorithm is given for the case when the region is rectangular and separation of variables can be applied. Algorithms are also suggested for the case when the region is nonrectangular. A numerical example in two space dimensions is given.

1. Introduction

The problem of solving a parabolic equation backwards in time is a classical example of an ill-posed problem. Consider the equation

$$\begin{cases} u_t = -Lu \quad \text{in } \Omega \times [0,T], \\ u = 0 \quad \text{on the boundary } \partial\Omega \times [0,T], \\ u(x,T) = w(x), \end{cases} \tag{1}$$

where L is a selfadjoint, elliptic operator with smooth coefficients (independent of t) and Ω is a bounded region in R^n with smooth boundary. Let A denote the evolution operator, which maps initial values $u(x,0)$ into final values $u(x,T) = w(x)$.

The ill-posed nature of the problem is equivalent to A^{-1} being unbounded, and any attempt to solve (1) backwards numerically (i.e. to compute $A^{-1}w$) will prove unstable. The problem is stabilized if we replace (1) by the following minimization problem in the Hilbert space $L^2(\Omega)$

$$\min_u \{||Au - w||^2 + \mu^2 ||u||^2\} , \tag{2}$$

(see e.g. [9],[10],[11],[15]) for some value of the parameter μ. (2) is usually called a <u>regularization method</u> [15] for the ill-posed problem (1).

To solve (2) numerically (1) is discretized in space and time by replacing the differential operators by finite difference (or finite element) approximations. Corresponding to (2) there will be a finite-dimensional least squares problem

$$\min_U \{||AU - W||^2 + \mu^2 ||U||^2\}, \tag{3}$$

where A is a matrix mapping the initial vector U into the final vector W.

Department of Mathematics
Linköping University
S-581 83 Linköping, Sweden

Typically A is the product of a sequence of sparse matrices (see section 2), and therefore A is a dense matrix. For this reason it has been assumed that the regularization method (2) cannot be used efficiently for the numerical solution of problems in two or more space dimensions [10].

It is the purpose of this paper to present an algorithm for the solution of the discretized problem (3). We show that the normal equations of (3) are equivalent to a sequence of sparse linear systems, and thus (3) can be solved without forming the dense matrix A. This algorithm is formulated for general problems (1) in Section 2. If the geometry is rectangular and the elliptic operator allows for variable separation then the sparse systems can be solved very efficiently using a method related to direct methods for separable elliptic problems [14]. This special case is described in Section 3. In Section 4 we discuss the case when the region is non-rectangular and we suggest two different iterative methods, which both utilize the sparsity of the problem. A numerical example is given in Section 5. The heat equation in two space dimensions is discretized on a 50×50 grid, and using the method of Section 3 it is solved backwards 50 time steps.

2. Factorization of the normal equations

First the differential equation (1) is discretized in the space variables in such a way that symmetry is preserved:

$$\frac{dU(t)}{dt} = \frac{1}{h^2} C\, U(t),$$

where $U(t)$ is a vector of unknowns at the time t, C is a symmetric, sparse matrix corresponding to the elliptic operator and h is a step length in the space discretization. Then the interval $[0,T]$ is divided into N subintervals of length k; put $t_\ell = \ell k$, $\ell = 0,1,2,\ldots,N$. Using the Crank-Nicholson scheme we get

$$(I - \frac{r}{2} C)U^{\ell+1} = (I + \frac{r}{2} C)U^{\ell}, \quad \ell = 0,1,2,\ldots,N-1,$$

where U^ℓ is an approximation of $U(t_\ell)$ and $r = k/h^2$. Putting

$$D = (I - \frac{r}{2} C)^{-1} (I + \frac{r}{2} C),$$

$$W = U^N, \quad U = U^0,$$

we can write (3)

$$\min_U \{||D^N U - W||^2 + \mu^2 ||U||^2\}, \tag{4}$$

where the norm is the Euclidean vector norm. Since the matrix D is symmetric the normal equations of (4) are

$$(D^{2N} + \mu^2 I)U = D^N W. \tag{5}$$

Of course these normal equations should not be formed explicitly, since already the matrix D is dense. However, the matrix polynomial $D^{2N} + \mu^2 I$ can be factorized into second degree real factors

$$\begin{cases} D^{2N} + \mu^2 I = \prod_{\ell=1}^{N} (D^2 + \theta_\ell D + \varphi I), \\ \theta_\ell = -2\,\mu^{1/N} \cos(\frac{2\ell-1}{2N}\pi), \\ \varphi = \mu^{2/N}. \end{cases}$$

Therefore, if we can solve linear systems of the form

$$(D^2 + \theta_\ell D + \varphi I)X = DY, \qquad (6)$$

(5) can be solved in N steps

U := W;
<u>for</u> ℓ := 1 <u>step</u> 1 <u>until</u> N <u>do</u>
U := $(D^2 + \theta_\ell D + \varphi I)^{-1}DU$;

The matrix D is a dense matrix, but (6) can be rewritten as a sparse linear system by multiplying by $(I - \frac{r}{2} C)^2$:

$$[(I + \frac{r}{2} C)^2 + \theta_\ell(I - \frac{r^2}{4} C^2) + \varphi(I - \frac{r}{2} C)^2]X = [I - \frac{r^2}{4} C^2]Y ,$$

and if we collect terms by powers of C we get

$$\begin{cases} [\alpha_\ell C^2 + \beta C + \gamma_\ell I]X = (I - \frac{r^2}{4} C^2)Y , \\ \alpha_\ell = r^2(1 - \theta_\ell + \varphi)/4 , \\ \beta = r(1 - \varphi) , \\ \gamma_\ell = 1 + \theta_\ell + \varphi. \end{cases} \qquad (7)$$

The matrix $\alpha_\ell C^2 + \beta C + \gamma_\ell I$ is definite (this follows from the fact that the linear factors of the polynomial $z^2 + \theta_\ell z + \varphi$ corresponding to (6) are complex), and therefore (7) can be solved iteratively, using e.g. the conjugate gradient method [12]. Note that when (7) is solved iteratively it is not necessary to form C^2 explicitly. This is important since normally C^2 is less sparse than C. In the next section we shall show that in the case when separation of variables if possible (7) can be solved efficiently using a direct method.

3. <u>The separable case</u>

We now consider the case when the elliptic operator is separable on a rectangular geometry in two dimensions:

$$\begin{cases} u_t = (a(x)u_x)_x + (b(y)u_y)_y + (c(x) + d(y))u, \\ \quad a(x), b(y) > 0, \text{ and } c(x), d(y) \leq 0 \text{ in } \Omega, \\ \Omega = \{(x,y) \mid x_* \leq x \leq x^*, \quad y_* \leq y \leq y^*\}, \end{cases} \tag{8}$$

If (8) is discretized on a rectangular grid (for simplicity we assume that the step lengths in both directions are equal to h) using the usual five point difference formula for selfadjoint operators [7, p. 363], and if the unknowns are ordered by vertical lines, the matrix C will be of the block tridiagonal form

$$C = \begin{bmatrix} B_1 & A_1 & & & O \\ A_1 & B_2 & A_2 & & \\ & A_2 & \ddots & \ddots & \\ & & \ddots & \ddots & A_{n-1} \\ O & & & A_{n-1} & B_n \end{bmatrix}, \tag{9a}$$

where the blocks are $q \times q$ matrices,

$$\begin{aligned} B_i &= B + b_i I, \quad B = B^T, \\ A_i &= a_i I, \end{aligned} \tag{9b}$$

and B is tridiagonal. The vector of unknowns is partitioned similarly

$$U = \begin{bmatrix} u_1 \\ u_2 \\ \vdots \\ u_n \end{bmatrix}, \qquad u_i = \begin{bmatrix} u_{i1} \\ u_{i2} \\ \vdots \\ u_{iq} \end{bmatrix},$$

where each u_i consists of the unknowns on a vertical line.

We shall now show how in this special case the normal equations (5) can be solved using the factorization algorithm of Section 2 and a variant of a method, which in the context of separable elliptic equations is called the matrix decomposition method [4].

Let Q_B denote the orthogonal matrix of eigenvectors of B and put

$$\Lambda = \text{diag}(\lambda_1, \lambda_2, \ldots, \lambda_q) = Q_B^T B Q_B. \tag{10}$$

Further let

$$Q = \text{blockdiag}(Q_B, Q_B, \ldots, Q_B).$$

Q is evidently orthogonal. Then the tranformation $Q^T C Q$ gives a block tridiagonal matrix, where all blocks are diagonal matrices

$$\hat{C} = Q^T C Q = \begin{bmatrix} \hat{B}_1 & A_1 & & & O \\ A_1 & \hat{B}_2 & A_2 & & \\ & A_2 & \ddots & \ddots & \\ & & \ddots & \ddots & A_{n-1} \\ O & & & A_{n-1} & \hat{B}_n \end{bmatrix} , \quad \hat{B}_i = \Lambda + b_i I, \; i = 1,2,\ldots,n. \tag{11}$$

Now if we multiply the normal equations (5) by Q^T and make the change of variables $\hat{U} = Q^T U$ we get

$$(\hat{D}^{2N} + \mu^2 I)\hat{U} = \hat{D}^N \hat{W} \tag{12}$$

where

$$\hat{D} = (I - \frac{r}{2}\hat{C})^{-1}(I + \frac{r}{2}\hat{C}) , \quad \hat{W} = Q^T W.$$

Using the same argument as in Section 2 we see that (12) can be solved for $\hat{U}$ in N steps, where in each step we must solve a system equivalent to (7):

$$(\alpha_\ell \hat{C}^2 + \beta\hat{C} + \gamma_\ell I)X = (I - \frac{r^2}{4}\hat{C}^2)Y. \tag{13}$$

But since $\hat{C}$ has the structure (11), $\hat{C}^2$ and the coefficient matrix in (13) are block pentadiagonal, where all blocks are diagonal matrices. If we reorder the unknowns by horizontal lines according to

$$X = \begin{bmatrix} x_1 \\ x_2 \\ \vdots \\ x_n \end{bmatrix} , \quad x_i = \begin{bmatrix} x_{i1} \\ x_{i2} \\ \vdots \\ x_{iq} \end{bmatrix} , \quad \bar{X} = \begin{bmatrix} \bar{x}_1 \\ \bar{x}_2 \\ \vdots \\ \bar{x}_q \end{bmatrix} , \quad \bar{x}_j = \begin{bmatrix} x_{1j} \\ x_{2j} \\ \vdots \\ x_{nj} \end{bmatrix} ,$$

we see that (13) is equivalent to q independent systems for the unknown vectors $\bar{x}_1, \bar{x}_2, \ldots, \bar{x}_q$

$$P_j^\ell \bar{x}_j = \bar{z}_j , \quad j = 1,2,\ldots,q , \tag{14}$$

where P_j^ℓ are pentadiagonal symmetric matrices and $\bar{z}_j$ are made up from the right hand side of (13) by the same reordering. An analogous reordering scheme in the block tridiagonal case is described in [4]. The derivation of the elements of P_j^ℓ is elementary but tedious and is therefore omitted. Since the coefficient matrix in (13) is positive definite so are the matrices P_j^ℓ and thus the systems (14) can be solved by a band-Cholesky algorithm.

We summarize the algorithm in the separable case.

1. Solve the eigenvalue problem for B.
2. $\hat{W} := Q^T W$.
3. $X := \hat{W}$;
   <u>for</u> ℓ := 1 <u>step</u> 1 <u>until</u> N <u>do</u>
   <u>begin</u>
   $Y := (I - \frac{r^2}{4} \hat{C}^2)X$;
   <u>for</u> j := 1 <u>step</u> 1 <u>until</u> q <u>do</u>
   $\bar{x}_j := (P_j^{\ell})^{-1} \bar{z}_j$;
   <u>end</u>;
4. U := QX.

If the coefficients b(y) and d(y) in (8) are constant, then the decomposition (10) is known explicitly and steps 2 and 4 can be computed very efficiently by the fast Fourier transform [6],[8]. If b(y) and d(y) are not constant we must compute the decomposition (10). Then steps 2 and 4 together require $2q^2n$ operations (multiplications and additions). Step 3 requires approximately 12qnN operations.

4. <u>Nonrectangular regions</u>

Suppose that the differential equation is defined as in (8) and that the region Ω is nonrectangular. Then suppose that Ω is imbedded into a rectangular region Ω_s (or splitted into several rectangles, the union of which is denoted Ω_s) [3]. Let C and C_s denote the discretization of the elliptic operator on Ω and Ω_s respectively, but assume that the extra unknowns introduced in the imbedding (or splitting) procedure are also accounted for in C so that both matrices are of the same order.

Now, since Ω is nonrectangular, we can not solve (7) using the algorithm of Section 3. Instead (7) can be solved iteratively. When a positive definite system of equations Hx = g is solved iteratively, the number of iterations needed for convergence is often proportional to the condition number of H, $\kappa(H)$. The rate of convergence can be accelerated if the equations and unknowns are scaled

$$G^{-T}HG^{-1}(G\,x) = G^{-T}g ,$$

where G is cosen so that the condition number $\kappa(G^{-T}HG^{-1})$ is smaller than $\kappa(H)$ and so that systems Gy = v can easily be solved. This technique is usually called <u>preconditioning</u> (see [1]).

One possible method for solving problems on nonrectangular regions is to use the factorization of Section 2 and to solve the systems (7) by the conjugate gradient with SSOR-preconditioning. This method can be implemented without forming C^2 explicitly [1].

However, it would be desirable to take advantage of the fact that in many cases the difference $C - C_s$ is small. It is not immediately obvious how the capacitance matrix technique [3] could be used efficiently here. Therefore we suggest a method, where (5) is solved iteratively using a preconditioning based on the matrix C_s.

Put

$$S_\mu^{-1} = (D_s^{2N} + \mu^2 I)^{-1} D_s^N, \quad D_s = (I - \frac{r}{2} C_s)^{-1}(I + \frac{r}{2} C_s).$$

If we make the change of variables $U = S_\mu^{-1} \hat{U}$ and multiply (5) by S_μ^{-1} we get the preconditioned system

$$S_\mu^{-1}(D^{2N} + \mu^2 I)S_\mu^{-1} \hat{U} = S_\mu^{-1} D^N W. \tag{15}$$

If C and C_s do not differ too much then $\kappa(S_\mu^{-1}(D^{2N} + \mu^2 I)S_\mu^{-1})$ will be small and the iterative solution of (15) e.g. by the conjugate gradient method will converge rapidly. Using this method a matrix-vector multiplication

$$S_\mu^{-1}(D^{2N} + \mu^2 I)S_\mu^{-1}X$$

must be performed in every iteration. $Y = S_\mu^{-1}X$ can be computed using the algorithm of Section 3. The multiplication $D^{2N}Y$ is equivalent to the forward solution of the parabolic equation 2N steps in time. Then in each step a system $(I - \frac{r}{2} C)X = Z$ must be solved. If the difference $C - C_s$ has low rank this can be done efficiently using a direct method for separable elliptic equations [13],[4] combined with the capacitance matrix method [3].

5. A numerical example

To illustrate the algorithm of Section 3 we solved the backwards heat equation on a square

$$\begin{cases} u_t = u_{xx} + u_{yy} \quad \text{in} \quad \Omega \times [0,2], \\ u = 0 \quad \text{on } \partial\Omega \times [0,2], \\ u(x,y,2) = e^{-4} \sin x \sin y, \\ \Omega = \{(x,y) \mid 0 \le x \le \pi,\ 0 \le y \le \pi\}. \end{cases}$$

This problem has the exact solution $u(x,y,0) = \sin x \sin y$. The equation was discretized in the space variables using the common 5-point approximation of the Laplacian on a square grid with $h = \pi/50$. Thus the number of unknowns was $49 \times 49 = 2401$. The number of time steps was 50.

The algorithm of Section 3 was programmed in Simula 67, but the most critical parts (e.g. the solution of pentadiagonal systems) were performed by Fortran subroutines. The program was run on a DEC-10 computer, which has the relative precision $\sim 1.5 \cdot 10^{-8}$.

Full machine precision was used in the computation of the vector of final values, and therefore relatively good accuracy could be obtained (see [2] and [5], where the stability of the problem is discussed).

With $\mu^2 = 10^{-7}$ the maximal absolute error was $0.7 \cdot 10^{-3}$. In Figure 1 we show a printout of the errors in the solution. Minus the logarithm (base 10) of the error rounded to the nearest power of ten is given for half the square (the upper half is omitted because of symmetry).

The CPU-time on the DEC-10 was approximately 64 seconds. We believe that it can be reduced significantly if the whole program is written in Fortran. The program did not take advantage of the fact that for this example the fast Fourier transform could have been used in Steps 2 and 4 of the algorithm.

6. Acknowledgement

I wish to thank Professor Gene H. Golub for valuable advice and Dr. Tommy Elfving for many helpful discussions.

References

1. Björck, Å., Elfving, T., Accelerated Projection Methods for Computing Pseudoinverse Solutions of Systems of Linear Equations, Linköping University, Department of Mathematics, Report LiTH-MAT-R-1978-5.
2. Buzbee, B.L., Carasso, A., On the Numerical Computation of Parabolic Problems for Preceding Times, Math. Comp. 27(1973), pp. 237-266.
3. Buzbee, B.L., Dorr, F.W., George, J.A., Golub, G.H., The Direct Solution of the Discrete Poisson Equation on Irregular Regions, SIAM J. Numer. Anal. 8(1971), pp. 722-736.
4. Buzbee, B.L., Golub, G.H., Nielson, C.W., On Direct Methods for Solving Poisson's Equations, SIAM J. Numer. Anal. 7(1970), pp. 627-656.
5. Carasso, A., Error Bounds in the Final Value Problem for the Heat Equation, SIAM J. Math. Anal. 7(1976), pp. 195-199.
6. Cooley, J.W., Tukey ,J.W., An Algorithm for Machine Calculation of Complex Fourier Series, Math. Comp. 19(1965), pp. 297-301.
7. Dahlquist, G., Björck, Å., Anderson, N., Numerical Methods, Prentice-Hall, Englewood Cliffs, 1974.
8. Hockney, R.W., A Fast Direct Solution of Poisson's Equation using Fourier Analysis, J. Assoc. Comput. Mach. 12(1965), pp. 95-113.
9. Miller, K., Least Squares Methods for Ill-posed Problems with a Prescribed Bound, SIAM J. Math. Anal. 1(1970), pp. 52-74.
10. Miller, K., Efficient Numerical Methods for Backward Solution of Parabolic Problems with Variable Coefficients, Improperly Posed Boundary Value Problems, Carasso & Stone (Eds.), Pitman Publishing, London, 1975.

11. Payne, L.E., Improperly Posed Problems in Partial Differential Equations, Society for Industrial and Applied Mathematics, Philadelphia, 1975.

12. Reid, J.K., On the Method of Conjugate Gradients for the Solution of Large Sparse Systems of Linear Equations, Large Sparse Sets of Linear Equations, J.K. Reid ed., Academic Press, New York, 1971, pp. 231-254.

13. Swartztrauber, P.N., A Direct Method for the Discrete Solution of Separable Elliptic Equations, SIAM J. Numer. Anal. 11(1974), pp. 1136-1150.

14. Swartztrauber, P.N., The Methods of Cyclic Reduction, Fourier Analysis and the FACR Algorithm for the Discrete Solution of Poisson's Equation on a Rectangle, SIAM Review 19(1977), pp. 490-501.

15. Tihonov, A.N., On the Solution of Incorrectly Posed Problems and the Regularization Method, Dokl. Akad. Nauk SSSR 151(1963), pp. 501-504.

```
4444444443333333333333333444444444444445555565565
4444444443333333333333333444444444444445555665565
4444444443333333333333333444444444444445556655565
4444444443333333333333343444444444444445556655565
4444444443333333333333343444444444444445556655565
4444444443333333333333343444444444444445556656565
4444444444333333333333343444444444444445556656565
4444444444333333333333344444444444444445556656565
4444444444333333333333444444444444444445556656565
5444444444334333333333444444444444444445556656565
5444444444344333334434444444444444444455556656565
5444444444344334433444444444444444444455556656566
5444444444444344334444444444444444444455557756566
5444444444444444444444444444444444444455557756566
5444444444444444444444444444444444444455557756566
5444444444444444444444444444444444444455557756566
5444444444444444444444444444444444444555557666566
5444444444444444444444444444444444445555557766566
5444444444444444444444444444444444455555557766566
5544444444444444444444444444444444455555557666666
5544444444444444444444444444444444455555557766666
5544444444444444444444444444444454555555558766666
5554444444444444444444444444444555555555567767676
5555554444444444444444444444555555555556567767676
6555555555555555555555555555555555555566668767677
```

Figure 1. Minus the logarithm (base 10) of the error rounded to the nearest power of ten. $\mu^2 = 10^{-7}$. The errors in the lower half of the square are shown.

Zur stochastischen Regularisierung nichtkorrekter Gleichungen in Hilberträumen

V. FRIEDRICH

Summary

Stochastical additional knowledge is used for regularization of ill-posed linear equations (2) in Hilbertspaces, which arise from indirect measurement problems. The unknown elements and the measurement error can be random or weak random variables. The optimal regularizator in the sense of mean square error (4) for Gaussian random elements is given. The minimal mean square error, which is available by linear bounded estimators, is approximated by α-families of bounded linear regularizators. Convergence theorems for decreasing measurement error are given without special assumptions about covariance operators for measurement errors.

1. Problemstellung und Ergebnisse für Gaußsche Elemente

Zur Schätzung des Objektzustandes x aus einer Meßgröße z und der Beziehung

$$z = Ax + y \tag{1}$$

mit y als Meßfehler sind im Falle nicht stetig invertierbarer Operatoren A (z.B. Fredholmscher Integraloperatoren, Faltungsoperatoren) zusätzliche Kenntnisse (bzw. Annahmen) über x und y notwendig, die zur Regularisierung der nichtkorrekten Gleichung

$$Ax = z \tag{2}$$

benutzt werden [8]. Für die Beurteilung von Näherungsverfahren zur Bestimmung von x ist angesichts des zufälligen Charakters der Meßfehler eine Mittelung über diese sinnvoll. Kann auch der Zuatand x selbst als Realisierung eines zufälligen Vektors (Elements) ξ aufgefaßt werden, so ist eine zweite Mittelung entsprechend der Wahrscheinlichkeit des Auftretens von x für ein Beurteilungskriterium in Betracht zu ziehen.

Bei der Übertragung bekannter Resultate zu besten Schätzvorschriften bezüglich des Kriteriums (4) auf die Betrachtung von (1) in Hilberträumen erwachsen Schwierigkeiten daraus, daß bereits im Falle Gaußverteilter zufälliger Elemente ξ und η in Hilberträumen letztere nukleare Kovarianzoperatoren besitzen. Die Nuklearität des Kovarianzoperators $C = \operatorname{cov}\eta$ des Meßfehlers reicht aber für eine Regularisierung von A^{-1} zu einem stetigen Operator nicht aus.

Diese Schwierigkeit kann formal durch einen Übergang zu schwachen zufälligen Meßfehlern [3] mit invertierbarem Kovarianzoperator überwunden werden, sie kann aber auch durch eine zusätzliche formale Regularisierung des nuklearen Kovarianzoperators C umgangen werden. Beide Möglichkeiten werden nachfolgend bezüglich ihrer Genauigkeit verglichen.

Sei A eine lineare Abbildung aus R^n in R^m. Seien ξ, η, $\zeta = A\xi + \eta$ zufällige Vektoren und bezeichnet $\mathcal{R}(z)$ eine Vorschrift, nach der z eine Näherung $\tilde{x} = \mathcal{R}(z)$ für x zugeordnet wird, so minimiert

$$\tilde{x}_{st} = \bar{x} + BA^*(ABA^* + C)^{-1}(z - A\bar{x}) \tag{3}$$

den mittleren quadratischen Fehler

$$v(\mathcal{R}) = E_{(\xi,\eta)}(\|\mathcal{R}(A\xi+\eta) - \xi\|^2) \tag{4}$$

über allen linearen Vorschriften $\mathcal{R}(z) = Rz + q$, wobei ξ und η als stochastisch unabhängige Vektoren vorausgesetzt werden, E für die Erwartungswertbildung steht, $E_{(\xi,\eta)} = E_\xi E_\eta$ genutzt und $E\eta = 0$ angenommen wird und die Bezeichnungen $\bar{x} = E\xi$, $B = \mathrm{cov}\,\xi$, $C = \mathrm{cov}\,\eta$ verwendet werden.

Sind ξ und η normalverteilte Vektoren, so stellt (3) bezüglich des Kriteriums (4) die beste Schätzvorschrift überhaupt dar, und das Verhältnis von trace $BA^*(ABA^*+C)^{-1}AB$ zur a-priori-Kenntnis $E_\xi(\|\xi-\bar{x}\|^2) = \mathrm{tr}\, B$ charakterisiert den mittleren erreichbaren Informationsgewinn durch Auswertung der Meßdaten unter Ausnutzung von (1) und der a-priori-Daten $\bar{x}$ und B. In jedem konkreten Anwendungsfall ist die Gegenüberstellung dieser Größen zweckmäßig, um die Wirksamkeit einer mathematischen Bearbeitung der Meßdaten beurteilen zu können.

Betrachtet man ξ und η als zufällige Elemente in Hilberträumen X und Y, so ist eine formale Anwendung von (3) nicht möglich, da Kovarianzoperatoren zufälliger Elemente mit der Eigenschaft $E_\eta(\|\eta\|^2) < \infty$ nicht stetig invertierbar sind und folglich auch ABA^*+C nicht stetig invertiert werden kann. Für Gaußsche zufällige Elemente ξ und η besitzt das Definitionsgebiet des Operators $(ABA^*+C)^{-1}$ nur das Maß 0 (bezüglich der Verteilung von ζ) [9]. In [5] wurde für Gaußsche Elemente ξ, η eine Erweiterung $\tilde{R}$ des Operators $R = BA^*(ABA^*+C)^{-1}$ angegeben, die für fast alle z (bezüglich des Maßes von ζ) anwendbar ist und die bestmögliche Schätzung von x bezüglich (4) darstellt. Zur Konstruktion von $\tilde{R}$, welche i.a. nicht mit einer Abschließung von R zusammenfällt, wurden Karhunen-Loeve-Zerlegungen nach den Eigensystemen der nuklearen Operatoren B und ABA^*+C benutzt.

Da $\tilde{R}$ i.a. ein unbeschränkter Operator ist, entsteht die Frage nach solchen beschränkten Operatoren R_α, die für $\alpha \longrightarrow 0$ die gleiche mittlere Genauigkeit wie $\tilde{R}$ garantieren, und nach dem durch eine zusätzliche Regularisierung eintretenden mittleren Genauigkeitsverlust.

Die gleichen Fragen entstehen auch bei der näherungsweisen Bestimmung linearer Funktionale (f,x) über dem unbekannten Zustand x. In den oben beschriebenen Fällen ist

$$\hbar(z) = (f,\tilde{x}_{st}) \tag{5}$$

die im Mittel beste lineare Schätzung bzw. die im Mittel beste Schätzung für (f,x) im Sinne des Kriteriums

$$v_f(\hbar) = E_{(\xi,\eta)}(|\hbar(A\xi+\eta) - (f,\xi)|^2) . \tag{6}$$

Die Wirksamkeit einer Funktionalschätzung erhält man durch Vergleich von (6) mit $E_\xi(\,|\,(f,\xi-\bar{x})\,|\,^2) = (f,Bf)$.

Nichtlineare Schätzvorschriften sind den linearen Vorschriften (3) bzw. (5) nur vorzuziehen, wenn wesentliche Abweichungen von der Normalverteilung vorliegen und wenn die Vorschriften in Beziehung zur nichtlinearen Regression bei bekannten Verteilungen für ξ und η gebracht werden können.

2. Beste lineare Schätzung für beliebige zufällige Elemente mit zweiten Momenten, ihre Approximation durch beschränkte Regularisatoren

Verzichtet man auf die Voraussetzung, daß ξ und η Gaußsche zufällige Elemente sind, so kann mit der in [5] benutzten Beweistechnik nicht mehr gezeigt werden, daß auch in diesem Fall die Erweiterung R für fast alle z anwendbar ist. Das Infimum der Schätzfehler (4) bzw. (6) über der Gesamtheit aller linearen stetigen Vorschriften $Rz + q$ bzw. $(h,z) + a$ ist jedoch leicht zu bestimmen, wobei die Forderung nach der Nuklearität von C - und für Funktionalschätzungen auch bezüglich B - fallengelassen werden können, so daß auch schwache Zufallsgrößen in die Betrachtung einbezogen werden können. Zur Vereinfachung sei im weiteren $\ker C = \{0\}$ vorausgesetzt.

Die günstigste Wahl von q bzw. a für vorgegebenes R bzw. h ist $a = (f,\bar{x}) - (h,A\bar{x})$ bzw. $q = \bar{x} - RA\bar{x}$, so daß

$$v_f(h) = (A^*h - f,\ B(A^*h - f)) + (h,\ Ch) \tag{7}$$

$$v(R) = \mathrm{tr}\,\{(RA - I)B(RA - I)^* + RCR^*\} \tag{8}$$

verbleibt. Unter Verwendung des beschränkten Operators $T = (ABA^*+C)^{-1/2}AB^{1/2}$ erhält man aus (7) und (8)

$$v_f(h) = (ABA^*+C)^{\frac{1}{2}}h - TB^{\frac{1}{2}}f\ ^2 + (f,(B-B^{\frac{1}{2}}T^*TB^{\frac{1}{2}})f) \tag{9}$$

$$v(R) = \mathrm{tr}\{(R(ABA^*+C)^{\frac{1}{2}}-B^{\frac{1}{2}}T^*)((ABA^*+C)^{\frac{1}{2}}R^*-TB^{\frac{1}{2}})\} + \mathrm{tr}\{B-B^{\frac{1}{2}}T^*TB^{\frac{1}{2}}\}. \tag{10}$$

Die an zweiter Stelle stehenden Ausdrücke sind jeweils von h bzw. R unabhängig.

Wird der Meßfehler als schwaches zufälliges Element mit positiv definiter Kovarianzoperator vorausgesetzt, so minimiert der lineare stetige Operator. $R_0 = BA^*(ABA^*+C)^{-1}$ bzw. das Funktional $h_0 = R_0^* f$ die Funktionale (8) bzw. (7) und eine weitere Regularisierung ist nicht erforderlich. Andernfalls erhält man über die Familie von Regularisatoren

$$R_\alpha = BA^*(ABA^*+C+\alpha I)^{-1} \tag{11}$$

und den folgenden Hilfssatz

Für eine auf $[0,\infty)$ nichtfallende Funktion beschränkter Variation g mit s = 0 als Stetigkeitspunkt gilt

$$\int_0^\infty \frac{\alpha^2}{(s+\alpha)^2}\, dg(s) \xrightarrow[\alpha\to 0]{} 0$$

den die Auswirkungen der formalen Regularisierung in (11) beschreibenden

<u>Satz 1:</u>

$$\lim_{\alpha\to 0} v_f(R_\alpha^* f) = \inf_h v_f(h) , \tag{12}$$

und

$$\lim_{\alpha\to 0} v(R_\alpha) = \inf_R v(R) , \quad \text{wenn } \operatorname{cov}\xi = B \text{ nuklear.} \tag{13}$$

Der Beweis nutzt die Identität

$$(ABA^*+C)^{\frac{1}{2}}R_\alpha^* - TB^{\frac{1}{2}} = -(ABA^*+C)^{-\frac{1}{2}}\alpha I(ABA^*+C+\alpha I)^{-1}(ABA^*+C)^{\frac{1}{2}}TB^{\frac{1}{2}} ,$$

die Zerlegung der Einheit E_s des Operators ABA^*+C und

$$\left\|(ABA^*+C)^{\frac{1}{2}}R_\alpha^* f - TB^{\frac{1}{2}}f\right\|^2 = \int_0^\infty \frac{\alpha^2}{(s+\alpha)^2} d(\|E_s(TB^{\frac{1}{2}}f)\|^2)$$

bzw.

$$\operatorname{tr}\left\{(R_\alpha(ABA^*+C)^{\frac{1}{2}} - B^{\frac{1}{2}}T^*)((ABA^*+C)^{\frac{1}{2}}R_\alpha^* - TB^{\frac{1}{2}})\right\} =$$

$$= \int_0^\infty \frac{\alpha^2}{(s+\alpha)^2}\, d\left\{\sum_{i=1}^\infty \|E_s(TB^{\frac{1}{2}}u_i)\|^2\right\}$$

mit $\{u_i\}$ als einem beliebigem vollständigen Orthonormalsystem in X.

Die Vorschriften $R_\alpha^* f$ und R_α können auch als Lösungen der regularisierten Variationsaufgaben

$$(h,(ABA^*+C)h) - 2(h,ABf) + \alpha\|h\|^2 \quad \text{bzw.}$$

$$\operatorname{tr}\{R(ABA^*+C)R^* - 2RAB + \alpha RR^*\}$$

aufgefaßt werden, die die i.a. nichtkorrekten Variationsaufgaben (7) und (8) ersetzen. Andere für die Schätzung von x bzw. (f,x) geeignete Vorschriften können natürlich auch aus einer (unvollständigen) Minimierung der Variationsaufgaben (7) bzw. (8) mit anderen Methoden erhalten werden (z.B. mit Iterationsverfahren oder Gradientenmethoden [1]).

Aussagen über eine bestimmte Konvergenzordnung der Differenzen

$$v_f(R_\alpha^* f) - \inf_h v_f(h) \qquad \text{bzw.} \qquad v(R_\alpha) - \inf_R v(R) \tag{14}$$

gegen Null, d.h. über die Geschwindigkeit, mit der der durch die zusätzliche Regularisierung von C auftretende mittlere Genauigkeitsverlust mit $\alpha \longrightarrow 0$ abnimmt, erfordern weitere Voraussetzungen. So gilt z.B.: Können eine Zahl $r > 0$ und ein Element w bzw. ein Hilbert-Schmidt-Operator W gefunden werden, so daß

$$ABf = (ABA^*+C)^{\frac{1}{2}+r}w \qquad \text{bzw.} \qquad AB = (ABA^*+C)^{\frac{1}{2}+r}W$$

erfüllt ist, so besitzen die Differenzen (14) die Konvergenzordnung $\mathcal{O}(\alpha^{2r})$.

3. Konvergenzuntersuchungen

Deterministisch motivierte Näherungsverfahren für die Gleichung (2) werden i.a. durch Konvergenzsätze gestützt. Analoge Aussagen können auch für die stochastisch begründeten Verfahren erhalten werden, wenn für den Kovarianzoperator des Meßfehlers $C = \sigma^2 N$ angenommen und $\sigma \longrightarrow 0$ betrachtet wird. [2]. Der Meßfehler kann hierbei auch als schwaches zufälliges Element vorausgesetzt werden. O.B.d.A. sei $N \leq I$.

<u>Satz 2:</u> Gilt für den Bildbereich $R_{B^{\frac{1}{2}}}$ des Operators $B^{\frac{1}{2}}$ die Beziehung $\ker(A) \cap R_{B^{\frac{1}{2}}} = \{0\}$, so ist

$$\inf_h v_f(h) \leq v_f(R_\sigma^* f) \xrightarrow[\sigma \to 0]{} 0 \qquad \text{und}$$

$$\inf_R v(R) \leq v(R_\sigma) \xrightarrow[\sigma \to 0]{} 0, \quad \text{falls B nuklear},$$

wobei $R_\sigma = BA^*(ABA^*+\sigma^2 I)^{-1}$ gewählt ist.

Zum Beweis werden die Operatorenungleichung

$$(I-R_\sigma A)B(I-R_\sigma A)^* + R_\sigma C R_\sigma^* \leq \sigma^2 B^{\frac{1}{2}}(B^{\frac{1}{2}}A^*AB^{\frac{1}{2}} + \sigma^2 I)^{-1}B^{\frac{1}{2}}$$

und ein Analogon des oben genannten Hilfssatzes benutzt sowie die Zerlegung der Einheit des Operators $B^{\frac{1}{2}}A^*AB^{\frac{1}{2}}$ verwendet.

Die Geschwindigkeit, mit der $v_f(R_\sigma^* f)$ gegen Null geht, hängt wesentlich vom betrachteten Funktional ab. Im speziellen Fall eines A-korrekten Funktionals f erhält man aus der Darstellung $f = A^*w$

$$v_f(R_\sigma^* f) \leq \sigma^2 \|w\|^2 \quad .$$

Für zufällige Meßfehler mit nuklearem Kovarianzoperator $C = \sigma^2 N$ ermöglicht die bei der deterministischen Regularisierung häufig getroffene Annahme einer Quelldarstellung des zu bestimmenden Elements auch Aussagen über die Konvergenzgeschwindigkeit des mittleren Schätzfehlers gegen Null für $\sigma \longrightarrow 0$. So erlaubt z.B. eine "zufällige Quelldarstellung" $\xi = A^*\omega$ mit $E\omega = 0$ und nuklearem Kovarianzoperator $W = \mathrm{cov}\,\omega$ über die Ungleichung $\|(A^*A+\alpha I)^{-\frac{1}{2}}A^*\| \leq 1$ für $\alpha = \sigma$ die Abschätzung

$$v((A^*A+\alpha I)^{-1}A^*) \leq \frac{1}{\alpha}(\alpha^2 \text{ tr } W + \sigma^2 \text{ tr } N) = \sigma(\text{tr } W + \text{tr } N) .$$

Analog führt eine Quelldarstellung $\xi = A^*A\tilde{\omega}$ unter entsprechenden Voraussetzungen an $\tilde{\omega}$ auf

$$v((A^*A+\sigma^{\frac{2}{3}}I)^{-1}A^*) \leq \mathcal{O}(\sigma^{\frac{2}{3}}) .$$

4. Bemerkungen zu Minimax-Schätzungen von Funktionalen und Elementen

Gehört x einer Menge $M_L = \{x \in X \mid \|Lx\| \leq 1\}$ zulässiger Elemente an, ohne daß eine Wahrscheinlichkeitsverteilung für x motiviert werden kann, so führt die Minimax-Schätzung

$$\max_{x \in M_L} E_\eta(|(h, Ax+\eta) - (f,x)|^2) \longrightarrow \min$$

zu analogen besten linearen Regularisatoren [7] .

Beste lineare Vorschriften für die Elementschätzung ergeben dagegen in den einfachsten noch auswertbaren Fällen Vorschriften, die als endlichdimensionale Operatoren stark oszillierende Anteile in den Näherungen auch völlig ausfiltern können, wobei eine gewisse Analogie zu der in [6] beschriebenen Vorgehensweise hervortreten.

Literatur

[1] BALAKRISHNAN,A.V. (1969). Communication theory. Mc.Graw Hill.

[2] BALAKRISHNAN,A.V. (1976), Applied funktional analysis. Springer-Verlag, New York - Heidelberg - Berlin.

[3] BENSOUSSAN,A. (1971), Filtrage optimal des systemes lineaires. Dunod, Paris.

[4] FRANKLIN,J.N. (197o). Well-posed stochastic extensions of ill-posed linear problems. J.Math.Anal.and Appl., 31, 682-716.

[5] FRIEDRICH,V. und UHLIG,A. (1978). Zur stochastischen Regularisierung linearer Gleichungen in Hilberträumen. Beiträge zur Numerischen Mathematik, 7, 33-48.

[6] HANSON,R.J. (1971). A numerical method for solving Fredholm integral eqations of the first kind using singular values. SIAM J. Num. Anal. 8, 616-622.

[7] PETROV,A.P. (1967). Ocenki linejnych funkcionalov dlja rešenija nekotorych obratnych zadač. Ž.vyč.mat. i mat.fiz. 7, 648-654

[8] TICHONOV,A.N. i ARSENIN,B.J. (1974). Metody rešenija nekorrektnych zadač. Nauka, Moskva.

[9] VACHANIJA.N.N. (1971). Verojatnostnyje raspredelenija v linejnych prostranstvach. Mecnereba, Tbilissi.

Prof. Dr. V. Friedrich, Sektion Mathematik, Technische Hochschule, DDR 901 Karl-Marx-Stadt

QUASI-LINEARIZATION AND THE IDENTIFICATION OF PARAMETERS IN PARTIAL DIFFERENTIAL EQUATIONS OF PARABOLIC TYPE

ILIO GALLIGANI(*)

ABSTRACT

Identification problems are fundamental problems of science, engineering and medicine. In such problems, characteristic properties of a system have to be determined by an "observed behavior" of the system itself. For example, in management and control studies of water resources, it is required to identify the transmissivity parameter of an aquifer when the underground water level has been measured and recorded at a number of wells drilled into the aquifer.

In this paper we describe the quasi-linearization method for solving the state-identification problem associated with a set of partial differential equations of parabolic type. For problems in two space dimensions, we analyse the main computational aspects of the method, especially when the parabolic equations exhibit the phenomena of "stiffness" and "numerical dispersion". Besides we illustrate an application of this method to the problem of identifying the transmissivity parameter in an underground water reservoir. This parameter-identification problem is improperly posed. Finally, we present the results of numerical experiments carried out on some test problems.

1. THE QUASI-LINEARIZATION METHOD FOR THE STATE IDENTIFICATION PROBLEM.

Let t denote time, $t \in [0,t^*]$, and let Ω be an open bounded and simply connected domain in $\mathbb{R}^m$ with piecewise smooth boundary Γ. The exterior normal ν shall exist at every point of Γ. Points of $\mathbb{R}^m$ are denoted by $x = (x_1 x_2 \ldots x_m)$.

Let be given the following set of n parabolic differential equations

$$(1) \quad y_t = \sum_{i=1}^{m} (A_i y_{x_i})_{x_i} + f \qquad \text{in } \Omega \times (0,t^*]$$

with the initial-boundary conditions:

(*) Istituto per le Applicazioni del Calcolo-CNR, Roma

(2) $y(0\ x) = h(x)$ in Ω

(3) $y(t\ x) = g(t\ x)$ on $\Gamma \times [0,t^*]$

where $y \equiv y(t\ x) = (y_k(t\ x))$, $f \equiv f(t\ x\ y) = (f_k(t\ x\ y_1(t\ x)\dots y_n(t\ x)))$, $h \equiv h(x) = (h_k(x))$ and $g \equiv g(t\ x) = (g_k(t\ x))$, $k = 1,2\dots n$, are real-valued sufficiently smooth vector functions of n components and $A_i \equiv A_i(t\ x\ y) = \mathrm{diag}\{a_{ik}(t\ x\ y)\}$ is a $n \times n$ diagonal matrix whose diagonal element $a_{ik}(t\ x\ y)$, $k=1,2,\dots,n$, is a real-valued sufficiently smooth function.
We have introduced the abbreviated notation for the derivatives of these vector functions at some point (t,x):

$$y_t = \left(\frac{\partial y_k(t\ x)}{\partial t}\right),\quad y_{x_i} = \left(\frac{\partial y_k(t\ x)}{\partial x_i}\right),\quad (A_i y_{x_i})_{x_i} = \left(\frac{\partial}{\partial x_i}\left(a_{ik}(t\ x\ y)\frac{\partial y_k(t\ x)}{\partial x_i}\right)\right),$$

$k = 1,2\dots n$ and $i = 1,2\ \dots m$.

We assume that the system (1)-(3) has an unique regular solution $y(t\ x)$. This solution depends on the initial-boundary data h and g; in order to display this dependence explicitly, we write $y(t\ x) = y(t\ x;u)$, where $u = (t\ g)$.

The *state-identification problem* associated with the system (1)-(3) is that of minimizing the functional

$$(4)\quad J(u) = \int_0^{t^*}\int_\Omega \|y(t\ x;u) - \tilde{y}(t\ x)\|^2 dtdx + \int_\Omega \|h(x) - \tilde{h}(x)\|^2 dx + \int_0^{t^*}\int_\Gamma \|g(t\ x) - \tilde{g}(t\ x)\|^2 dtds$$

where $\tilde{y}(t\ x)$, $\tilde{h}(x)$ and $\tilde{g}(tx)$ are measurements of $y(t\ x)$, $h(x)$ and $g(t\ x)$, respectively.

We assume that these measurements are sufficiently smooth maps obtained, for example, by deterministic or stochastic interpolators.

In order to solve this non linear identification problem we use the *quasi-linearization method*, which reduces this problem to a sequence of linear state identification problems. This gives rise to the following iterative process.
At each outer iteration $\mu = 1,2,\dots$, compute a vector $u_p^{(\mu)}$ which minimizes the functional:

$$(5)\quad J(u^{(\mu)}) = \int_0^{t^*}\int_\Omega \|y^{(\mu)}(t\ x;u^{(\mu)}) - \tilde{y}(t\ x)\|^2 dtdx + \int_\Omega \|v^{(\mu)}(x) - \tilde{h}(x)\|^2 dx \\ \int_0^{t^*}\int_\Gamma \|w^{(\mu)}(t\ x) - \tilde{g}(t\ x)\|^2 dtds$$

along a trajectory $y^{(\mu)}(t\,x;u^{(\mu)})$ of the linear system:

$$(6)\quad y_t^{(\mu)} = \sum_{i=1}^{m} (A_i^{(\mu-1)} y_{x_i}^{(\mu)})_{x_i} + \sum_{i=1}^{m} (B_i^{(\mu-1)} y^{(\mu)})_{x_i} + C^{(\mu-1)} y^{(\mu)} + s^{(\mu-1)} \qquad \text{in } \Omega \times (0,t^*]$$

$$(7)\quad y^{(\mu)}(0\;x) = v^{(\mu)}(x) \qquad \text{in } \Omega$$

$$(8)\quad y^{(\mu)}(t\;x) = w^{(\mu)}(t\;x) \qquad \text{on } \Gamma \times [0,t^*]$$

where $A_i^{(\mu-1)} = \mathrm{diag}\{a_{ik}^{(\mu-1)}\}$, $B_i^{(\mu-1)} = (b_{ikj}^{(\mu-1)})$ and $C = (c_{kj}^{(\mu-1)})$ are $n \times n$ matrices with elements

$$a_{ik}^{(\mu-1)} = a_{ik}(t\,x\,y^{(\mu-1)}),\quad b_{ikj}^{(\mu-1)} = \frac{\partial y_k^{(\mu-1)}}{\partial x_i} \cdot \frac{\partial a_{ik}^{(\mu-1)}}{\partial y_j} \quad \text{and}$$

$$c_{kj}^{(\mu-1)} = \frac{\partial f_k^{(\mu-1)}}{\partial y_j}\;;\; f_k^{(\mu-1)} = f_k(t\,x\,y^{(\mu-1)}),\quad s_k^{(\mu-1)} \equiv s_k^{(\mu-1)}(t\;x) =$$

$$= f_k^{(\mu-1)} - \sum_{i=1}^{m} (B_i^{(\mu-1)} y^{(\mu-1)})_{x_i} - C^{(\mu-1)} y^{(\mu-1)} \quad \text{and } u^{(\mu)} = (v^{(\mu)} w^{(\mu)}).$$

The initial guess $y^{(0)}(t\;x)$ of the vector $y(t\;x)$ is supposed to be given everywhere.

Under weak assumptions, the functional (5) may be represented as a quadratic functional on a suitable (convex and closed) Hilbert space $\mathcal{H}$. Thus, the functional (5) has an unique minimum $u_p^{(\mu)} = (v_p^{(\mu)} w_p^{(\mu)})$, which satisfies the characterization equation:

$$(9)\quad (u^{(\mu)},\; \nabla J(u_p^{(\mu)}))_{\mathcal{H}} = 0 \qquad \text{for all } u^{(\mu)} \in \mathcal{H}$$

where

$$(10)\quad \nabla J(u_p^{(\mu)}) = \begin{pmatrix} z^{(\mu)}(0\,x\,;u_p^{(\mu)}) + v_p^{(\mu)} - \tilde{h} \\ -\dfrac{\partial z^{(\mu)}(t\,x;u_p^{(\mu)})}{\partial \nu} + w_p^{(\mu)} - \tilde{g} \end{pmatrix}$$

is the gradient of $J(u^{(\mu)})$ in $u_p^{(\mu)}$ ($(\cdot,\cdot)_{\mathcal{H}}$ denotes inner product in $\mathcal{H}$). The k-th component ($k=1,2,\dots,n$) of $\dfrac{\partial z^{(\mu)}}{\partial \nu} \equiv \dfrac{\partial z^{(\mu)}(t\,x;u_p^{(\mu)})}{\partial \nu}$ is $\sum_{i=1}^{m} a_{ik}^{(\mu-1)} \dfrac{\partial z_k^{(\mu)}}{\partial x_i} \cos(\nu, x_i)$ and $z^{(\mu)} \equiv z^{(\mu)}(t\,x) \equiv z^{(\mu)}(t\,x;u_p^{(\mu)})$ is the solution of the adjoint system to (6)-(8):

$$(11)\quad -z_t^{(\mu)} = \sum_{i=1}^{m} (A_i^{(\mu-1)} z_{x_i}^{(\mu)})_{x_i} + \sum_{i=1}^{m} ((B_i^{(\mu-1)})^T z^{(\mu)})_{x_i} + (C^{(\mu-1)})^T z^{(\mu)}$$
$$+ y^{(\mu)}(t\, x; u_p^{(\mu)}) - \tilde{y}(t\, x) \qquad \text{in } \Omega \times [0,t^*)$$

$$(12)\quad z^{(\mu)}(t^* x) = 0 \qquad \text{in } \Omega$$

$$(13)\quad z^{(\mu)}(t\, x) = 0 \qquad \text{on } \Gamma \times [0,t^*]$$

$(B_i^{(\mu-1)})^T$ and $(C^{(\mu-1)})^T$ are the tranposes of the matrices $B_i^{(\mu-1)}$ and $C^{(\mu-1)}$, respectively.
It is assumed that $z^{(\mu)}$ is sufficiently differentiable.

Now, at each outer iteration μ, the vector $u_p^{(\mu)}$ is determined as the limit point of a weakly convergent sequence generated by the following *conjugate gradients method* $(\lambda = 0,1\ldots)$:

$$(14)\quad \begin{cases} u^{(\mu\,\lambda+1)} = u^{(\mu\,\lambda)} + \bar{\rho}_\lambda q^{(\mu\,\lambda)} \\ q^{(\mu\,\lambda+1)} = -\nabla J(u^{(\mu\,\lambda+1)}) + \bar{\sigma}_\lambda q^{(\mu\,\lambda)} \end{cases}$$

where $u^{(\mu\,0)} = u_p^{(\mu-1)}$, $q^{(\mu\,0)} = -\nabla J(u_p^{(\mu-1)})$, $\bar{\sigma}_\lambda = \|\nabla J(u^{(\mu\,\lambda+1)})\|^2 / \|\nabla J(u^{(\mu\,\lambda)})\|^2$

and $\bar{\rho}_\lambda$ is chosen so that $J(u^{(\mu\,\lambda)} + \bar{\rho}_\lambda q^{(\mu\,\lambda)}) = \inf_{\rho_\lambda} J(u^{(\mu\,\lambda)} + \rho_\lambda q^{(\mu\,\lambda)})$.

It is possible to obtain an explicit expression for $\bar{\rho}_\lambda$.
Here the superscript λ stands for inner iteration number. This iterative process realizes the infimum $u_p^{(\mu)}$ of $J(u^{(\mu)})$ when $\|\nabla J(u^{(\mu\,\lambda+1)})\| \leq \text{eps}$ for a given eps > 0.

2. A PARAMETER ESTIMATION PROBLEM OF GROUNDWATER HYDROLOGY.

The above quasi-linearization method has been adapted to the problem of identifying the transmissivity parameter in an underground water reservoir.
We shall consider the case of a two dimensional inhomogeneous isotropic aquifer extending over a region $\Omega \in \mathbb{R}^2$ with boundary $\Gamma = \Gamma_0 + \Gamma_1$. The equation of flow valid for this aquifer is:

$$(15)\quad \frac{\partial p(t\, x)}{\partial t} = \operatorname{div}(\sigma(x) \operatorname{grad} p(t\, x)) + \theta(t\, x)$$

where $\sigma(x)$ denotes the unknown transmissivity parameter, $\theta(t\, x)$ is a known function related to the production rate of the wells drilled into the aquifer and $p(t\,x)$ is the piezometric head. The initial-boundary conditions associated with (15) are:

(16) $\quad p(0\,x) = h(x) \qquad \text{in } \Omega$

(17) $\quad p(t\,x) = g(t\,x) \qquad \text{on } \Gamma \times [0,t^*]$

Let be given measured values $\tilde{p}(t\,x)$, $\tilde{h}(x)$ and $\tilde{g}(t\,x)$ of the functions $p(t\,x)$, $h(x)$ and $g(t\,x)$, respectively; then it is required to determine the functions $\sigma(x)$, $h(x)$ and $g(t\,x)$ which, under the constrains (15)-(17), will "best" fit the measured values.

In order to solve this parameter identification problem, it is convenient to think the parameter $\sigma(x)$ as a dependent variable and to put

$$y(t\,x) \equiv \begin{pmatrix} y_1(t\,x) \\ y_2(t\,x) \end{pmatrix} = \begin{pmatrix} p(t\,x) \\ \sigma(x) \end{pmatrix}$$

Thus the equation (15) may be written in the form:

$$(18) \qquad \begin{cases} \dfrac{\partial y_1(t\,x)}{\partial t} = \operatorname{div}(y_2(t\,x)\operatorname{grad} y_1(t\,x)) + \vartheta(t\,x) \\ \dfrac{\partial y_2(t\,x)}{\partial t} = \varepsilon\,\Delta_2\, y_2(t\,x) \end{cases} \qquad \text{in } \Omega \times (0,t^*]$$

$$(19) \qquad \begin{cases} y_1(0\,x) = h(x) \\ y_2(0\,x) = \sigma(x) \end{cases} \qquad \text{in } \Omega$$

(20) $\quad y_1(t\,x) = g(t\,x) \qquad \text{on } \Gamma \times [0,t^*]$

where $\varepsilon > 0$ is allowed to go to zero.
By using the quasi-linearization method, we have finally to find a sequence of vectors $\{u_p^{(\mu)}\}$, $\mu = 1,2\ldots$ such that each $u_p^{(\mu)}$ extremizes the functional:

$$J(u^{(\mu)}) = \int_0^{t^*}\int_\Omega \left\{\left|y_1^{(\mu)} - \tilde{p}\right|^2 + \left|y_2^{(\mu)} - y_1^{(\mu-1)}\right|^2\right\} dt\,dx +$$

$$+\int_\Omega \left\{\left|v_1^{(\mu)} - \tilde{h}\right|^2 + \left|v_2^{(\mu)} - v_2^{(\mu-1)}\right|^2\right\} dx + \int_0^{t^*}\int_\Gamma \left\{\left|w_1^{(\mu)} - \tilde{g}\right|^2 + \left|w_2^{(\mu)} - w_2^{(\mu-1)}\right|^2\right\} dt\,ds$$

along a trajectory $y^{(\mu)} \equiv y^{(\mu)}(t\,x)$ of a linear parabolic system.

In order to evaluate the computational performance of this method some numerical studies have been carried out.

The domain Ω is composed by a rectangle R with eventually some lateral rectangular subregions removed: the sides of R are parallel to the axes x_1 and x_2. In the removed subregions "small" values (10^{-6} was used) of $\sigma(x)$ and $\vartheta(t\,x)$ are assigned. The parameter ε is taken $\varepsilon = 10^{-3}$. On R a rectangular network of 21×19 mesh-points is overlapped.

Two studies concern the estimation of $\sigma(x)$ in a problem with $p(t\,x)$, $\vartheta(t\,x)$, $h(x)$, $g(t\,x)$ and $\sigma(x)$ given exactly: $p(t\,x)$ and $\sigma(x)$ are slowly varying functions with values included into (1.,2.) and (0.5,1.) respectively. In the first study $\tilde{p}(t\,x) = p(t\,x)$, $\tilde{h}(x) = h(x)$, $\tilde{g}(t\,x) = g(t\,x)$ and $y^{(o)}(t\,x)$ is obtained by adding a white noise to the truth values of $p(t\,x)$ and $\sigma(x)$.

In the second study the previous values of $y^{(o)}(t\,x)$ are used but the measured values of $p(t\,x)$ are changed by subtracting a white noise from the exact values.

In both studies $\sigma(x)$ was determined with three significant digits after 3 and 5 outer iterations (and 7(=4+2+1) and 13=(6+3+2+1+1) inner iterations) in 280 and 620 sec. of IBM 370/168 computer time, respectively.

A further study is related to the numerical simulation of a three-year (15-days time steps) pumping test conducted on a realistic aquifer with 48 wells distributed in a domain composed by a square of 12 Km width with some lateral rectangular subregions removed. We suppose that the history of the production rate of the aquifer is known and, at a fixed time $t = 0$, the piezometric and transmissivity maps are available: it means that the functions $\vartheta(t\,x)$, $h(x)$ and $\sigma(x)$ are assigned. We put $g(t\,x) = 0$. With these data we simulate the measures $\tilde{p}(t\,x)$ of the piezometric head by integrating the system (15)-(17). Now, ignoring the exact value of $\sigma(x)$ we solve the identification problem with an initial guess $y^{(o)}(t\,x)$ equal to a constant and compare the computed value of the transmissivity with the exact function $\sigma(x)$. These two values are different. Only near the wells these two values are in good agreement. In the mesh-points near the boundary of the aquifer the computed values of the transmissivity are a little different from the initial guess values. This phenomenon is related to the question of "non unicity" in the solution of this example. Indeed a necessary condition for existence and uniqueness of a solution of the above identification problem is that, in addition to the piezometric head, the transmissivity must be known along a curve Γ_o crossed by all streamlines in the aquifer. This is because $\sigma(x)$ satisfies a first order linear hyperbolic equation in two independent variables which, under appropriate smoothness conditions, has a solution of class C^1 when "initial" values of $\sigma(x)$ (of class C^1) are assigned on a suitable regular curve Γ_0: a characterization of Γ_0 is that the characteristics of the hyperbolic equation through each point of Γ_0 must cover completely the domain Ω. Since this condition on Γ_0 is difficult to satisfy, in hydrological applications much attention has been directed in attempting to bypass the uniqueness requirement. However, when the necessary condition on Γ_0 is lacking, the problem of identifying the transmissivity parameter is improperly posed and in general yields multiple solutions.

Generally, the reconstruction by a set of distributed observations of the transmissivity and piezometric maps is obtained by using a particular Kriging technique. Since the hydrological data are usually few, unevenly scattered and affected by instrument arrors, the use of a stochastic interpolator, instead of a deterministic interpolator of the type [1], is widely warranted. In a Kriging technique the event to reconstruct, i.e. the piezometric at a prefixed time or the transmissivity, is considered as a random field described in terms of its mean and covariance functions [2],[3]. In this way the required smoothness conditions on the data of the above identification problem make sense.

3. COMPUTATIONAL COMPLEXITY OF THE QUASI-LINEARIZATION METHOD.

The most expensive (with respect to the computing-time) steps of the quasi-linearization method are those related to the numerical integration of the system (6)-(8) and of the adjoint system (11)-(13). Besides the solution of these systems can be expected to exhibit either numerical dispension or some oscillatory behavior. Therefore a deeper analysis of these steps is useful.

To this end, we will consider the parabolic system (6)-(8), in which henceforth, for the sake of simplicity, the superscripts μ and μ-1 will be dropped. (The analysis of the adjoint system (11)-(13) is identical: now the system must be solved backward from $t = t^*$).

The solution of (6)-(8) may exhibit either numerical dispension or oscillatory behavior when the set of first-order differential equations $y_t - \sum_{i=1}^{m} (B_i y)_{x_i} = 0$ is of hyperbolic-type and the physical dispension term A_i is very small when compared with the advection term B_i (i=1,2..m): briefly we will write $B_i >> A_i$.

In this paper we will limit the analysis of the system (6)-(8) to the extreme cases $B_i = 0$ and $B_i >> A_i$ with $m = 2$ and Ω composed by a rectangle with sides parallel to the axes x_1 and x_2.

a) Case $B_i = 0$, $i = 1,2$

Consider $\Omega_{\Delta x}$, $\Gamma_{\Delta x}$ and $T_{\Delta t}$ along with Ω,Γ and $[0,t^*]$. Let $T_{\Delta t}$ be the set of points $\{t_\tau\}$, $t_{\tau+1} - t_\tau = \Delta t : \tau = 0,1\ldots, t_o = 0$.
Let us construct an alternating direction difference scheme for system (6)-(8) of the form:

$$(21)\quad \begin{cases} (I - \frac{\Delta t}{2}(D_1^\tau + H_1^\tau))\varphi^{\tau+1/2} = (I + \frac{\Delta t}{2}(D_2^\tau + H_2^\tau))\varphi^\tau + \frac{\Delta t}{2}\, s^\tau \\ (I - \frac{\Delta t}{2}(D_2^\tau + H_2^\tau))\varphi^{\tau+1} = (I + \frac{\Delta t}{2}(D_1^\tau + H_1^\tau))\varphi^{\tau+1/2} + \frac{\Delta t}{2}\, s^\tau \end{cases}$$

Here φ^0 and s^τ are restrictions of $v(x)$ and $s(t_\tau x)$, respectively, on a rectilinear network overlapping Ω with mesh-spacings Δx_1 and Δx_2. The p th order matrix $D^\tau = D_1^\tau + D_2^\tau$ consists of the elements which represent the spatial differencing in (6) at time $t = t_\tau$: therefore these elements are inversely proportional to the square of the mesh-spacings Δx_1 and Δx_2. The p th order matrix $H^\tau = H_1^\tau + H_2^\tau$ consists of the elements which are independent upon the mesh-spacings.
We assume that $D_i^\tau + (D_i^\tau)^T$ (i=1,2) is negative definite: this condition is satisfied by many conventional difference schemes.

For a fixed discretization of the spatial variables of (6) at time $t = t_\tau$, the *transition matrix* $M_2^\tau N_1^\tau M_1^\tau N_2^\tau$, where $M_i^\tau = (I - \frac{\Delta t}{2}(D_i^\tau + H_i^\tau))^{-1}$, $N_i^\tau = I + \frac{\Delta t}{2}(D_i^\tau + H_i^\tau))$, (i=1,2) has a matrix power series development about $\Delta t = 0$ ($\frac{\Delta t}{2}\|D_i^\tau + H_i^\tau\| < 1$, i=1,2) that agrees through at least linear terms with the expansion of $\exp(\Delta t(D^\tau + H^\tau))$.
In order to have stability for this scheme, it is sufficient that the matrices $\prod_{\ell=0}^{\tau}(M_2^\ell N_1^\ell M_1^\ell N_2^\ell)$ be uniformly bounded for all $\Delta t < \overline{\Delta t}$ and $\tau\cdot\Delta t < \tilde{t}$, where $\tilde{t}$ is a fixed time-station.

When the matrices A_1 and A_2 are independent of time, by using a theorem on the uniform boundedness of sparse matrices proved in [5], it can be shown that, if $\Delta t/\Delta x_1^2$ and $\Delta t/\Delta x_2^2$ are arbitrary positive constants, the difference scheme (21) is stable.
Thus, if we continually refine the time step size Δt, and hence also the spatial mesh-spacings Δx_1 and Δx_2 (in such way that $\Delta t/\Delta x_i^2$ is an arbitrary positive constant, i=1,2), and require more and more steps to reach a fixed time-station $\tilde{t}$, the computed solution is always bounded.

b) Case $B_i >> A_i$, i=1,2

In this case, on each interval $[t_{\tau-1}, t_{\tau+1}]$, $\tau=1,2\ldots(t_o = 0)$, we split the system (6)-(8) into two systems ($y(t_o x) = v(x)$):

$$(22)\quad \begin{cases} \psi_t - \sum_{i=1}^{2} (B_i\psi)_{x_i} = 0 & \text{in } \Omega \times (t_{\tau-1}, t_{\tau+1}] \\ \psi(t_{\tau-1} x) = y(t_{\tau-1} x) & \text{in } \bar{\Omega} = \Omega + \Gamma \end{cases}$$

$$(23)\quad \begin{cases} \varphi_t = \sum_{i=1}^{2} (A_i \varphi_{x_i})_{x_i} + C\varphi + s & \text{in } \Omega \times (t_{\tau-1}, t_{\tau+1}] \\ \varphi(t_{\tau-1} x) = \psi(t_{\tau+1} x) & \text{in } \Omega \\ \varphi(t\, x) = w(t\, x) & \text{on } \Gamma \times [t_{\tau-1}, t_{\tau+1}] \end{cases}$$

$$y(t_{\tau+1} x) = \varphi(t_{\tau+1} x) \qquad \text{in } \bar{\Omega}$$

By hypothesis the system (22) is of hyperbolic-type. We therefore can trasform this system into the *normal form*, by determining n real non trivial couples $(\gamma_\ell\ \rho_\ell)$ which satisfy the characteristic equation $(\ell = 1,2\ldots n)$:

$$\det(I - \gamma_\ell B_1 - \rho_\ell B_2) = 0$$

Each differential equation of the system written in the normal form contains directional derivatives in the planes $(t\ x_1)$ and $(t\ x_2)$. Let $(t_{\tau-1}, x_1+\gamma_\ell\Delta x_1, x_2)$ and $(t_{\tau-1}, x_1, x_2+\rho_\ell\Delta x_2)$ be the intersections of the plane t = const through $(t_{\tau-1}, x_1, x_2)$ (more briefly, plane $P_{\tau-1}$) with the lines through $(t_{\tau+1}, x_1, x_2)$ in the planes $(t\ x_1)$ and $(t\ x_2)$ having the characteristic directions $1/\gamma_\ell$ and $1/\rho_\ell$, respectively. Then, we choose the time step size $2\Delta t$ and the couples $(\gamma_\ell\ \rho_\ell)$ in such a way that the polygonal in the plane $P_{\tau-1}$ with vertices $(t_{\tau-1}, x_1+\gamma_\ell\Delta x_1, x_2)$ and $(t_{\tau-1}, x_1, x_2+\rho_\ell\Delta x_2)$, $\ell = 1,2\ldots n$, contains the intersection of the characteristic cone at $(t_{\tau+1}, x_1, x_2)$ with the plane $P_{\tau-1}$. Besides, in order to satisfy the Courant condition, the time step size $2\Delta t$ must be limited by $2\Delta t \leq \min_{1\leq\ell<n}\{|\gamma_\ell|\cdot\Delta x_1, |\rho_\ell|\cdot\Delta x_2\}$.

By using the values of the restriction on $\Omega_{\Delta x}$ of $\psi(t\,x)$ at time $t=t_{\tau-1}$, we approximate the directional derivatives which appear in the normal system with the Courant-Isaacson-Rees scheme [4]. By solving, at each mesh-point of $\Omega_{\Delta x}$, a simultaneous system of n equations, we determine the restriction on $\Omega_{\Delta x}$ of $\psi(t\ x)$ at time $t = t_{\tau+1}$.

Successively, the system (23) is solved with the alternating direction difference scheme of case a).

The numerical performance of this splitting-up method has been established by carrying out experimental calculations on some sample problems of the form of those which describe the mass transport of a miscible fluid in a flow (n = 1) and simulate the spatial structuration of a prey-predator population in a aquatic ecosystem (n = 2).

REFERENCES

[1] Akima,H.: A Method of Bivariate Interpolation and Smooth Surface Fitting for Irregularly Distributed Data Points. *ACM Trans. Math. Software 4,2* (1978) 148-164.

[2] Delfiner,P., Delhomme,J.P.: Optimum Interpolation by Kriging. In: *Dispaly and Analysis of Spatial Data* (J.C.Davis-M.J.McCullagh, ed.) J.Wiley, London (1975).

[3] Drozdov,O.A., Sepelevskii,A.A.: The Theory of Interpolation in a Stochastic Field of Meteorological Elements and its Application to Meteorological Map and Network Rationalization Problems. *Trudy N.I.U. GUGSM 1,18* (1946).

[4] Forsythe,G.E., Wasow,W.R.: *Finite Difference Methods for Partial Differential Equations*. J. Wiley, New York (1960).

[5] Galligani,I.: Un teorema sulle matrici sparse stabili. *Consiglio Naz. Ricerche, Pubbl. Ist. Appl. Calcolo 140* (1978).

Über die Behandlung Inverser Probleme bei streng zeichenfesten Kernen

Klaus Glashoff
Universität Hamburg
Institut für Angewandte Mathematik
Bundesstrasse 55
D-2000 Hamburg 13

1. Einführung.

Wir betrachten die Integralgleichung

(1.1) $$z(s) = \int_T K(s,t)u(t)dt;$$

hierbei ist K(s,t) eine reellwertige Funktion auf S × T, wobei S = (a,b) und T = (c,d) beschränkte Intervalle der reellen Achse sind; z(s) ist eine vorgegebene Funktion. Genauere Angaben über K(s,t) und z(s) folgen später; gesucht ist u(t).

Um eine Näherungslösung von (1.1) zu berechnen, kann man wie folgt vorgehen:

Man wählt eine "große" Zahl M > 0 und betrachtet die folgende Optimierungsaufgabe:

(1.2) Minimiere $$\max_{s\in S} \left|z(s) - \int_T K(s,t)u(t)dt\right|$$

über alle u mit

(1.3) $$|u(t)| \leq M, \qquad t \in T.$$

In dieser Arbeit soll dieses Optimierungsproblem genauer untersucht werden; wir wollen insbesondere auf das sog. "Bang-Bang-Prinzip" hinsteuern, das folgendes besagt: Es gibt eine eindeutige Lösung u_M des Problems (1.2)-(1.3) mit der Eigenschaft

$$u_M(t) = M\varepsilon(-1)^r \text{ für } t \in [t_{r-1}, t_r), \quad r = 1,\dots,n+1$$

für ein $\varepsilon = +1$ oder -1, ein $n \geq 1$ und eine Zerlegung

$$c = t_o < t_1 < \dots t_n < t_{n+1} = d$$

des Intervalls T. Dies erlaubt dann die Anwendung spezieller, rasch konvergenter Verfahren zur Lösung der Aufgabe (1.2)-(1.3), worauf wir zum Schluß eingehen.

Um solche Resultate zu gewinnen, muß man gewisse Annahmen über den Kern K(s,t) machen; nämlich die, daß es sich hierbei um einen <u>streng zeichenfesten</u> Kern handelt: Es gebe eine Folge von Zahlen $\varepsilon_m = +1$ oder -1 derart, daß

(1.4) $$\varepsilon_m \det \{K(s_i, t_k)\} > 0$$

gilt für alle $a < s_i < \dots < s_m < b$, $c < t_1 < \dots t_m < d$. Kerne dieser Art haben "variationsvermindernde" Eigenschaften, die in die Herleitung unserer Resultate eingehen; Beweise werden in [2] gegeben.

Die Eigenschaft (1.4) ist bei vielen konkreten Problemen erfüllt. Wenn etwa (1.1) den Wärmeleitungsvorgang in einem Stab beschreibt mit Anfangstemperatur u und Endtemperatur z, dann kann man für die zugehörige Green'sche Funktion die Eigenschaft (1.4) herleiten (Karlin [9]); dies trifft auch für große Klassen von Green'schen Funktionen bei parabolischen Anfangsrandwertaufgaben zu.

2. <u>Das endliche Bang-Bang-Prinzip</u>

Wir setzen voraus, daß durch (1.1) ein linearer, beschränkter Operator

$$K : L_\infty(T) \to C(\bar{S})$$

definiert wird, der außerdem stetig bezüglich der schwachen Topologie auf $C(\bar{S})$ und der schwach-stern-Topologie auf $L_\infty(T)$ ist.

Mit der Definition

$$B_M = \{u \in L_\infty(T) \;/\; |u(t)| \leq M \quad \text{f. ü. in } T\}$$

läßt sich das Problem (1.2)-(1.3) dann in der folgenden Gestalt schreiben:

$$\text{(2.1)} \qquad \underset{y \in K(B_M)}{\text{Minimiere}} \; ||z - y||_\infty \;;$$

die <u>Existenz</u> einer Lösung u_M folgt mit Standardschlüssen; vgl. [5].

Bei (2.1) handelt es sich um ein restringiertes Approximationsproblem, für das der folgende Chrakterisierungssatz gilt (vg. Holmes [7], §22).

<u>Satz 1.</u> y_M ist genau dann Lösung von (2.1), wenn es ein Maß $d\alpha$ gibt mit

$$\int_S \{z(s) - y_M(s)\}\, d\alpha(s) = ||z - y_M||_\infty$$

$$\int_S y_M(s)\, d\alpha(s) = \max_{y \in K(B)} \int_S y(s)\, d\alpha(s) \;.$$

Wenn wir die Funktion $\lambda(t)$ folgendermaßen definieren:

$$\text{(2.2)} \qquad \lambda(t) = \int_S K(s,t)\, d\alpha(s) \;, \qquad t \in T,$$

dann gilt mit $y_M = Ku_M$

$$\int_T u_M(t)\lambda(t)\,dt = \max_{u \in B_M} \int_T u(t)\lambda(t)\,dt$$

(Maximumprinzip!). Hieraus folgt sofort

$$u_M(t) = M \operatorname{sgn} \lambda(t) \qquad \text{für } \lambda(t) \neq 0 \;;$$

es muß nun noch gezeigt werden, daß $\lambda(t)$ nur endlich viele Nullstellen auf T besitzt. Einen entsprechenden Satz haben wir in [2] bewiesen:

<u>Satz 2.</u> Sei $\min_{u \in B_M} ||z - Ku||_\infty > 0$. Wenn $K(s,t)$ streng zeichenfest ist, dann hat λ aus (2.2) nur endlich viele Nullstellen; jede Lösung u_M von (1.2)-(1.3) ist daher endlich bang-bang (wie üblich folgt hieraus unmittelbar die Eindeutigkeit von u_M).

3. Numerische Behandlung

Zur numerischen Lösung von (1.2)-(1.3) empfiehlt es sich, in zwei Stufen vorzugehen. In der ersten Stufe bestimmt man die Struktur der Lösung u_M, d.h. die Anzahl der "Sprünge" zwischen +M und -M, sowie das Vorzeichen von $u_M(0)$; dies geschieht am besten mit Hilfe der Linearen Optimierung. In der zweiten Stufe führt man die Koordinaten $t_1,\dots,t_n$ der Sprünge als Variable ein und löst das zugehörige nichtlineare Approximationsproblem.

Stufe 1. Man wähle eine feste Unterteilung

$$c = \tau_0 < \tau_1 < \dots < \tau_N < \tau_{N+1} = d$$

von T und definiere für i = 1, ... , N+1

$$u_i(t) = \begin{cases} 1 & \text{für } t \in [\tau_{i-1},\tau_i) \\ 0 & \text{sonst} . \end{cases}$$

Dann löse man das lineare Approximationsproblem

$$\underset{|\alpha_i|\leq M}{\text{Minimiere}} \quad \|z - \sum_{i=1}^{N+1} \alpha_i K u_i\|_\infty \quad .$$

Diese Aufgabe läßt sich in üblicher Weise als lineare Optimierungsaufgabe (mit unendlich vielen Nebenbedingungen) umformulieren; effektive Lösungsalgorithmen für solche "Semi-infiniten" Probleme werden in dem Buch von Glashoff und Gustafson [4] angegeben. Ein ausführlich durchgerechnetes Problem (Bestimmung der Randdaten bei einer Wärmeleitungsaufgabe) findet man bei Glashoff und Gustafson [3].

Stufe 2. Aus den Ergebnissen von Stufe 1 muß die Anzahl n der Sprünge der optimalen Lösung u_M geschätzt werden (möglicherweise muß man die erste Phase der Rechnung mit größerem N wiederholen). Außerdem braucht man Näherungen für die Koordinaten der Sprünge sowie den Wert $u_M(0)$.

Man betrachtet nun das nichtlineare Approximationsproblem (wobei wir annehmen, daß $u_M(0) = -M$ gilt):

$$\underset{c=t_0<t_1<\dots<t_{n+1}=d}{\text{Minimiere}} \quad \max_{s\in\overline{S}} \left|z(s) - M \sum_{i=1}^{n+1} (-1)^i \int_{t_{i-1}}^{t_i} K(s,t)dt\right| .$$

Solche Probleme der nichtlinearen gleichmäßigen Approximation kann man etwa mit Linearisierungsalgorithmen lösen und äußerst effektiv auch mit dem von Hettich [6] beschriebenen Newton-Verfahren.

Man überlegt sich leicht, daß für die Approximationsfamilie

$$F(t_1,\dots,t_n;s) = M \sum_{i=1}^{n+1} (-1)^i \int_{t_{i-1}}^{t_i} K(s,t)\,dt$$

die lokale Haar'sche Bedingung erfüllt ist, vgl. Collatz und Krabs [1], wenn K(s,t) als streng zeichenfest vorausgesetzt wird. Für die numerische Behandlung sind also sehr günstige Voraussetzungen vorhanden.

Literatur

[1] L. Collatz und W. Krabs.: Approximationstheorie. B.G. Teubner, Stuttgart 1973

[2] K. Glashoff.: Restricted approximation by strongly sign regular kernels: the finite bang-bang-principle. J. Approximation Theory (to appear).

[3] K. Glashoff und S.-Å. Gustafson.: Numerical treatment of a parabolic boundary-value control problem, J. Opt. Th. Appl. 19 (1976), 645-663.

[4] K. Glashoff und S.-Å. Gustafson.: Einführung in die lineare Optimierung. Wissenschaftliche Buchgesellschaft, Darmstadt 1978.

[5] K. Glashoff und N. Weck.: Boundary control of parabolic differential equations in arbitrary dimensions: supremum-norm problems. SIAM J. Control and Optimization, Vol. 14, No. 4, 662-681, 1971

[6] R. Hettich.: A Newton-method for nonlinear Chebyshev approximation, Approx. Theory, Proc. int. Colloqu. Bonn 1976, Springer Lecture Notes in Math. 556, 222-236, 1976.

[7] R.B. Holmes.: A course on optimization and best approximation, Lecture Notes in Mathematics 257, Springer Verlag, Berlin - Heidelberg - New York, 1972

[8] A. Karafiat.: The problem of the number of switches in parabolic equations with control. Ann. Pol. Math. XXXIV, 289-316, 1977

[9] S. Karlin.: Total positivity, Vol. I, Stanford University Press, Stanford, California, 1968.

OPTIMAL CONTROL OF COEFFICIENTS IN LINEAR ELLIPTIC EQUATIONS
II. NECESSARY OPTIMALITY CONDITIONS

MANFRED GOEBEL *)

SUMMARY

In the note necessary optimality conditions are derived for a control system which behaviour is described by the (weak) solutions of a linear elliptic boundary value problem of the first kind. The control function occurs in all coefficients of the given differential operator and for fixed control the corresponding boundary value problem satisfies the FREDHOLM alternative.

0. INTRODUCTION

In our note we shall consider an optimal control problem for a linear elliptic boundary value problem of first kind. The problem is particularly characterized by occuring of control function in all coefficients and the right-hand side of the differential equation. Such control problems are closely connected with inverse problem in elliptic equations. However, they also arise directly in many branches of sciences and engineering.

In the first part of this paper /2/ we were working on existence assertions on optimal control and now we are going to derive necessary optimality conditions from a general method for deducing maximum principles to optimal control developed by von WOLFERSDORF and myself (cf. /1,3/). Up to date such conditions for coefficients control problems in elliptic equations have been already proved by ZOLEZZI /7/, GOGODZE /4/ and various special cases among others by LIONS, RAITUMS, NENACHOV, GORSCHAKOV, CHAVENT and myself; regarding precise references we refer the reader to /1/. In all this papers the unique solvability of the corresponding boundary value problem for every admissible control plays an essential role. But, we shall describe necessary optimality conditions under the more natural assumption that the elliptic boundary value problem only satisfies the FREDHOLM alternative for every fixed control. Under such a condition control problems for elliptic equations with the right-hand side function as control parameter were first considered by MIERSEMANN /5/.

---

*) Dozent Dr. M. Goebel, Sektion Mathematik, Bergakademie Freiberg DDR - 92 Freiberg

1. PROBLEM STATEMENT, ASSUMPTIONS

Let Ω be an open bounded connected set in the n-dimensional real space R^n, $n \geq 2$. $\partial\Omega$ denotes its boundary and $x = \{x_1, \ldots, x_n\}$ an arbitrary element of $\partial\Omega \cup \Omega$. $(.|.)$ denotes the scalar product in R^n and $|.|$ its norm. $L^\infty(\Omega)$, $L^2(\Omega)$ and $H_0^1(\Omega)$ are the usual BANACH spaces of real-valued functions defined on Ω. Their norms are denoted by $\|.\|_\infty$, $\|.\|$ and $\|.\|_0$, respectively. $(.,.)$ is the scalar product in $L^2(\Omega)$. $H^{-1}(\Omega)$ is the dual space of $H_0^1(\Omega)$ and $\langle .,. \rangle$ the corresponding generalized scalar product.

We begin the description of the control problem considered here with the set of control functions, which is given by

(1) $$V = \{u \in L^\infty(\Omega) \mid \xi_1(x) \leq u(x) \leq \xi_2(x) \quad \text{a.e.} \quad x \in \Omega\}$$

where $\xi_{1,2} \in L^\infty(\Omega)$ are fixed. Let $\tilde{I}$ be an open interval with

$$\tilde{I} \supset [\operatorname{ess}\inf_{\Omega} \xi_1(x), \operatorname{ess}\sup_{\Omega} \xi_2(x)].$$

We assume that the given functions

$$d(x,u) = a_{ij}(x,u),\ a_i(x,u),\ b_i(x,u),\ c(x,u),\ i,j = 1(1)n,$$

defined on $\Omega \times \tilde{I}$ satisfy following conditions:

(2) (a) $d(\cdot,u)$ is measurable on Ω for every $u \in \tilde{I}$.

(b) $d(x,.)$ is continuously differentiable on $\tilde{I}$ for almost every $x \in \Omega$; the derivation is denoted by $d_u(x,.)$.

(c) There exists a real number d such that $|d(x,u)|,\ |d_u(x,u)| \leq d \quad \forall u \in \tilde{I}$, a.e. $x \in \Omega$.

(d) There exists a real-valued function $\chi = \chi(u)$ defined on a certain neighbourhood $\mathcal{S} \subset R^1$ of $u = 0$ with $\chi(u) \longrightarrow 0$ if $u \longrightarrow 0$ and $|d_u(x,u) - d_u(x,v)| \leq \chi(u-v) \quad \forall u,v \in \tilde{I}$ with $u-v \in \mathcal{S}$, a.e. $x \in \Omega$.

Moreover, let $f_i(x,u)$, $i=0(1)n$, be given functions defined on $\Omega \times \tilde{I}$, too, having the properties (3a,b) written down in (2a,b) for $d(x,u)$; the corresponding derivations are denoted by $f_{iu}(x,.)$.

Further we assume:

(3) (c) There exists an element $\eta \in L^2(\Omega)$ such that $|f_i(x,u)|,\ |f_{iu}(x,u)| \leq \eta(x) \quad \forall u \in \tilde{I}$, a.e. $x \in \Omega$.

Now we introduce the following abbrevations:

(4) $$A(x,u) = ((a_{ij}(x,u))),\quad x \in \Omega,\quad u \in \tilde{I},$$

$$(5)\qquad \left.\begin{aligned} A(u) &= A(.,u(.)),\\ a(u) &= a(.,u(.)) = \{a_1(.,u(.)),\dots a_n(.,u(.))\},\\ b(u) &= b(.,u(.)) = \{b_1(.,u(.)),\dots b_n(.,u(.))\},\\ c(u) &= c(.,u(.)),\quad f_o(u) = f_o(.,u(.)),\\ f(u) &= f(.,u(.)) = \{f_1(.,u(.)),\dots f_n(.,u(.))\} \end{aligned}\right\}\ u\in V.$$

In these notations the state equation of our control problem reads as follows:

$$(6)\qquad L(u)y = f_o(u) - \operatorname{div} f(u) \text{ in } \Omega\ ,\ y=0 \text{ on } \partial\Omega,$$

$$(7)\qquad L(u)y = -\operatorname{div}\{A(u)\nabla y + a(u)y\} + (b(u)|\nabla y) + c(u)y.$$

Assuming

$$(8)\qquad (A(x,u)\xi|\xi)\geq\alpha|\xi|^2\ \forall\xi\in R^n,\ \forall u\in\tilde{I},\quad \text{a.e. } x\in\Omega\ ,\ \alpha=\text{const.}>0$$

this means that the state is given by solutions from the elliptic boundary value problem (6), where all coefficients and the right-hand side are NEMYZKI operators defined by functions introduced above.

Finally we have to describe the cost functional. It is chosen as:

$$(9)\qquad \begin{aligned} &J(y,u) = J_1(y) + J_2(u),\\ &J_1(y) = \int_\Omega g(x,y,\nabla y)dx,\quad J_2(u) = \int_\Omega h(x,u)dx, \end{aligned}$$

where the functions $g(x,l)$, $l=\{l_o,\dots l_n\}$, and $h(x,u)$ defined on $\Omega\times R^{n+1}$ and $\Omega\times R^1$, respectively, satisfy the following assumptions:

(10) (a) $g(.,l)$ is measurable on Ω for every $l\in R^{n+1}$.

(b) $g(x,.)$ is continuously differentiable with respect to $l_o,\dots l_n$ in R^{n+1} for almost every $x\in\Omega$; the partial derivations are denoted by $g_{l_i}(x,.)$, $i=0(1)n$.

(c) There exist $\alpha_1\in L^1(\Omega)$, $a_2\in L^2(\Omega)$ and $\beta_{1,2}\in R^1$ such that

$$\left.\begin{aligned}|g(x,l)|&\leq\alpha_1(x)+\beta_1|l|^2\\ |g_{l_i}(x,l)|&\leq\alpha_2(x)+\beta_2|l|\end{aligned}\right\}\ \forall\, l\in R^{n+1},\ \text{a.e. } x\in\Omega.$$

(11) (a) $h(.,u)$ is measurable on Ω for every $u\in R^1$.

(b) $h(x,.)$ is continuously differentiable in R^1 for almost every $x\in\Omega$; $h_u(x,.)$ denotes the derivation.

(c) For every $\lambda>0$ exists an element $\alpha_\lambda\in L^1(\Omega)$ such that

$$|h(x,u)|,\ |h_u(x,u)|\leq\alpha_\lambda(x)\ \forall u\in R^1 \text{ with } |u|\leq\lambda,\ \text{a.e. } x\in\Omega.$$

With the cost functional the control problem treated here is completely described. For the sake of brevity we denote this by problem (P):

(P) $\quad \min J(y,u)$ subject to constraints (1,6).

In the next section we shall settle the problem (P) into framework of our theory developed in /1,3/ and show that its hypotheses are satisfied here. Various supplementary remarks to the needed assumptions listened above can be found in /1/.

2. PRELIMINARY THEOREMS

As usual, for fixed $u \in V$ an element $y \in H_0^1(\Omega)$ is called a (weak) solution of the boundary value problem (6) if and only if

(12) $\quad \mathbb{A}(u)y = F(u)$,

where $\mathbb{A}(u) \in \mathcal{L}(H_0^1(\Omega), H^{-1}(\Omega))$ is a FREDHOLM operator induced by the bilinear form

(13) $$L_u(y,z) = \int_\Omega \{(A(u)\nabla y \mid \nabla z) + (a(u) \mid \nabla z)y + (b(u) \mid \nabla y)z + c(u)yz\}\, dz$$

defined on $H_0^1(\Omega) \times H_0^1(\Omega)$ and the right-hand side given by

(14) $$\langle F(u), y\rangle = \int_\Omega \{f_0(u)z + (f(u) \mid \nabla z)\}\, dx, \quad y \in H_0^1(\Omega)$$

belongs to $H^{-1}(\Omega)$.

Thus the problem (P) with (12) instead (6) is such one as considered in /1,3/, where we have to put $X = H_0^1(\Omega)$, $Y = H^{-1}(\Omega)$, $U = L^\infty(\Omega)$ and $T(y,u) = \mathbb{A}(u)y - F(u)$. We shall prove that the hypotheses from /1,3/ are fulfilled here. For this goal we introduce the sets

$$\tilde{V} = \{u \in L^\infty(\Omega) \mid u(x) \in \tilde{I} \quad \text{a.e.} \quad x \in \Omega\}$$
$$\mathcal{S}_{\bar{u}} = \{u \in L^\infty(\Omega) \mid \|u - u\|_\infty < \varrho\}$$

where $\bar{u}$ is an arbitrary, but fixed, element belonging to V and ϱ is such a fixed number that $\mathcal{S}_{\bar{u}} \subset \tilde{V}$. Further we use the following abbrevations ($y,z \in H_0^1(\Omega)$, $v \in \mathcal{S}_{\bar{u}}$, $u \in L^\infty(\Omega)$):

(15) $$\begin{aligned} A'(v) &= A_u(\cdot,v(\cdot)) = ((a_{iju}(\cdot,v(\cdot)))), \\ a'(v) &= a_u(\cdot,v(\cdot)) = \{a_{1u}(\cdot,v(\cdot)),\ldots a_{nu}(\cdot,v(\cdot))\}, \\ b'(v) &= b_u(\cdot,v(\cdot)) = \{b_{1u}(\cdot,v(\cdot)),\ldots b_{nu}(\cdot,v(\cdot))\}, \\ c'(v) &= c_u(\cdot,v(\cdot)), \quad f_0'(v) = f_{ou}(\cdot,v(\cdot)), \\ f'(v) &= f_u(\cdot,v(\cdot)) = \{f_{1u}(\cdot,v(\cdot)),\ldots f_{nu}(\cdot,v(\cdot))\}, \\ g_i(y) &= g_{l_i}(\cdot,v(\cdot), y(\cdot)), \quad i = 0(1)n, \\ g(y) &= \{g_1(y),\ldots g_n(y)\}. \end{aligned}$$

The bilinear form

(16) $$L_{v,u}(y,z) = \int_\Omega \{(A'(v)\nabla y \mid \nabla z) + (a'(v) \mid \nabla z)y + (b'(v) \mid \nabla y)z + c'(v)yz\}\, u\, dx$$

is important for the next considerations.

<u>Lemma 1:</u> Under the assumptions (2) the operator $\mathbb{A}: L^{\infty}(\Omega) \longrightarrow \mathcal{L}(H_0^1(\Omega), H^{-1}(\Omega))$ is continuously FRECHET-differentiable in $\mathcal{J}_{\bar{u}}$. The FRECHET derivation $\mathbb{A}'(v)$ at the point $v \in \mathcal{J}_{\bar{u}}$ satisfies

$$\langle [\mathbb{A}'(v)u]y,z \rangle = L_{v,u}(y,z) \quad \forall u \in L^{\infty}(\Omega),\ \forall y,z \in H_0^1(\Omega),$$

where $L_{v,u}(y,z)$ is defined by (16).

<u>Proof:</u> By (2) it follows that the NEMYZKI operator $d(v) = d(.,v(.))$ is continuously FRECHET-differentiable in $\mathcal{J}_{\bar{u}}$ as an operator acting in $L^{\infty}(\Omega)$; its FRECHET derivation at the point $v \in \mathcal{J}_{\bar{u}}$ is given by $d'(v) = d_u(.,v(.))$ (cf. /6, part X, §1/,/1/).

Obviously, the bilinear form $L_{v,u}(y,z)$ is continuous and bounded over $H_0^1(\Omega) \times H_0^1(\Omega)$ for every fixed $v \in \mathcal{J}_{\bar{u}}$ and $u \in L^{\infty}(\Omega)$ and, therefore, a linear bounded operator $K(v)u$ acting on $H_0^1(\Omega)$ into $H^{-1}(\Omega)$ is induced by $L_{v,u}(y,z)$. It is rather easy to see, that $K(v)u$ is linear with respect to u and that the inequality

$$\| K(v)u \|_{\mathcal{L}(H_0^1(\Omega),H^{-1}(\Omega))} \leq \text{const.} \| u \|_{\infty} \quad \forall u \in L^{\infty}(\Omega)$$

is valid. This means $K(v) \in \mathcal{L}(L^{\infty}(\Omega), \mathcal{L}(H_0^1(\Omega),H^{-1}(\Omega)))$.

Using the continuous properties of $d'(v)$ mentioned above it can be shown that the map $v \longrightarrow K(v)$ is a continuous one from $\mathcal{J}_{\bar{u}} \subset L^{\infty}(\Omega)$ into $\mathcal{L}(L^{\infty}(\Omega), \mathcal{L}(H_0^1(\Omega),H^{-1}(\Omega)))$. Hence the proof will be finished, if the relation $K(v) = \mathbb{A}'(v)$ is proved.

Let δ be a real positive number, such that $v+u \in \mathcal{J}_{\bar{u}}$ for every $u \in L^{\infty}(\Omega)$ with $\|u\|_{\infty} < \delta$. Considering that $\mathbb{A}'(v)$ is given by (13) and $K(v)u$ by (16) obtain the estimate

$$\begin{aligned}
&\| \mathbb{A}(v+u) - \mathbb{A}(v) - K(v)u \|_{\mathcal{L}(H_0^1(\Omega),H^{-1}(\Omega))} \\
&\leq \text{const.} \Big\{ \sum_{i,j=1}^{n} \| a_{ij}(v+u) - a_{ij}(v) - a_{ij}'(v)u \|_{\infty} + \\
&\sum_{i=1}^{n} (\| a_i(v+u) - a_i(v) - a_i'(v)u \|_{\infty} + \\
&\| b_i(v+u) - b_i(v) - b_i'(v)u \|_{\infty}) + \\
&\| c(v+u) - c(v) - c'(v)u \|_{\infty} \Big\} \quad \forall u \in L^{\infty}(\Omega) \text{ with } \|u\|_{\infty} < \delta
\end{aligned}$$

Because $d(v)$ has the FRECHET derivation $d'(v)$ the assertions of Lemma 1 are proved.

<u>Lemma 2:</u> Under the assumptions (3) the map $F\colon L^\infty(\Omega) \longrightarrow H^{-1}(\Omega)$ is continuously FRECHET-differentiable in $\mathcal{S}_{\bar{u}}$. For its FRECHET derivation $F'(v)$ at the point $v \in \mathcal{S}_{\bar{u}}$ the representation

$$\langle F'(v)u, z\rangle = \int_\Omega \{f_0'(v)z + (f'(v)|\nabla z)\}\, u\, dx \quad \forall u \in L^\infty(\Omega), \forall z \in H_0^1(\Omega)$$

is valid.

Of course, in the proof omitted in this paper we have to use that by (3) the NEMYZKI operators $f_i(v) = f_i(.,v(.))$, $i = 0(1)n$, are continuously FRECHET-differentiable in $\mathcal{S}_{\bar{u}}$ as operators from $L^\infty(\Omega)$ into $L^2(\Omega)$ and that their FRECHET derivations at $v \in \mathcal{S}_{\bar{u}}$ are given by $f_i'(v) = f_{iu}'(.,v(.))$, $i = 0(1)n$, (see /6,1/).

Let $\{\mathring{y},\mathring{u}\} \in H_0^1(\Omega) \times L^\infty(\Omega)$ be an arbitrary pair with $\mathring{u} \in V$. Then from the results obtained in the above lemmata it directly follows that the operator $T(y,u) = \mathbb{A}(u)y - F(u)$ is continuously FRECHET-differentiable in a sufficiently small neighbourhood of $\{\mathring{y},\mathring{u}\}$, if only the hypotheses (2,3) are fulfilled. Moreover, its partial FRECHET derivations $\mathring{T}_y = T_y(\mathring{y},\mathring{u})$ and $\mathring{T}_u = T_u(\mathring{y},\mathring{u})$ are given by

$$(17) \qquad \mathring{T}_y y = \mathbb{A}(\mathring{u})y, \quad \mathring{T}_u u = [\mathbb{A}'(\mathring{u})u]\mathring{y} - F'(\mathring{u})u \quad \forall \{y,u\} \in H_0^1(\Omega) \times L^\infty(\Omega).$$

<u>Lemma 3:</u> Under the assumptions (10,11) the functionals J_1 and J_2 defined on $H_0^1(\Omega)$ and $L^\infty(\Omega)$ by (9) are continuously FRECHET-differentiable at every point $\mathring{y} \in H_0^1(\Omega)$ and $\mathring{u} \in L^\infty(\Omega)$, respectively. Their corresponding FRECHET derivations $J_1'(\mathring{y})$ and $J_2'(\mathring{u})$ are given by

$$\langle J_1'(\mathring{y}), y\rangle = \int_\Omega \{g_0(\mathring{y})y + (g(\mathring{y})|\nabla y)\}\, dx \quad \forall y \in H_0^1(\Omega),$$

$$\langle J_2'(\mathring{u}), u\rangle_\infty = \int_\Omega h_u(x,\mathring{u}(x))u(x)\, dx \quad \forall u \in L\ (\Omega).$$

(In the last formula $\langle .,.\rangle_\infty$ denotes the scalar product between $L^\infty(\Omega)$ and $L^\infty(\Omega)^*$.)

<u>Proof:</u> Again using /6, part X, §1/ it is clear by (10) that the CARATHEODORY functions $g_{l_i}(x,l)$, $i = 0(1)n$, define continuous maps from $(L^2(\Omega))^{n+1}$ into $L^2(\Omega)$. Therefore with $\mathring{y} \in H_0^1(\Omega)$ we have $g_i(\mathring{y}) \in L^2(\Omega)$, $i = 0(1)n$, and by the structure theorem of $H^{-1}(\Omega)$ the term $J_1'(\mathring{y}) = g_0(\mathring{y}) - \operatorname{div} g(\mathring{y})$ belongs to $H^{-1}(\Omega)$. It is also easy to see that $\mathring{y} \longrightarrow J_1'(\mathring{y})$ is a continuous map from $H_0^1(\Omega)$ into $H^{-1}(\Omega)$. In order to prove that $J_1'(\mathring{y})$ gives the FRECHET derivation of J_1 at the point $\mathring{y} \in H_0^1(\Omega)$ we consider

$$\tilde{J}_1(z_0,\dots z_n) = \int_\Omega g(x,z_0,\dots z_n)\,dx,\ z_i \in L^2(\Omega).$$

By (10) it follows that $\tilde{J}_1$ has the partial FRECHET derivations

$$(\tilde{J}_{1z_i}(\tilde{z}_0,\dots\tilde{z}_n),z_i) = \int_\Omega g_{1_i}(x,\tilde{z}_0,\dots\tilde{z}_n)z_i dx,\ z_i \in L^2(\Omega),$$

$(i = 0(1)n)$ at an arbitrary fixed point $\{\tilde{z}_0,\dots\tilde{z}_n\} \in (L^2(\Omega))^{n+1}$. Because $Py = \{y, y_{x_1},\dots y_{x_n}\}$ defines a linear bounded operator acting from $H_0^1(\Omega)$ into $(L^2(\Omega))^{n+1}$ the functional $J_1(y) = \tilde{J}_1(Py)$ is FRECHET-differentiable at every point $\mathring{y} \in H_0^1(\Omega)$, where its FRECHET derivation is given by

$$\int_\Omega \left\{g_{1_0}(x,\mathring{y},\nabla\mathring{y})y + \sum_{i=1}^{n} g_{1_i}(x,\mathring{y},\nabla\mathring{y})y_{x_i}\right\} dx$$

for every $y \in H_0^1(\Omega)$. Considering (15) this proves the first part of the lemma. The second part we leave out here.

3. NECESSARY OPTIMALITY CONDITIONS

Summarizing the results of the previous section we can establish that our problem (P) fulfils the hypotheses needed in /1,3/, if we assume (2,3,8,10,11). Hence we are able to formulate necessary optimality conditions for the solution of (P) basing on these papers. Let $\{\mathring{y},\mathring{u}\} \in H_0^1(\Omega) \times L^\infty(\Omega)$ be a fixed optimal solution of (P). Then $N(L(\mathring{u}))$ and $N(L(\mathring{u})^*)$ denote the finite-dimensional null spaces of $L(\mathring{u})$ and $L(\mathring{u})^*$, respectively, where the operator $L(\mathring{u})^*$ adjoint to $L(\mathring{u})$ is given by

(18) $\quad L(\mathring{u})^*y = -\operatorname{div}\left\{A(\mathring{u})^*\nabla y + b(\mathring{u})y\right\} + (a(\mathring{u})|\nabla y) + c(\mathring{u})y$

and its corresponding bilinear form induces the operator $\mathbb{A}(\mathring{u})^*$ being the adjoint one to $\mathbb{A}(\mathring{u})$. In (18) $A(\mathring{u})^*$ denotes the adjoint matrix to $A(\mathring{u})$. In the following we must distinguish two cases in dependence on the dimension q of the subspace $N(L(\mathring{u})^*)$. If $q = \dim N(L(\mathring{u})^*) > 0$, then $\{z_j\}_{j=1(1)q}$ denotes a fixed basis belongig to $N(L(\mathring{u})^*)$. We introduce the abbrevation ($y,z \in H_0^1(\Omega)$, $u \in L^\infty(\Omega)$, $v \in V$, $\mu_0,\dots \mu_q \in R^1$)

(19) $\quad L_{v,u}(y,z;\mu_0,\dots\mu_q) = L_{v,u}(y,\tilde{z}) - \langle F'(v)u,\tilde{z}\rangle + \mu_0\langle J_2'(v),u\rangle_\infty$

$\qquad \tilde{z} = \mu_0 z + \mu_1 z_1 + \dots + \mu_q z_q.$

The main result is the following theorem.

<u>Theorem:</u> Let $\{\overset{\circ}{y},\overset{\circ}{u}\} \in H_0^1(\Omega) \times L^\infty(\Omega)$ be an optimal solution of (P). Supposing (2,3,8,10,11) the following necessary conditions are fulfilled:

(o) $$L(\overset{\circ}{u})\overset{\circ}{y} = f_0(\overset{\circ}{u}) - \operatorname{div} f(\overset{\circ}{u}) \quad \text{in } \Omega, \quad \overset{\circ}{y} = 0 \quad \text{on } \partial\Omega, \quad \overset{\circ}{u} \in V$$

(i) $$\int_\Omega \{g_0(\overset{\circ}{y})z + (g(\overset{\circ}{y})|\nabla z)\}\, dx = 0 \quad \forall z \in N(L(\overset{\circ}{u}))$$

(ii) Let $z_0 \in H_0^1(\Omega)$ be an arbitrary fixed solution of the adjoint boundary value problem

(20) $$L(\overset{\circ}{u})^* z = - g_0(\overset{\circ}{y}) + \operatorname{div} g(\overset{\circ}{y}) \quad \text{in } \Omega, \quad z = 0 \quad \text{on } \partial\Omega .$$

If $q = 0$, then

$$L_{\overset{\circ}{u},u-\overset{\circ}{u}}(\overset{\circ}{y},z_0;1,0,\dots 0) \geq 0 \quad \forall u \in V.$$

If $q > 0$ and $\xi_2(x) - \xi_1(x) \geq \text{const.} > 0$ a.e. $x \in \Omega$ (cf. (1)), then there exist q+1 real numbers $\mu_0,\dots\mu_q$ with $\mu_0 \geq 0$ and $|\mu_0| + \dots + |\mu_q| > 0$ such that

$$L_{\overset{\circ}{u},u-\overset{\circ}{u}}(y,z_0;\mu_0,\dots\mu_q) \geq 0 \quad \forall u \in V.$$

These assertions immediately follow by results decribed in /1,3/. (It must be noted that the adjoint states in /1/ and /3/, here given by the solutions of (20), differ in their sign.)

The optimality conditions in (ii) are integral inequalities. Applying a lemma by KRASNOSELSKI from these we can imply local conditions, too. Namely, if we introduce the notation

$$L_{v,u}(y,z;\mu_0,\dots\mu_q) = \int_\Omega H_v[y,z;\mu_0,\dots\mu_q](x)u(x)\, dx$$

(cf. (19)), then under the above assumptions the integral conditions are equivalent to

$$H_{\overset{\circ}{u}}[\overset{\circ}{y},z_0;\mu_0,\dots\mu_q](x)(u - \overset{\circ}{u}(x)) \geq 0$$
$$\forall u \in [\xi_1(x), \xi_2(x)] \quad \text{a.e. } x \in \Omega$$

with the corresponding parameters $\mu_0,\dots\mu_q$. Hereby and by the maximum principle for solutions of elliptic equations one can conclude bang-bang properties for the optimal control in various particular cases. We refer to /1/.

In some ways the above theorem gives final results. But, in the case $q > 0$ a result of preliminary character is of special interest. It reads as follows:

Supposing that the assumptions of the theorem are fulfilled, that the functions $h_j^o(.)$, $j = 1(1)q$, given by the defining relations

$$L_{\mathring{u},u}(\mathring{y}, z_j) - \langle F'(\mathring{u})u, z_j \rangle = \int_\Omega h_j^o(x)u(x)\, dx, \quad j = 1(1)q,$$

are linearly independent in $L^1(\Omega)$ and, finally, that there exists at least one $u \in \operatorname{int} V$ with

$$(21) \qquad \int_\Omega h_j^o(x)\, (u(x) - \mathring{u}(x))\, dx = 0, \quad j = 1(1)q,$$

the inequality

$$L_{\mathring{u},u-\mathring{u}}(\mathring{y}, z_o; 1,0,\ldots 0) \geq 0$$

is valid for <u>every</u> solution $z_o \in H_o^1(\Omega)$ of (20) and for every $u \in V$ satisfying the condition (21). In other word, this is a variant of the normal case.

Lastly we remark that we can obtain similar results described above for coefficients control problems in elliptic equations with smooth-convex cost functional; we refer to /1/.

BIBLIOGRAPHY

/1/ GOEBEL, M.: Zur Theorie optimaler Steuerprobleme im BANACH-Raum und deren Anwendung auf elliptische Systeme. Diss. B, Bergakademie Freiberg 1978.

/2/ GOEBEL, M.: Optimal control of coefficients in linear elliptic equations. I. Existence of optimal controls, to appear.

/3/ GOEBEL, M., v. WOLFERSDORF, L.: Optimale Steuerprobleme bei Noetherschen Operatorgleichungen III. Math. Nachr. <u>82</u> (1978), 77-85.

/4/ GOGODZE, I.K.: Necessary optimality conditions in elliptic control problems with integral constraints. Soobš. AN Gruz. SSR <u>81</u>, No. 1 (1976), 17-20; in Russian.

/5/ MIERSEMANN, E.: Lineare elliptische Zustände mit nichtlinearem Zielfunktional. ZAMM <u>54</u>, (1974), 19-25.

/6/ ZABREJKO, P.P. and others: Integral equations. Izd. Nauka, Moscow 1968; in Russian.

/7/ ZOLEZZI, T.: Necessary conditions for optimal controls of elliptic or parabolic problems. SIAM J. Control, vol. 10, Nr. 4 (1972), 594-607.

HADAMARD'S AND OPTIMIZATION-THEORETICAL INSTABILITY.

A. GÖPFERT +)

Summary. One of the most famous results of the optimization theory is the Rockafellar stability theorem for a pair of dual problems, f.i. written in the Fenchel form with conjugate functionals. Instability is corresponding to a duality gap or nontaking of one of the optimal values. Further on f.i. an initial value problem for an elliptic differential equation is generally incorrectly stated in the sense of Hadamard, that is especially, the solution has an instable behaviour, if it is existing and the dates are disturbed. It is pointed out, done with problems of approximation type, that the just mentioned instability is producing also an optimization-theoretical instability. Furthermore some other results and open questions, connected with these problems, are given.

Introduction

An important and in many cases open problem in connection with control problems is, to characterize the set of states (reachable set), which is given by the admissible controls. We are especially interested in the case, that the set of states is dense in a function-space. In the following we will give a class of control-problems which is leading to such a problem. As a realization we state a problem of approximation-type, founded in the linear theory of elasticity, where in the proof we have to use the jump-relations of H. Weyl and W.D. Kuprade for potentials of the theory of elasticity in a wider sense. Then we discuss this problem as an usual optimization problem. Therefore it must be possible to characterize the set of nonreachable states by the optimization-theoretical instability of the dual problem. As background of this unstable dual problem we are finding out an incorrectly stated problem in the sense of Hadamard for the strong elliptic system of partial differential equations of the linear theory of elasticity.

+) Prof. Dr. A. Göpfert, Technische Hochschule Leuna-Merseburg, Sektion Mathematik-Rechentechnik, 42 Merseburg, Geusaer Str.

1.

If A is a linear continuous operator, which is mapping the Hilbert space $\mathfrak{X}$ into the Hilbert space $\mathfrak{Y}$, and if the kernel of the adjoint operator A^* is only $\{o\}$, then the range R(A) is dense in $\mathfrak{Y}$. If not, there would be an element $h \in \mathfrak{Y} \ominus R(A)$, $h \neq o \in \mathfrak{Y}$, with the property

$$0 = \langle h, Ax \rangle_{\mathfrak{Y}} = \langle A^* h, x \rangle_{\mathfrak{X}} \quad \forall x \in \mathfrak{X} , \tag{1}$$

and therefore $A^* h = o$, that is $h = o$. Let b be an element from $\mathfrak{Y}$. If $b \notin R(A)$, then b is not reachable by the set of controls .

2.

As direct realization of 1. let us now consider a simply connected domain D in R^3, the boundary ∂D of which may be decomposed in the set $(\partial D)_1$, having a representation by real-analytical functions, and the remaining part $(\partial D)_2$, belonging piecewise to the class $C^{2,\lambda}$, $0 < \lambda \leq 1$. Let Γ be a closed surface of the class $C^{2,\lambda}$, contained completely in the interior of D which divides D (see fig. 1). We now make the assumption, that D is an elastic body, obeying the classical linear theory of elasticity. Hence, the dilatation-vector $\vec{u}$ is satisfying the system of elliptic differential equations

$$\Delta^* \vec{u} = \Delta \vec{u} + \frac{\lambda + \mu}{\mu} \text{ grad div } \vec{u} = 0 \tag{2}$$

in int D; μ and λ are the moduls of elasticity of Lamé, $\lambda + 2\mu > \frac{4}{3}\mu$, $\mu > 0$. To this elliptic system of differential equations there exists [7] the Green's tensors G_I, G_{II}, belonging to the following two boundary value problems:

$$\text{(I)} \quad \vec{u}(\partial D) = \vec{\varphi}(Q), \quad Q \in \partial D, \ \vec{\varphi} \text{ continuous} \tag{3a}$$

$$\text{(II)} \quad T\vec{u}(\partial D) = \vec{\psi}(Q), \quad Q \in \partial D, \ \vec{\psi} \text{ hölderian continuous,} \tag{3b}$$

where T is the vector differential operator with the following property: in a point of the surface ∂D with the normal n, T means the stress-vector

$$T\vec{u} = 2\mu \frac{\partial \vec{u}}{\partial n} + \lambda n \operatorname{div} \vec{u} + \mu (n \times \operatorname{rot} \vec{u}). \tag{4}$$

The solution of the first boundary value problem is existing and uniquely determined, the second has 6 (linearly independent) eigen-solutions $\vec{a}_i(P)$ (translation and rotation of a rigid body). Therefore the solution of the second problem does not exist for arbitary $\vec{\psi}$, the conditions for solubility are

$$\int_{\partial D} \vec{a}_i(Q) \quad \vec{\psi}(Q) \, d\sigma_Q = 0, \qquad i = 1,2,\ldots,6. \tag{5}$$

The columns of the Green's tensors are satisfying the following conditions relative to P:

$$\Delta^*_P G_I (P,Q) = 0, \; P \in \operatorname{int} D, \; P \neq Q \tag{6a}$$

$$G_I(P,Q) = 0, \; Q \in D, \; P \in \partial D \tag{6b}$$

$$\mu \Delta^*_P G_{II}(P,Q) = \sum_{i=1}^{6} \vec{a}_i(P) \, \vec{a}_i{}^T(Q) \tag{7a}$$

$$T_P G_{II}(P,Q) = 0, \; Q \in \operatorname{int} D, \; P \in \partial D. \tag{7b}$$

The representations of the solutions with the Green's tensors $G_I(P,Q)$, $G_{II}(P,Q)$ read as follows:

$$\text{(I):} \qquad \vec{u}(P) = \int_{\partial D} T_Q G_I(P,Q) \quad \vec{\varphi}(Q) \, d\sigma_Q \tag{8a}$$

$$\text{(II):} \qquad \vec{u}(P) = \int_{\partial D} G_{II}(P,Q) \quad \vec{\psi}(Q) \, d\sigma_Q + \sum_{i=1}^{6} c_i \vec{a}_i(P). \tag{8b}$$

Let $\mathbb{X}$ be the triple product of spaces $L_2(\partial D)_1$, $\mathbb{Y}$ the triple product of $L_2(\Gamma)$ and let Ax be the restriction of the solution of the boundary value problems to Γ, where the boundary values may be zero along $(\partial D)_2$. Then $\vec{u}$ in (8a) or (8b) appear as generalized solutions of the boundary value problems. We can now state the following

<u>Theorem:</u> The kernel of A^* is only the zero-element from $\mathbb{Y}$.

<u>Proof.</u> Let us consider the first boundary value problem. If, in opposition to the theorem, there would be a function $\vec{h} \neq o$, $\vec{h} \in L_2^3(\Gamma)$ in ker A^*, it follows that

$$0 = \left\langle A^{*}h,\ x\right\rangle_{\mathfrak{X}} = \int_{(\partial D)_1} \int_{\Gamma} \vec{h}(P)\ T_Q G_I(P,Q)\ \vec{\varphi}(Q) d\tau_P d\sigma_Q \qquad (9)$$

for all $x \in \mathfrak{X}$; therefore almost everywhere on $(\partial D)_1$ we have

$$T_Q\ \vec{F}(Q) = \int_{\Gamma} \vec{h}(P)\ T_Q G_I(P,Q) d\tau_P = 0, \qquad (10)$$

and because of the integral is a continuous vector-function for Q on $(\partial D)_1$, (10) holds for all Q of $(\partial D)_1$.
Now we place the operator T before the integral and, having in mind, that the lines of G_I relative to Q are satisfying the homogeneous first boundary value problem ($Q \neq P$), we can complete (10) to the following Cauchy initial value problem:

$$\left.\begin{aligned} \vec{F}\ (Q) &= 0 \\ T_Q\ \vec{F}\ (Q) &= 0 \end{aligned}\right\} \quad Q \in (\partial D)_1 \qquad (10a)$$
$$\Delta^{*}_Q\ \vec{F}\ (Q) = 0, \qquad Q \in (\text{int } D) \setminus \Gamma$$

On the one hand, this is an incorrectly stated problem in the sense of Hadamard, on the other hand, because of our hypotheses, we can use the theorem of Cauchy-Kowalewski: the solution of (10a, 10b) is existing and uniquely determined. $\vec{F}$ is a solution. Hence $\vec{F}$ is vanishing in that part of ∂D, that is lying outside of the closed surface Γ (that is, in int D_a, fig. 1). We have to conclude h = o. On that account we are considering $\vec{F}(Q)$ as a "generalized" elastic potential of the simple layer: the kernel is not the fundamental solution S (Tensor of Somigliana [7]) from Δ^{*}, but G_I (G_I-S is regular in int D), furthermore the density function $\vec{h}$ is only an element in L_2^3 (Γ). We are seeking jump-relations along Γ. H.Weyl [7] introduced the operator N instead of T:

$$N\vec{u} = \frac{\lambda+3\mu}{(\lambda+2\mu)\,2\mu^2}\,\frac{\partial}{\partial n}\vec{u} + \frac{(\lambda+\mu)(\lambda+3\mu)}{4\mu^3(\lambda+2\mu)}\,n\,\mathrm{div}\,\vec{u} + \frac{\lambda+\mu}{\lambda+3\mu}\left(n \times \mathrm{rot}\,\vec{u}\right). \qquad (11)$$

N has especially the property, that $N_Q S(P,Q)$ for $Q \rightarrow P$ has componentwisely singularities only like $\frac{\partial}{\partial n_Q}\frac{1}{r}$, while $T_Q S$ is possessing a strong singularity, and this is corresponding (in connection with jump-relations) to the behaviour of potentials of the usual Laplace equation. Therefore (see chapter 3) we can show, that the classical jump-relations of W.D. Kupradse and H. Weyl [4, 7] are also valid for L_2-density functions: for the elastic potential $\vec{F}$ of the simple-layer type are valid for $\vec{h} \in L^3(\Gamma)$, (let $\overline{D} = \partial D \cup D_a \cup D_i \cup \Gamma$, see fig. 1) the jump-relations:

$$\lim_{\substack{Q \to P_o, Q \in D_a \\ Q \to P_o, Q \in D_i}} N\vec{F}(Q) = N\vec{F}(P_o) \pm \frac{2\pi}{\mu} \vec{h}(P_o) \tag{12}$$

for almost all $P_o \in \Gamma$.

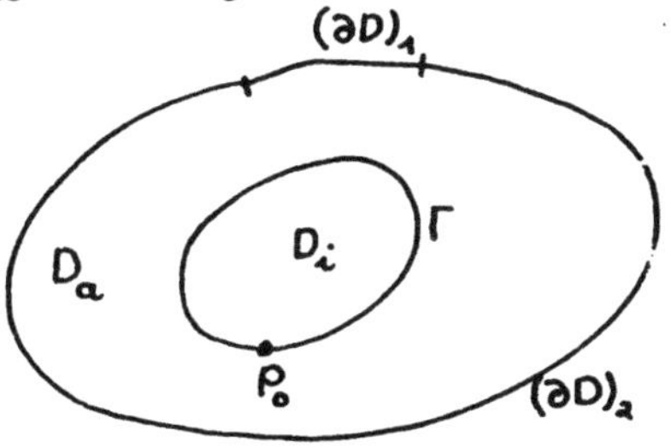

Figure 1

Taking the limes for $Q \in D_a$, because of $\vec{F} = 0$ in D_a, we get the weakly singular integral equation system of Fredholm type along Γ

$$0 = N\vec{F}(P_o) + \frac{2\pi}{\mu} h(P_o). \tag{13}$$

By iteration it follows that $\vec{h}$ is continuous. The simple layer potential with continuous density function is going continuously through the surface Γ, hence $\vec{F} = 0$ also (uniqueness of the first boundary value problem) in D_i. The difference of the jump-relations (12) then gives $\vec{h} = o$.

Finally we are considering the second boundary value problem. For the proof we state $Y = L_2^3(\Gamma) \ominus \{a_1, \ldots a_6\}_{|\Gamma}$, where $\{\ldots\}_{|\Gamma}$ is the linear hull H of the eigen-functions, restricted to Γ. Then with the analogous procedure with $\vec{h} \neq o$, $\vec{h} \in Y$ we get the initial value problem

$$\Delta^*_Q F_{II} = 0 \quad \text{for } Q \in (\text{int } D) \setminus \Gamma \tag{14a}$$

$$\vec{F}_{II}(Q) = T\vec{F}_{II}(Q) = 0 \quad \text{for } Q \in (\partial D)_1 \tag{14b}$$

with $\vec{F}_{II} = \vec{F} - \vec{l}$, where $\vec{F}$ is formed likewise $\vec{F}$ in (9), but with the kernel-function G_{II}, and $\vec{l}$ is an element of H . Again we obtain the same singularity-behaviour, $\vec{l}$ gives no effect relative to the jump-relations. Therefore we can conclude: $\vec{h} = o \in Y$. To each solution of the boundary value problem II we can add eigen-functions, so supplementary we can drop the restriction of $\mathbf{Y}$ by adding $\mathbf{H}$.

Together with chapter 1 our theorem also implies, that we are able to approximate (in a closed domain, lying in int D_i) each possible displacement-field by variation of the displacements or stress-vectors along a (small) part of the boundary $(\partial D)_1$, because the displacements in D_i are determined by the displacements on Γ and these we can L_2-approximate. There is an interesting question: is it possible, that the set the solutions of an elliptic boundary value problem, restricted to a closed surface Γ in D, is a dense subset in the sense of the norm of the Banach-space of continuous functions over Γ ? Some weaks ago G. Wanka +), Merseburg, had found the proof for the Dirichlet boundary value problem related to the Laplacian equation in R^n. He had used the balayage-principle, because instead of $h \in R(A)^{\perp}$ he had to take a Borel-measure as element of the annihilator of R(A) and for such potentials there are no jump-relations. It is also of interest, that there are such results [2] of approximation for problems of the Laplacian equation, if the domain for the solutions is constructed by subsets with sufficiently regular boundaries, and the solutions have to satisfy transition relations along the boundaries between domains with different parameters $\varepsilon_i > 0$.

In accordance with figure 2 let $D_0, D_1 \dots D_m$ be of class $C^{2,\lambda}$, let the Hilbert space $\mathbf{X}$ be given on the boundary of D_1, the Hilbert space $\mathbf{Y}$ on Γ (closed or not, $\Gamma \cap D_i = \emptyset$). Then we consider the solutions of the following boundary value problem

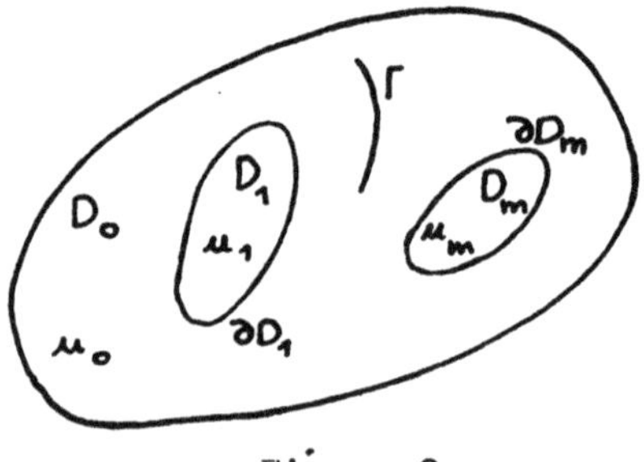

Figure 2

$$\Delta u(P) = 0 \text{ for } P \in D_i,\ u \in C^1(\overline{D}_i),\ i = 0,\dots,m$$

$$u_{|\partial D} = f,\ f \in C^{1,\alpha} \quad (\partial D),$$

$$u \in C^2(D_i),\ i = 0,\dots,m \qquad (u \equiv u_j \quad \text{in } D_j)$$

+) International Congress for Optimization, Eisenach, Nov. 1978 with the contact conditions

$$u_0|_{\partial D_k} = u_k|_{\partial D_k} \quad , \; k = 1,\dots, m,$$

$$\left(\varepsilon_0 \frac{\partial u_0}{\partial n} - \varepsilon_k \frac{\partial u_k}{\partial n}\right)\Big|_{\partial D_k} = g_k \quad , \; g_k \in C^{0,\alpha}(\partial D_k)\,, \; \varepsilon_i > 0, \; i = 0,1,\dots,m.$$

The solution exists uniquely in the mentioned classes [6] . The problem is taken from the stationary theory of heat. If we are controlling in Υ along Γ with g_k as the control function. $\underline{k}$ fixed, we are also getting a result of density [2] . It is an open question, if it is possible to control along Γ with the contact function $g_{\underline{k}}$, where $g_{\underline{k}}$ is zero on a given part of $\partial D_{\underline{k}}$.

3.

The proof of the jump-relations (12) of W.D. Kupradse/H.Weyl but for density functions $\vec{h} \in L_1^3(\Gamma)$ can be done analogously to such proofs in Günter [3] . Let P_o be a Lebesgue-point of the density function $\vec{h}(Q)$ in the potential

$$\vec{F}(Q) = \int_\Gamma \vec{h}(P) G_I(P,Q) d\tau_P,$$

let Q be a point, lying on the normal in P_o relative to Γ , then we have (see Fig.3)

$$\begin{aligned} \vec{F}(Q) &= \int_\Gamma \vec{h}(P) N_Q G_I(P,P_o) d\tau_P \\ &+ \vec{h}(P_o) \int_\Gamma [N_Q G_I(P,Q) - N_Q G_I(P,P_o)] \, d\tau_P \\ &+ \int_\Gamma (\vec{h}(P) - \vec{h}(P_o)) N_Q G(P,Q) d\tau_P \\ &- \int_\Gamma (\vec{h}(P) - \vec{h}(P_o)) N_Q G(P,P_o) d\tau_P. \end{aligned} \qquad (15)$$

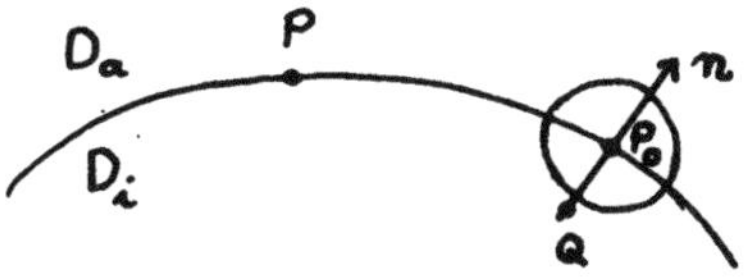

Figure 3

The first term in (15) is the direct value $N\vec{F}(P_o)$; according to the theorem of Fubini, it is summable. The second term is giving, analogous to regular density functions, the actual jump-terms, the third and fourth integral are vanishing (if $Q \to P_o$). We now have to split the surface Γ using round P_o a Ljapunov-ball and to treat then both the last integrals componentwisely. As we have just said, the orders of singularities in that case are not stronger than that of the usual potential theory, so we can use the lemma of Fichera [s. 3] and obtain the postulated result.

In the realization of the situation in chapter 1, spoken about in the last chapters, it cannot hold $R(A) = \mathbf{Y}$, obviously. Hence, if we state for $b \in \mathbf{Y} \setminus R(A)$ the problem

$$\| b - Ax \| = \text{Inf!}, \quad x \in \mathbf{X} , \tag{16}$$

there is no $\bar{x} \in \mathbf{X}$ with $\| b - A\bar{x} \| = 0$. Therefore the dualproblem to (16) in the sense of the optimization theory has to be instable exactly then, if $b \in \mathbf{Y} \setminus R(A)$, that is, if b is not reachable. We shall see, that just this instability in the sense of optimization theory we can interprete by an instable initial value problem in the sense of Hadamard. The dual to (16) problem has the form

$$\left\{ \langle b, w^* \rangle_{\mathbf{Y}} \mid A^* w^* = o, \ \| w^* \|_{\mathbf{Y}} \leqq 1 \right\} = \text{Sup!} , \tag{17}$$

and this is a result of the general duality theorem of the approximation theory [5] . Obviously we see by our premises, that the maximal value of (17) is zero, and so there is no gap of duality between (17) and (16). Hence, the optimization-theoretical instability must exactly correspond to the fact, that in the primal problem there is no argument $\bar{x}$, at which the objective function is achieving its infimum. And now: the problem (17) is instable in the sense of Rockafellar, (s. f.i. [1]), if a general local disturbation z^* of the restrictions of the dual problem (17) has the consequence, that the directional derivatives of the maximal value function of the disturbed problem (17) at the point $z^* = 0$ are bounded above. The disturbation z^* is related to both of the conditions (17); but the disturbation is to choose sufficient small, so that disturbation of the norm-condition in (17) has no effect, only of substance is the disturbation of the operator equation $A^* w^* = o$ to

$$A^* w^* = z_1^*, \quad z_1^* \in U(o) \subset \mathbf{X} . \tag{18}$$

Explicitely the last terms have the form, remembering (1) and (9):

$$T_Q \vec{F}(Q) = z_1^*, \quad z_1^* \in U(o) \subset \mathbf{X} . \tag{19}$$

Together with (10a), (10b):

$$\vec{F}(Q) = 0, \qquad \Delta_Q^* F(Q) = 0, \tag{20}$$

we have (relative to our spaces) with (19) and (20) a disturbed incorrectly stated initial value problem in the sense of Hadamard (because of the differential system in (20) is elliptic).
The instable behaviour of this produces such w^*, that in dependence of b the mentioned set of directional derivatives of the maximal value function of the disturbed problem (17) is not bounded above.

Bibliography

[1] Göpfert, A., Mathematische Optimierung in allgemeinen Vektorräumen, Leipzig, Teubner 1973

[2] Göpfert, A., Wanka, G., Zur Approximation durch Lösungen partieller Differentialgleichungen, XXI. Int. Wiss. Koll. 1976, Reihe A1, 61-64

[3] Günter, N.M., Die Potentialtheorie und ihre Anwendung auf Grundaufgaben der mathematischen Physik, Leipzig, Teubner 1957

[4] Kupradse, W.D., Randwertaufgaben der Schwingungstheorie und Integralgleichungen, DVW Berlin 1956

[5] Tichomirov, W.M., Einige Fragen der Theorie der Näherungen, Moskau, Mir, 1976 (russ.)

[6] Wanka, G., Wanka, J., Die erste Randwertaufgabe der Potentialtheorie mit Kopplungsbedingungen, Wiss. Z. KMU 25 (Heft 1) 1976, 79-87.

[7] Weyl, H., Das asymptotische Gesetz der Eigenschwingungen eines beliebig gestalteten elastischen Körpers, Rend. Circ. Mat. di Palermo 39, 1915, 1-50.

NUMERICAL TREATMENT OF ABEL INTEGRAL EQUATIONS

RUDOLF GORENFLO [1)]

Summary: Several numerical methods for treating first kind Abel integral equations are surveyed: (a) quadrature methods, (b) fitting and smoothing techniques, (c) estimation in function spaces. The methods (b) and (c) offer the possibility of using extra information on the shape of the solution (e.g. non-negativity, monotonicity, convexity) often available when physical measurements are to be evaluated. The author proposes a method of piecewise linear interpolation capable of using such shape information via quadratic optimization under linear inequality constraints.

§ 0 Introduction

For $s,t \in [0,1]$, $\alpha \in (0,1)$ consider Abel's integral equation

$$(1) \quad (A\varphi)(t) := \int_{s=0}^{t} \frac{\varphi(s)\,ds}{(t-s)^{\alpha}} = \phi(t).$$

The problem (1) of determining φ when ϕ is given is ill-posed in customary pairs of function spaces; let us simply remark that $A: L_2(0,1) \to L_2(0,1)$ or $C[0,1] \to C[0,1]$ does not have a bounded inverse. Formally (see text books on integral equations or on operator calculus) the solution is given by

$$(2) \quad \varphi(s) = \frac{\sin(\alpha\pi)}{\pi} \frac{d}{ds} \int_{t=0}^{s} \frac{\phi(t)\,dt}{(s-t)^{1-\alpha}}$$

$$= \frac{\sin(\alpha\pi)}{\pi} \left\{ \frac{\phi(0)}{s^{1-\alpha}} + \int_{t=+0}^{s} \frac{d\phi(t)}{(s-t)^{1-\alpha}} \right\}.$$

Of particular importance is the case $\alpha = 1/2$ already studied by N. H. Abel (1825, tautochrone problem).

The second, Stieltjes integral form of the solution offers us an intuitive visualization of illposedness. If $\phi(t) = \varepsilon$ for $0 < a_1 < t \le a_2 < 1$ and 0 else, then (see Fig. 1)

$\pi\varphi(s)/\sin(\alpha\pi) = 0$ for $0 \le s \le a_1$,

$\varepsilon(s-a_1)^{\alpha-1}$ for $a_1 < s \le a_2$, and

$\varepsilon\{(s-a_1)^{\alpha-1} - (s-a_2)^{\alpha-1}\}$ for $a_2 < s \le 1$.

1) Fachbereich Mathematik
Freie Universität Berlin
Arnimallee 2-6
D - 1000 Berlin 33

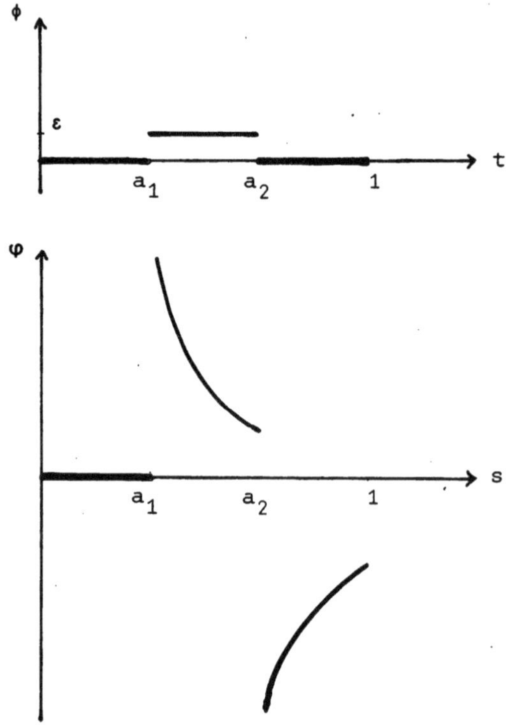

Fig. 1

For $\alpha = 1/2$ the substitutions $x^2 = 1-t$, $r^2 = 1-s$, $\phi(t) = G(x)$, $\varphi(s) = g(r)$ yield the equation (for $0 \leq r \leq 1$, $0 \leq x \leq 1$)

$$(3) \quad 2 \int_{r=x}^{1} \frac{g(r)\, rdr}{\sqrt{r^2-x^2}} = G(x)$$

which is often used in the evaluation of spectroscopic measurements (see [1] and [8]).

$$G(x) = \int_{y=-\bar{y}}^{\bar{y}} g(r)\, dy, \text{ with } r = \sqrt{x^2+y^2} \text{ and } \bar{y} = \sqrt{1-x^2},$$

is obtained by side-on measurement of an intensity $g(r)$ of radiation radially distributed in the unit circle $\{(x,y) \in \mathbb{R}^2 \mid x^2+y^2 \leq 1\}$, integration being carried out along the straight line connecting the peripheral points $(x,-\bar{y})$ and $(x,\bar{y})$.

Equation (1) is also interesting as a model equation for more general problems, e.g. for equations

$$(4)\quad \int_0^t \frac{k(t,s)\varphi(s)\,ds}{(t-s)^\alpha} = \phi(t)$$

or even for nonlinear equations

$$(5)\quad \int_0^t \frac{k(t,s,\varphi(s))ds}{(t-s)^\alpha} = \phi(t).$$

A theorem of Kowalewski [10] may be quoted. If in (4) k and $\partial k/\partial s$ are continuous for $0 \le s \le t \le 1$, $k(t,t) \neq 0$ for $0 \le t \le 1$, and

$\frac{d}{dt}\int_0^t \frac{\phi(s)\,ds}{(t-s)^{1-\alpha}}$ is continuous for $0 \le t \le 1$, then equation (4) has exactly one solution $\varphi \in C\,[0,1]$.

Theorems on existence, uniqueness, and smoothness of solutions of (5) can be found in Branca's papers [3] and [4].

§ 1 Quadrature Methods

Since a long time physicists have applied naive interpolatory quadrature methods for numerical solution of (1) and (3). Usually they lead to triangular systems of linear equations. From the large literature on this subject we only quote [1], [2], and [8]. A rigorous theory of convergence of some of these methods (for successive refinement of the computational grid) has been worked out in this decade by R.S. Anderssen and R. Weiss ([16] and [17]). They investigate "product integration" methods for equation (4).

For $N \in \mathbb{N}$ let $h = 1/N$, $t_i = ih$, $\phi_i = \phi(t_i)$, and try to approximate φ by a step-function $\tilde{\varphi}$ equal to a value $\tilde{\varphi}_{i+\frac{1}{2}}$ in each interval (t_i, t_{i+1}).

Putting $k_{i,j+\frac{1}{2}} = k(t_i, t_{j+\frac{1}{2}})$ and $W_\lambda = \int_{(\lambda-1)h}^{\lambda h} s^{-\alpha} ds$ leads to a triangular system of linear equations

$$\sum_{j=0}^{i-1} \tilde{\varphi}_{j+\frac{1}{2}}\, k_{i,j+\frac{1}{2}}\, W_{i-j} = \phi_i,\quad i = 1,2,\ldots,N.$$

If $|k(t,t)| \geq k_o > 0$ for $t \in [0,1]$, k and $\partial k/\partial s$ are L(ipschitz)-continuous in s and in t for $0 \leq s \leq t \leq 1$, and if φ is L-continuous, then $\tilde{\varphi}$ converges uniformly to φ as $N \to \infty$. If further φ' is L-continuous then convergence is $O(h)$. If still further $\varphi(0) = \varphi'(0) = 0$ and if k satisfies some additional smoothness requirements then convergence is $O(h^{2-\alpha})$. Compare [16].

By trapezoidal product integration higher order of convergence is obtainable (compare [17]). Here in (4) for $t = t_i$ the left hand numerator $k(t_i,s)\varphi(s)$ is replaced by

$$((t_{j+1}-s)k_{ij}\,\tilde{\varphi}_j+(s-t_j)k_{i,j+1}\,\tilde{\varphi}_{j+1})/h \text{ for } t_j \leq s \leq t_{j+1}.$$

Now if k and φ are sufficiently smooth and if $\alpha \in [\alpha_1,1]$ the method is $O(h^2)$-convergent. The constant $\alpha_1 \approx 0.2117$ stems from the technique of proof.

Branca ([3] and [4]) analyzes a piecewise linear interpolation method for the nonlinear equation (5) and shows it to be $O(h^2)$-convergent under appropriate smoothness assumptions. By a piecewise quadratic interpolation technique he obtains an $O(h^3)$-convergent method for the classical Abel equation (1) in the special case $\alpha = 1/2$.

Brunner [5] has developed a scheme of "implicit interpolation" for (4). He tries to approximate by piecewise polynomials of degree $\leq m$ which are continuously stuck together at the nodal points of a partition $0 = \xi_0 < \xi_1 < \ldots < \xi_N = 1$.

The polynomials are determined by collocation at an appropriate set of equidistant points in each subinterval. For $N \to \infty$ and regular refinement of the partition $O(H^m)$ can be attained as oder of convergence in the nodes ξ_j, but only $O(H)$ elsewhere; $H = \max\,(\xi_{k+1} - \xi_k)$.

§ 2 Evaluation of Measurements

The methods in § 1 are very useful if ϕ or G is smooth enough and known very accurately (e.g. given by explicit formula). Spectroscopists however often experience a lot of trouble when using quadrature methods for evaluating their measurements via equation (3). Noise is amplified, the more the finer the computational grid is taken. Even worse, shape properties (for example non-negativity) which the true solution must have for physical reasons are often not reproduced by the numerical computation. A kind of regularization is required.

As a model of the measuring procedure let us assume that there are given linear functionals α_j for $j = 1,2,\ldots,J$, acting on G, but perturbed by unknown random errors ρ_j (noise), so that as available information we have values

(6) $a_j = \langle\alpha_j, G\rangle + \rho_j$, $j = 1,2,\ldots,J$.

In the simplest case we have

(7) $\langle\alpha_j, G\rangle = G(x_j)$, $0 \leq x_1 < x_2 < \ldots < x_J \leq 1$,

but in a more realistic model we have

$\langle\alpha_j, G\rangle = \int_0^1 w_j(x) G(x)\, dx$ with non-negative weight functions w_j.

We want the approximate solution $\tilde{g}$ to share one or more of the following properties with the exact solution g: smoothness, non-negativity, boundedness (by a known constant), monotonicity, convexity, unimodality, etc. With linearly independent basis functions g_n and functions G_n. calculated from the g_n via (3) take

(8)
$$g(r) \approx \tilde{g}(r) = \sum_{n=0}^{N} c_n g_n(r)$$
$$G(x) \approx \tilde{G}(x) = \sum_{n=0}^{N} c_n G_n(x).$$

With the positive weights γ_j (most simply all $\gamma_j = 1$) fit $\tilde{G}$ to the data $a_j = \langle\alpha_j, G\rangle + \rho_j$ by minimizing

(9) $$Q(c) := \sum_{j=1}^{J} \gamma_j \left(a_j - \langle\alpha_j, \tilde{G}\rangle\right)^2$$

(or another measure of deviation). N may be smaller or larger than J+1, the particular choice N=J+1 corresponds to interpolation ($Q(c) = 0$). Here c denotes the vector with components c_n. The quadratic form Q is strictly positive definite if the matrix $(\langle\alpha_j, G_n\rangle \mid j = 1,2,\ldots,J;\ n = 0,1,\ldots,N)$ has rank N+1 (this, of course, requires $N+1 \leq J$).

Gorenflo and Kovetz in [9] and Minerbo and Levy in [14] take polynomials or polynomials multiplied by a suitable function of r as functions g_n. and N small, thus achieving "smoothness" (in a colloquial sense: oscillations suppressed). Disadvantages: low accuracy, choice of larger values N leads to unwanted oscillations of $\tilde{g}$. By forcing $\tilde{g}(r) \geq 0$ at a finite set of points (this leads to a quadratic optimization problem, see (9)) Gorenflo and Kovetz obtain functions $\tilde{g}$ "approximately non-negative".

A very interesting variant has been proposed and numerically tested by Larkin [11] for evaluation of inaccurate measurements. In an appropriate function space $\tilde{F}$ (representing the set of candidates for the exact solution), he determines an element by a maximum likelihood estimation technique. As an example he treats the problem of this section with $\langle \alpha_j, G \rangle$ of the form

$$\int_{x_j}^{x_{j+1}} G(x)\, dx \text{ and looks for non-negative solutions.}$$

He takes $\tilde{F} = \{g \in L_1(0,1) \mid g(x) \geq 0 \text{ almost everywhere}\}$.

After deriving conditions for the measured values to be compatible with a non-negative solution g, he arrives at the problem of minimizing

$$\int_0^1 \tilde{g}(r) \log \tilde{g}(r)\, r\, dr \text{ under the restriction that } \tilde{g} \text{ should pro-}$$

duce the values a_j.

This problem can be treated by the Lagrange multiplier method.

If the compatibility conditions are not met, complications arise which require statistical techniques of correcting the a_j so as to fulfill these conditions.

As a second method Larkin proposes an ansatz $\tilde{g}(r) = (h(r))^2$ so that certainly $\tilde{g}(r) \geq 0$. To obtain smoother solutions than the ones usually obtainable by his first method he now minimizes

$$\int_0^1 (h'(r))^2\, r\, dr \text{ under the restrictions that g produces the values}$$

a_j and the (physically motivated) boundary conditions $h'(0) = h(1) = 0$.

The Euler-Lagrange technique leads to a nonlinear generalized eigenvalue problem for which he gives an iterative method.

§ 3 Continuous Piecewise Linear Ansatz

The author (with some students) is presently experimenting with continuous piecewise linear approximating functions $\tilde{g}$, thereby using the well-known "hat" (or "roof") functions as basis functions g_n. With this ansatz non-negativity or monotonicity or convexity can be tested by looking at a finite set of nodes so that the constraints of the minimization problem - see (9) - consist of a finite set of linear inequalities. It has a unique solution if Q is strictly positive definite. It should be stressed that in case (7) the points at which G is measured can be chosen independently of the nodal points of the ansatz.

Theoretical Justification and Numerical Tests

Sufficiently strong inequality conditions force the approximate solution to lie in a compact subset of $C[0,1]$ or $L_2(0,1)$ (independently of N and J). For appropriate compactness criteria look at those of Helly, Arzelà and Ascoli, Rellich, and at embedding theorems. See also Tikhonov and Arsenin ([15], Theorems 1 and 2 of chapter I) for conditions for uniqueness and continuous dependence of quasi-solutions (in compact subsets) on the data . Compare also Eckhardt ([6] and [7]).

We have carried out some numerical case studies in order to test the method's ability to suppress unwanted oscillations caused by perturbing G. We present results for the pair

$$G(x) = \frac{3\pi}{16}(1-x^2)^2, \quad g(r) = \frac{1}{2}(1-r^2)^{3/2}.$$

G(x) was used (according to (7)) at 31 equidistant points of [0,1] (step-length 1/30), whereas the nodes of the ansatz were taken as 11 equidistant points of [0,1] (step-length 1/10).

One computation (a) was with exact values $G(x_j)$, the other one (b) with perturbed values $G^*(x_j) = G(x_j) + 0.05 \sin(100\, x_j)$.

The errors $|\tilde{g}(r) - g(r)|$, multiplied by 1000, are given in Table 1. Intensive further test computations taking account also of extra conditions such as monotonicity etc. are planned.

r	(a)	(b)
0.0	1.50	107
0.1	1.13	22.2
0.2	1.16	4.50
0.3	1.03	2.78
0.4	0.928	0.395
0.5	0.626	0.832
0.6	0.543	0.503
0.7	0.439	0.414
0.8	0.371	0.366
0.9	4.84	4.87
1.0	$1.75 \cdot 10^{-7}$	$1.75 \cdot 10^{-7}$

Table 1

Conclusions:

Cleverly devised quadrature methods are very good for problems with smooth right hand sides known to a high precision but of a rather low value for evaluation of inaccurate measurement data. In this latter case one should better use Larkin's methods or the method proposed in § 3. These methods can, of course, be generalized to other types of first kind integral equations, but it is worthwhile to study in detail their application to Abel integral equations: effective methods in this area find a high resonance among computing **physicists** (in particular among spectroscopists).

Much remains to be done. The theoretical foundations should be laid deeper as hinted at in this paper, Larkin's ideas should be made applicable to other kinds of a-priori extra information (he only considered non-negativity), the method of § 3 should be thoroughly investigated from both, the theoretical and computational viewpoints, and the problem of how best to choose the functionals α_j (measuring points in case (7)) and the nodes of the continuous piecewise linear ansatz should be studied. For analyzing this last problem appropriate criteria of optimality have to be developed (ideas of the book [13] might prove useful in this respect).

Acknowledgement: The author is indebted to Dipl.-Math. W. Neun for writing computer programs and for carrying out several numerical case studies.

Bibliography:

[1] BIELSKI,A., KACZMAREK,W., KUBRYCHT,J., WOLNIKOWSKI,J.: On the Determination of the Radial Intensity Distribution in Cylindrical Plasma. Acta Physica Polonica 33 (1968), 701-709.

[2] BRACEWELL, R.N.: Strip Integration in Radio Astronomy. Australian Journal of Physics 9 (1956), 198-217.

[3] BRANCA, H.W.: The Nonlinear Volterra Equation of Abel's Kind and Its Numerical Treament. Computing 20 (1978), 307-324.

[4] BRANCA, H.W.: Die nichtlineare Volterra-Integralgleichung vom Abelschen Typ und ihre numerische Behandlung. Dissertation Köln 1976.

[5] BRUNNER, H.: Global Solution of the Generalized Abel Integral Equation by Implicit Interpolation. Math. of Comp. 28 (1974), 61-67.

[6] ECKHARDT, U.: Zur numerischen Behandlung inkorrekt gestellter Aufgaben. Computing 17 (1976), 193-206.

[7] ECKHARDT, U.: Incorrectly Posed Problems. ISNM 31 (1976), 57-63.

[8] EDELS,H., HEARNE,K., YOUNG,A.: Numerical Solutions of the Abel Integral Equation. J. Math. and Phys. 41 (1962), 62-75.

[9] GORENFLO,R., KOVETZ,Y.: Solution of an Abel Type Integral Equation in the Presence of Noise by Quadratic Programming. Numerische Mathematik 8 (1966), 392-406.

[10] KOWALEWSKI, G.: Integralgleichungen. De Gruyter, Berlin und Leipzig 1930.

[11] LARKIN, F.M.: Estimation of a Non-Negative Function. BIT 9 (1969), 30-52.

[12] LINZ, P.: Applications of Abel Transforms to the Numerical Solution of Problems in Electrostatics and Elasticity. MRC Technical Summary Report Nr. 826, Nov. 1967.

[13] MICCHELLI, Ch.A.and RIVLIN, T.J. (eds): Optimal Estimation in Approximation Theory. Plenum Press, New York & London 1977.

[14] MINERBO, G.N. and LEVY, M.E.: Inversion of Abel's Integral Equation by Means of Orthogonal Polynomials. SIAM J. Numer. Anal. 6 (1969), 598-616.

[15] TIKHONOV, A.N. and ARSENIN, V.Y.: Solutions of Ill-Posed Problems. Translated from the Russian. Wiley, New York, 1977.

[16] WEISS, R. and ANDERSSEN, R.S.: A Product Integration Method for a Class of Singular First Kind Volterra Equations. Numer. Math. 18 (1972), 442.456.

[17] WEISS, R.: Product Integration for the Generalized Abel Equation. Math. of Comp. 26 (1972), 177-190.

ON UNIQUENESS OF SOLUTIONS FOR INVERSE PROBLEMS OF POTENTIAL THEORY

by

V. M. ISAKOV

Institute of Mathematics, Sibirian Branch
USSR Academy of Sciences, Novosibirsk 630090

In this talk we will outline some recent progress we have made in the investigation of uniqueness of solutions of the inverse problem for the heat equation and the wave equation

In general form the inverse problem of potential theory is formulated in the following manner: The value of the gravitational field is given outside a given domain. It is required to find the domain and the density which create the field. It is connected with important problems in geophysics and celestical mechanics. This problem has been extensively studied by P.S. Novikov, A.N. Tikhonov, V.K. Ivanov, M.M. Lavrentiev, L.N. Sretensky, A.I. Prylepko, etc. [4]. Theorems on uniqueness, stability and existence of solutions have been obtained; the conditions of the latter ones being rather strict due to incorrectness of the problem.

In the present paper we outline several results which are natural analogies of the inverse problem in potential theory for classical non-stationary equations of mathematical physics. Only questions on uniqeness which play an important role in studying the problems according to A.N. Tikhonov [5] are considered.

Let E_n be an n-dimensional Euclidean space with elements $x = (x_1, \dots, x_n)$, $|x|$ the distance between 0 and x. Let further E_{n+1} be the space of points $X = (t,x)$, $t \in E_1$, and D a cylinder $(0,T) \times G$ in E_{n+1}, where $G \subset E_n$ is a domain. Denote by $E(0,T)$ the set $\{X : 0 < t < T\}$, by $\partial_x D_j$ the side boundary $\partial D_j \setminus \partial E(0,T)$ of a set $D_j \subset E(0,T)$ and by χ_{D_j} the characteristic function of D_j. Fix the density

$$\mu \in C^{(0,\lambda)}(D),\ \mu > 0 \text{ in } D\,, \qquad (1)$$

where $\lambda \in (0,1)$. Consider the bounded subsets

$$D_j \subset \bar{D}_j \subset ([0,T] \times G),\ j = 1,2\,,$$

which satisfy the following conditions:

$$D_j \cap \{t = \theta\},\ (D \setminus \bar{D}_j) \cap \{t = \theta\} \text{ are domains in } E_n \text{ for} \qquad (2)$$

$\theta \in (0,T)$;

the boundary $\partial_x D_j$ is in a neighbourhood of each of its points equal to the graph of the function

$$x_i = f(t,x_1, \dots , x_{i-1},x_{i+1}, \dots , x_n),\ f \in C^{(1,\lambda)}(E_n). \tag{3}$$

In the following the non-empty cylinder $(0,T) \times G_o$ in $D \setminus (\bar{D}_1 \cup \bar{D}_2)$ is of interest, where G_o is a ball in E_n.

Theorem 1: If the sets

$$(D \setminus (\bar{D}_1 \cup \bar{D}_2)) \cap \{t = \theta\} \text{ are connected at } \theta \in (0,T) \tag{4}$$

and the u^j are (generalized) solutions of the Cauchy problem

$$u^j_t - \Delta u^j = \mu \chi_{D_j} \text{ on } D,\ u^j = 0 \text{ on } \{0\} \times G,\ j = 1,2 \ , \tag{5}$$

satisfying

$$u^1 = u^2 \text{ on } (0,T) \times G_o \ , \tag{6}$$

then $D_1 = D_2$.

Here we will outline the proof of Theorem 1. The following lemma holds.

Lemma 1: $\bar{D}_1 \cap \{t = 0\} = \bar{D}_2 \cap \{t = 0\}$.

Denote by D^- the cylinder $(-T,0) \times G_1$, where G_1 is the interior of the cross-section $\bar{D}_1 \cap \{t = 0\}$. Using properties of thermal potentials, condition (1), condition (2) and Lemma 1 it follows that

$$u = u^2 - u^1 \in C^{(1,2)}(K_j), \tag{7}$$

where $K_1 = \overline{D_1 \setminus \bar{D}_2}$, $K_2 = \overline{D_2 \setminus \bar{D}_1}$ are compact subsets of $E(0,T)$. Using (5), (6) and (7) we obtain the following statemant

Lemma 2: There exists a $t_o \in (0,T)$ such that

$$\mu u(\chi_{D_2} - \chi_{D_1}) \leqq 0 \text{ on } E(0,t_o).$$

While proving Theorem 1 one may assume that

$$D_1 \cap E(0,t) \neq D_2 \cap E(0,t) \tag{8}$$

for any $t > 0$. Due to (5) we have

$$u_t - \Delta u = \mu(\chi_{D_2} - \chi_{D_1}) \text{ on } D,\ u = 0 \text{ on } \{0\} \times G_1 \tag{9}$$

From equality (6) due to conditions (2), (4) we obtain that

$$u = 0 \text{ on } D\setminus(\bar{D}_1 \cup \bar{D}_2) \tag{10}$$

We multiply both parts of equation (9) by u and integrate over the domain E(0,t):

$$\int_0^t d\tau \int_G (u_\tau - \Delta u)u\,dx = \int_0^t d\tau \int_G \mu(\chi_{D_2} - \chi_{D_1})u\,dx.$$

The left-side integral due to (9), (10) is greater or equal to

$$\frac{1}{2}\int_G u^2(t,x)\,dx \,.$$

From Lemma 2 it follows that u = 0 on $E(0,t_0)$, which contradicts assumption (8). Therefore $D_1 = D_2$.

Conditions (2), (4) are of topological character and hold, in particular, for domains D_j which are starlike (in a certain sense). The complete proof of Theorem 1 is given in [1] in a more general form. Uniqueness theorems for solutions of a variational inequality of parabolic type have been essentially used [3] .

We now consider the wave operator

$$\Box = \frac{\partial^2}{\partial t^2} - \Delta \quad \text{in } E_4 \,.$$

We suppose that the density μ satisfies the condition

$$\mu \in C^2(E_4),\; 0 < \varepsilon < \mu < \varepsilon^{-1} \text{ on } E_4 \,. \tag{11}$$

In Theorem 2 we consider the following domains

$$D_j = \{X : |x| < d_j(t,\, |x^{-1}|x)\} \quad , \tag{12}$$

where $d_j \in C^3(E_1 \times \partial B(0;1))$, $|(d_j)_t| < 1$ and $0 < d_j < R$.

Theorem 2: Let μ be a density satisfying (11) and D_j, j = 1, 2 , domains satisfying (12). If u^j are solutions of the Cauchy problem

$$\Box u^j = \mu\chi_{D_j} \quad \text{on } (0,T) \times E_3 \,,$$

$$u^j = u^j_t = 0 \quad \text{on } \{0\} \times E_3 \,,$$

j = 1,2 , then from

$$u^1 = u^2 \quad \text{on} \quad (0,T) \times \{X : R < x < R + \varepsilon\}$$

it follows that

$$D_1 \cap \{0 < t,\; t - |x| < T - R\} = D_2 \cap \{0 < t,\; t - |x| < T - R\} .$$

The proof of Theorem 2 is similar to that of Theorem 1. For more general hyperbolic equations it was given in [2] .

In the case of the E_2 we can apply another method to obtain the uniqueness theorem for minimal restrictions on domains and densities. Let D be a domain

$$\{(t,x) : t - |x| < T - R,\ t < T\}$$

on the plane $E_2 = \{(t,x)\}$. We consider the density

$$\mu \in L_1(D),\ \varepsilon_1 < \mu < \varepsilon_1^{-1} \quad \text{on} \quad D \cap \{t^2 + x^2 < R_1^2\}, \tag{13}$$

where $\varepsilon_1 > 0$ and μ only denpends on R_1. We introduce the following domains

$$D_j = \{(t,x) : x_j^-(t) < x < x_j^+(t),\ t \in E_1\},$$

where the functions x_j^+, x_j^- , which determine the side boundary of D_j, satisfy the conditions:

$$x_j^+,\ x_j^- \in C(E_1),\ -R < x_j^- < -\varepsilon < \varepsilon < x_j^+ < R \text{ for some } \varepsilon \in (0,R), \tag{14}$$

any straight line of the family $t \pm x = \text{const.}$ crosses tho graph of x_j^+, x_j^- in only one point (15)

<u>Theorem 3:</u> Let μ and D_j, $j = 1,2$, satisfy conditions (13) - (15). If

$$u_{tt}^j - u_{xx}^j = \mu \chi_{D_j} \quad \text{in } D \tag{16}$$

and

$$u^1 = u^2 \quad \text{on} \quad (-\infty,T) \times \{x : R < |x| < R + \varepsilon\}$$

then $D_1 \cap D = D_2 \cap D$.

Let us consider an exemple illustrating the importance of condition (15) of Theorem 3. Denote by $\Delta_1^{\pm}$, $\Delta_2^{\pm}$ rectangular triangulars with $\pi/4$ as one of the angles, and perpendiculars of length 1 which lie on straight lines $\{x = \pm 2\}$, $\{t = 2\}$, $\{t = 6\}$ in such a way that

$$\Delta_1^+ \subset \{t < 2,\ 0 < x < 2\},\ \Delta_2^+ \subset \{2 < t < 3,\ 2 < x\},$$

$$\Delta_1^- \subset \{5 < t < 6,\ x < -2\},\ \Delta_2^- \subset \{6 < t,\ -2 < x\}.$$

Define $D_1 = \{(t,x) : t - |x| < T - 4,\ |x| < 2\}$ having assumed $T > 10$. Define

$$D_2 = (D_1 \cup \Delta_2^+ \cup \Delta_1^-) \setminus (\bar{\Delta}_1^+ \cup \bar{\Delta}_2^-).$$

Let the density μ satisfy condition (13) and, besides, $\mu = 1$ on $(0,7) \times (-4,4)$. The function

$$u^j(t,x) = \int_{D_j \cap \{\tau - t + |y-x| < 0\}} \mu(\tau,y)\, d\tau dy$$

is a solution of equation (16). The relation $u^1 = u^2$ holds on

$$(-\infty,T) \times \{x : 4 < |x|\} ,$$

but, nevertheless, $D_1 \neq D_2$.

A more general version of Theorem 2 and further references can be found in [2] .

REFERENCES

[1] ISAKOV, V. M.: Uniqueness theorems for inverse problems of thermal potentials. Sib. Mat. Z. 17 (1976), pp. 259 - 272

[2] ISAKOV, V. M.: On the inverse problem of potential theory for hyperbolic equations. Dokl. AN SSSR 237 (1977), pp. 257 - 259

[3] LIONS, J. L.: Contrôle optimal de systèmes gouvernés par des équations aux dérivées partielles. Paris 1968

[4] PRYLEPKO, A. I.: Inverse problems of potential theory. Matemat. Zametki 14 (1973), pp. 755 - 765 (Russian)

[5] TIKHONOV, A. N., IVANOV, V. K. and LAVRENT'EV, M. M.: Improperly posed problems. In Partial Differential Equations, M., Nauka, 1970, pp. 224 - 238 (Translation into English)

ON SOME INVERSE PROBLEMS IN VARIATIONAL AND QUASI-VARIATIONAL INEQUALITIES

Reinhard Kluge [1)]

This paper deals with inverse problems in form of minimum problems for parametric operator equations, evolution equations, variational inequalities and quasi-variational inequalities (parts 2 and 4). Besides we give existence results for quasi-variational inequalities (part 2), some resultsabout subgradients of functionals and addition theorems for multivalued mappings (part 1). The solution of the inverse problems and the existence results for quasi-variational inequalities are based on implicit mapping theorems for parametric relations (part 3). The results can be applied to problems of determination of coefficients in partial differential equations and of constraints in inequalities.

Contents

1. On some mappings and functionals with one argument

1.1. Let B be a real reflexive Banach space with dual space B (with norm $\|.\|$ in both spaces). By $(.,.)$ we denote the pairing between B^* and B and by $\xrightarrow{s}$ and $\xrightarrow{w}$ the strong and the weak (sequential) convergence (in both B and B^*), respectively. Let ς denote the strong ($\varsigma = s(=1)$) or the weak ($\varsigma = w(=0)$) convergence in B , respectively. Let $\{C_n\}$ be a sequence of nonempty subsets $C_n \subset B, n=1,2,\ldots$ We use the limits $\varsigma\text{-}\overline{\text{Lim}}\, C_n$ and $\varsigma\text{-}\underline{\text{Lim}}\, C_n$ in the usual sense [10]. Let $T: B \rightarrow 2^{B^*}$ be a possible multivalued mapping from B into B^*. By G(T) we denote the graph of T in $B^* \times B$, by D(T) and R(T) the domain and the range of T, respectively. If D(T)=B holds then we write $T \in (B \rightarrow 2^{B^*})$ and in the case of single-valued mappings $T \in (B \rightarrow B^*)$, too.
A functional $g: B \rightarrow R^1$ is called proper, if $g(b) > -\infty$ for $b \in B$ and $g(b) \not\equiv +\infty$. We denote the set $\{b \in B : g(b) < +\infty\}$ by $D_e(g)$. The set

1) ZIMM der AdW der DDR. DDR-108 Berlin, Mohrenstraße 39

$E(g)=\{[r,b]\in R^1\times B: r\geq g(b)\}$ is called the epigraph of g.If $C\subset B$ is a nonempty subset of B,we denote by I_C the indicator function of C:

$$I_C(b) = \begin{cases} 0 & \text{if } b\in C \\ +\infty & \text{if } b\notin C \end{cases}$$

Obviously,I_C is proper and $D_e(I_C)=C$ holds.

A proper functional $f:B\to R^1$ is called ϱ-lower (ϱ-upper)semicontinuous,if $b_n \xrightarrow{\varrho} b$ in B implies $f(b)\leq \underline{\lim}\, f(b_n)$ $(f(b)\geq \overline{\lim}\, f(b_n))$.

f is called ϱ-continuous, if f is ϱ-lower and ϱ-upper semicontinuous.

Let $C\subset B$ be convex.The proper functional $f:C\to R^1$ is called quasi-convex (convex) if for $a,b\in C$ and $t\in[0,1]$ $f(ta+(1-t)b)\leq \max\{f(a), f(b)\}$ $(f(ta+(1-t)b)\leq tf(a)+(1-t)f(b))$ holds.

The mapping $T:B\to 2^{B^*}$ is called monotone if $(b^*-c^*,b-c)\geq 0$ for all pairs $[b^*,b]$, $[c^*,c]\in G(T)$.T is called maximal monotone if $G(S)\supset G(T)$ for another monotone mapping $S:B\to 2^{B^*}$ implies $G(S)\subset G(T)$.T is called (ϱ,γ)-generalized pseudomonotone if $b_n\in B$, $b_n^*\in Tb_n$, $b_n\xrightarrow{\varrho} b$, $b_n^*\xrightarrow{\gamma} b^*$ and $\overline{\lim}\,(b_n^*,b_n-b)\leq 0$ imply $b^*\in Tb$ and $(b_n^*,b_n)\to(b^*,b)$.

T is (w,w)-generalized pseudomonotone if and only if T is generalized pseudomonotone in the sense of Browder-Hess [2] .

1.2. We define the subgradient $\partial f:B\to 2^{B^*}$ of the proper functional $f:B\to R^1$ by $\partial f(a)=\{a^*\in B^*: f(b)\geq f(a)+(a^*,b-a) \text{ for all } b\in D_e(f)\}$.

Lemma 1. 1. ∂f is monotone; $\partial f(a)$ is convex and w-closed for $a\in D(\partial f)$.
2. The problems $f(a)=\inf\limits_{b\in B} f(b)$ and $o\in\partial f(a)$ are equivalent.

Theorem 2.([12]).If $f:B\to R^1$ is proper,convex and s-lower semicontinuous, then ∂f is maximal monotone.

Corollary 3.Under the conditions of Theorem 2 ∂f is (w,w)-generalized pseudomonotone.

For the proof we remark that a maximal monotone mapping is generalized pseudomonotone [2] .

We give a generalization of Corollary 3.

Theorem 4. Let $f:B\to R^1$ be proper and w-lower semicontinuous.Then ∂f is monotone and generalized pseudomonotone.

Proof.See Corollary 19 of section 1.3 in [7].

Corollary 5.Let f be proper,quasi-convex and s-lower semicontinuous.Then ∂f is monotone and generalized pseudomonotone.

Proof.See Corollary 20 of section 1.3 in [7].

Corollary 6. Let $C\subset B$ be nonempty and w-closed.Then ∂I_C is monotone and generalized pseudomonotone.

Proof. I_C is proper and w-lower semicontinuous.

The mapping $T:B \longrightarrow 2^{B^*}$ is called hemicontinuous in $b \in B$ if for each $c \in B$ the mapping $t \longrightarrow T((1-t)b+tc)$ is upper semicontinuous from $[0,1] \subset R^1$ into the weak topology of B^*.

Theorem 7. ([13]). Let T be single-valued, monotone and hemicontinuous and $C \subset D(T)$ be nonempty, convex and closed. Then the sum $T+\partial I_C$ is maximal monotone.

Corollary 8. Under the conditions of Theorem 7 the sum $T+\partial I_C$ is monotone and generalized pseudomonotone.

We give now a generalization of Corollary 8.

Theorem 9. Let $T:B \longrightarrow 2^{B^*}$ be monotone and hemicontinuous, Tb be w-closed for $b \in D(T)$ and D(T) be open or Tb be bounded for $b \in D(T)$. Then the sum $T+\partial I_C$ is monotone and generalized pseudomonotone.

Proof. See Corollary 30 of section 1.3 in [7].

Remark. Under the conditions of Theorem 9 the equation

$$b^* \in Tb + \partial I_C b \tag{1}$$

with $b^* \in B$ is equivalent to the variational inequality

$$(Tb - b^*, c-b) \geq 0 \text{ , } c \in C. \tag{2}$$

Let B_i be B-spaces, i=1,2. We denote the convergence ς in B_i by ς_i, i=1,2. In $B_2 \times B_1$ we introduce the (coordinatewise) convergence $\varsigma=(\varsigma_2,\varsigma_1)$. The single-valued mapping $T:B_1 \longrightarrow B_2$ is called $(\varsigma_1 \longrightarrow \varsigma_2)$-continuous if $b_n \xrightarrow{\varsigma_1} b$ in D(T) implies $Tb_n \xrightarrow{\varsigma_2} Tb$ in B_2.
We give a further example for (ς, β)-generalized pseudomonotone mappings.

Theorem 10. Let $C \subset B$ be nonempty and ς-closed, $T:B \longrightarrow B^*$ with $D(T) \supset C$ be $(\varsigma \longrightarrow \beta)$-continuous with $\max\{\varsigma,\beta\}=s$. Then the sum $T+\partial I_C$ is (ς,w)-generalized pseudomonotone.

Proof. See Corollary 28 of section 1.3 in [7].

Remark. Under the conditions of Theorem 10 the equation (1) is equivalent to the variational inequality (2) with not necessarily convex constraints.

Remark. Let $f:B \longrightarrow R^1$ be proper, $T:B \longrightarrow 2^{B^*}$, $b^* \in B^*$ and $C \subset B$ be nonempty. Then the following problems (3)-(5) are equivalent:

$$f(a) = \inf_{b \in C} f(b) \tag{3}$$

$$g(a) = \inf_{b \in B} g(b) \tag{4}$$

wit. $g = f + I_C$ and

$$o \in \partial g(a) \ . \tag{5}$$

Furthermore, the problems (6)-(8) are equivalent, too:

$$(Tb-b^*, c-b) \geq 0 \ , \ c \in C, \tag{6}$$

$$(Tb-b^*, c-b) \geq I_C(b)-I_C(c) \ , \ c \in B \tag{7}$$

and

$$b^* \in Tb + \partial I_C b \ . \tag{8}$$

2. Inverse problems and quasi-variational inequalities

We consider here two fundamental problems- the inverse problems by means of minimum problems with constraints and the quasi-variational inequalities. This considerarions give us the motivation for our next investigations about implicit mapping theorems in part 3.

2.1. Let Y and V be two further B-spaces with convergence notions σ and τ, respectively.

Theorem 11. Let $G \subset Y \times V$ be (σ, τ)-closed and the functional $J \in (G \rightarrow R^1)$ be (σ, τ)-lower semicontinuous. Furthermore, let there exist an element $[x_o, u_o] \in G$ such that the set $\{[y,v] \in G : J(y,v) \leq J(x_o, u_o)\}$ is (σ, τ)-compact. Then the solution set SS(J,G) of the minimum problem

$$J(x,u) = \inf_{[y,v] \in G} J(y,v) \tag{9}$$

is nonemty, (σ, τ)-closed and (σ, τ)-compact.

Proof. See Theorem 1 in [4].

We remark that for G there exist a mapping $S: V \longrightarrow 2^Y$ such that $G = G(S)$ holds. In our considerations S denotes the solution operator of some parametric relations. Then Theorem 11 gives a sufficient condition for the existence of "optimal" parameters (of the "components" u of the solutions $[x,u]$ in (9)) in inverse problems for these parametric relations.

For a general discussion of the conditions related to J in Theorem 11 see [4]. In part 3 we give some sufficient conditions for the (σ, τ)-closedness of G(S) for some parametric relations (implicit mapping theorems).

2.2. Let be $T: Y \times U \longrightarrow 2^{Y^*}$, $U \subset V$ and $V=Y$. The problem

$$y^{*} \in T(y,y) \ , \ y \in U \ , \tag{10}$$

with $y^{*} \in Y^{*}$ is called a quasi-variational inequality (Q.V.I.). We give an example.Let be $C \in (U \longrightarrow 2^{Y})$ and $A: Y \times U \longrightarrow 2^{Y^{*}}$.We consider the problem

$$y \in C(y) \ , \ (A(y,y)-y^{*}, z-y) \geq 0, \ z \in C(y) \ . \tag{11}$$

(11) is equivalent to (10) with $T(y,y)=A(y,y) + \partial_{y} I_{C(y)} y$ .So (11) is a Q.V.I.

Instead of (10) we consider the parametric extension

$$y^{*} \in T(y,v) \ , v \in U. \tag{12}$$

For $v \in U$ we denote the set of all solutions of $y[v]$ (12) by Sv; $y[v] \in Sv$.It is very easy to see that the set of all solutions of (10) is the same as the set of all fixed points of S in U.

To prove the existence of solutions of (10) we can use some fixed point theorems for mappings in a reflexive Banach space.

Theorem 12. Let $U \subset Y$ be nonemty,convex and closed and $S \in (U \longrightarrow 2^{U})$. Then the following assertions hold:

1. Let G(S) be (w,w)-closed,Sd be convex for $d \in U$ and S(U) be bounded. Then the set of all fixed points FP(S,U) of S in U is nonempty,w-closed and w-compact.
2. Let G(S) be (s,w)-closed,S be single-valued and S(U) be s-compact. Then FP(S,U) is nonemty,s-closed and s-compact.If furthermore G(S) is (w,w)-closed, then FP(S,U) is w-closed ,too.

Proof.For the first assertion see Theorem 13 and for the second assertion see Theorem 16 in section 1.1 of [7].

Remark.The condition D(S)=U in Theorem 12 is connected with existence theorems for (12) for each fixed parameter value $v \in U$.See for instance [1, 2, 6] .Besides of this,in Theorem 12 the closedness properties of G(S) play an important role.In the next part (section 3.2) we give some sufficient conditions for such properties.

3. Implicit mapping theorems for parametric equations and inequalities

3.1. Parametric relations with nonparametric constraints

We formulate the condition (i):

(i) Let $\Lambda_0 : Y \longrightarrow 2^{Y^{*}}$ be monotone,$U \subset V$ be w-closed and $A(\cdot,v): Y \rightarrow Y^{*}$ be monotone and hemicontinuous with $D(A(\cdot,v)) \supset D(\Lambda_0)$ for each $v \in U$.Furthermore,let $\Lambda_0 + A(\cdot,v)$ be maximal monotone for $v \in U$ and $A(y,\cdot)$ be a) $(w \longrightarrow s)$-continuous or b) $(w \longrightarrow w)$-continuous for each $y \in D(\Lambda_0)$.

We consider the parametric equation

$$\Lambda_0 y + A(y,v) \ni y^* , v \in U. \tag{13}$$

Theorem 13. Let the condition (i) be satisfied. Then G(S) of (13) is (w,w)-closed in the case a) and (s,w)-closed in the case b).

Proof. By Lemma 6 and Corollary 7 in section 1.2 of [7] the mapping $\Lambda_0 + A$ is ((w,w),s)-generalized pseudomonotone in the case a) and ((s,w),w)-generalized pseudomonotone in the case b). The closedness properties of G(S) then follow from assertion 3 of Lemma 3 in section 1.2 of [7].

We give some special cases of the parametric problem (13).

α) Operator equations

Let be $A(\cdot,v) \in (Y \longrightarrow Y^*), v \in U$ and

$$A(y,v) = y^* \quad , \ v \in U. \tag{14}$$

For problem (14) Theorem 13 in the case a) is proved in [8] for parametric elliptic partial differential equations and in [9] for parametric pseudoparabolic partial differential equations (inverse problems to determine coefficients in the partial differential equations, which depend from the gradient of the solutions, i.e. inverse problems to determine material laws in the nonlinear elasticity and in the nonlinear viscoelasticity - inverse problems to determine the type of nonlinearity of the partial differential equations.

β) Evolution equations

Let A be as in (14), $\Lambda_0 : Y \longrightarrow Y^*$ be single-valued and

$$\Lambda_0 y + A(y,v) = y^* , \ v \in U. \tag{15}$$

For problem (15) Theorem 13 in the case a) is proved in [11] for parabolic equations (inverse problems to determine coefficients, which dependfrom the gradient of the solutions). For the abstract case (15) see [5, 6], too.

γ) Variational inequalities

Let $X \subset Y$ be nonempty, convex and closed, $A(\cdot,v)$ with $D(A(\cdot,v)) \supset X$ be single-valued for $v \in U$ and

$$(A(y,v) - y^*, z - y) \geq 0, z \in X, v \in U. \tag{16}$$

For the proof of the closedness properties of G(S) in (16) for single-valued S see [3] and in the general case a) see [5, 6].

3.2. Implicit mapping theorems for problems with parametric constraints

We formulate the condition (ii) :

(ii) Let $U \subset V$ be τ-closed, $C \in (U \longrightarrow 2^Y)$ satifies the condition
$$\sigma - \overline{\mathrm{Lim}}\, C(v_n) \subset C(v) \subset \alpha - \underline{\mathrm{Lim}}\, C(v_n) \text{ for } v_n \xrightarrow{\tau} v \text{ in } U, \quad (17)$$
$A(.,v): Y \longrightarrow Y^*$ with $D(A(.,v)) \supset C(v)$ for $v \in U$, A be $((\sigma,\tau) \longrightarrow \varrho)$-continuous with $\max\{\varrho, \min\{\sigma,\alpha\}\}=s$ or $A(.,v)$ is monotone and hemicontinuous, $C(v)$ is convex for $v \in U$ and A is $((\alpha,\tau) \longrightarrow \varrho)$-continuous.

We consider the parametric variational inequality
$$(A(y,v)-y^*, z-y) \geq 0 \ , \ z \in C(v) \ , \ v \in U \ . \quad (18)$$

<u>Theorem 14</u>. Let the condition (ii) be satisfied. Then $G(S)$ of (18) is (σ,τ)-closed.

<u>Proof</u>. By Corollary 15 of section 1.3 in [7] the mapping $A+\partial I_C$ is $((\sigma,\tau),\varrho)$-generalized pseudomonotone. The closedness properties of $G(S)$ for (18) then follows from assertion 3 of Lemma 3 in section 1.2 of [7].

<u>Remark</u>. Let the condition (ii) be satisfied. Then from Theorem 14 follow the following implications:

1. $\sigma=s$, $\tau=w$, $\varrho=w,s$ and $\alpha=s,w$. $\Longrightarrow$ $G(S)$ of (18) is (s,w)-closed.
2. $\sigma=w$, $\tau=w$, $\varrho=s$ and $\alpha=s,w$. $\Longrightarrow$ $G(S)$ of (18) is (w,w)-closed.
3. $\sigma=s$, $\tau=w$, $\varrho=s$, $C(v)$ is convex and $A(.,v)$ is monotone and hemicontinuous. $\Longrightarrow$ $G(S)$ of (18) is (w,w)-closed.

<u>4. Inverse problems for quasi-variational inequalities</u>

We consider the space B with convergence notions ϱ and σ and the space V with convergence notion τ. Let be $U \subset V, K \in (U \rightarrow 2^B)$, $Z=G(K), C \in (Z \longrightarrow 2^B), C(.,v) \in (K(v) \longrightarrow 2^{K(v)})$ for $v \in Z$ and let be $A: B \times Z \longrightarrow B^*$.

We consider the parametric quasi-variational inequality (pQ.V.I.):

(pQ.V.I.) Find $[b,v] \in Z$ such that $b \in C(b,v)$ and
$$(A(b,b,v)-b^*, c-b) \geq 0, \ c \in C(b,v), \ v \in U \ . \quad (19)$$

We denote the solution operator of (19) by S. Besides of (pQ.V.I.) we consider the parametric variational inequality (pV.I.):

(pV.I.) Find $[b,d,v] \in B \times Z$ such that $b \in C(d,v)$ and

$$(A(b,d,v)-b^{*}, c-b) \geqslant 0, \quad c \in C(d,v), \quad v \in U. \tag{20}$$

We denote the solution operator of (20) by S_0.

Lemma 15. If $G(S_0)$ is $(\varsigma, \sigma, \tau)$-closed, then $G(S)$ is (α, τ)-closed with $\alpha = \max\{\varsigma, \sigma\}$.

Proof. The proof of this Lemma is the same as for parametric quasi-minimum problems in [7].

Lemma 16. Let U be τ-closed, $K \in (U \longrightarrow 2^B)$ satisfies the condition

$$\sigma - \overline{\mathrm{Lim}}\ K(v_n) \subset K(v) \quad \text{for } v_n \xrightarrow{\tau} v \text{ in } U. \tag{21}$$

Then $Z=G(K)$ is (σ, τ)-closed.

Proof. See Lemma 2 of section 2.3 in [7].

On the base of the Lemmata 15 and 16 we can prove the closedness condition's of $G(S_0)$ by the results of section 3.2 and solve the inverse problem for the parametric quasi-variational inequality (19) by Theorem 11 in an obvious way (see for inverse problems in parametric quasi-minimum-problems [7]).

References

[1] Browder,F.E.,Nonlinear operators and nonlinear equations of evolution in Banach space.Proc.Symp.Pure Math.Vol. XVIII.Part 2.Providence 1976.

[2] Browder,F.E.,Hess,P.,Nonlineare mappings of monotone type in Banach spaces.J.Funct.Anal. 11 (1972) ,251-294.

[3] Kluge,R.,Optimal control with minimum problems and variational inequalities.Lecture Notes in Computer sciences 27 (1975) ,377-382.

[4] Kluge,R., Über eine Klasse von Minimumproblemen mit Nebenbedingungen.Math.Nachr. 78 (1977),37-64.

[5] Kluge,R.,On some parameter determination problems and quasi-variational inequalities.In:Theory of nonlinear operators.Abhandl. der AdW der DDR.N6.Akademie-Verlag,Berlin 1978.Ed.R.Kluge.S.129-139.

[6] Kluge,R.,Nichtlineare Variationsungleichungen und Extremalaufgaben. VEB Deutscher Verlag der Wissenschaften.Berlin 1979.

[7] Kluge,R.,On some parameter determination problems in variational and quasi-variational inequalities.I.Math.Nachr.

[8] Kluge,R.,Langmach,H.,On some problem of determination of functional parameter in partial differential equations.In:Distributed Parameter Systems.Modelling and Identification.IFIP Working Conf. Lect. Notes in Control and Information Sciences 1,Springer-Verlag,Berlin-Heidelberg-New York 1978,p.298-309.

[9] Kluge,R.,Langmach,H.,On the determination of some rheologic properties of mechanical media.In:Theory of nonlinear operators.Abhandl. der AdW der DDR. N6.Akademie-Verlag,Berlin 1978.Ed. R.Kluge.S.141-158.

[10] Kuratowski,K.,Topology.I.Academic Press,New York 1966.

[11] Langmach,H.,On the determination of functional parameters in some parabolic differential equation.In:Theory of nonlinear operators. Abhandlungen der AdW der DDR.N6.Akademie-Verlag ,Berlin 1978.Ed. R.Kluge.S.175-184.

[12] Rockafellar,R.T.,On the maximal monotonicity of subdifferentials mappings.Pacif.J.Math. 33 (1970),209-216.

[13] Rockafellar,R.T.,On the maximality of sums of nonlinear monotone operators.Trans.Amer.Math.Soc. 149 (1970),75-88.

Uniqueness and Continuous Dependence of the Null Solution in the Cauchy Problem for a Nonlinear Elliptic System

by

R.J. Knops Heriot-Watt University, Edinburgh

L.E. Payne Cornell University, Ithaca, N.Y.

Introduction. In this paper we investigate the Cauchy problem for a class of nonlinear elliptic systems using a method which was employed by Payne [4], [5] to study questions of uniqueness and continuous dependence in the Cauchy problem, first for the Laplace equation, and later for a more general second order elliptic equation. This method relies on logarithmic convexity arguments which have been employed by several authors to analyze a variety of properly and improperly posed problems. The interested reader is referred to a recent survey on the subject by Payne [6]. The major aim of this paper is to establish a fundamental differential inequality which will permit the application of logarithmic convexity techniques. The derivation of uniqueness and continuous dependence of the null solution (as well as other results) from the convexity property is by now standard and is therefore only sketched.

The nonlinear system which we consider is not covered by the results of Trytten [10] and Schaefer [8], [9], since they did not consider coupling in the highest order derivatives as occurs for instance in the system of equations of nonlinear elasticity and in other equations of continuum physics. It should be emphasized, however, that we establish uniqueness and continuous dependence results only for the null solution. Thus our results must be regarded as a first step in what will hopefully be a more general theory for fully nonlinear systems.

In Section 2 we set out the notation and basic equations and indicate the relation to nonlinear elasticity. Section 3 is devoted entirely to the proof of the fundamental inequality, while the last section discusses uniqueness and continuous dependence results.

2. Notation and Basic Formulation.

Let D be a domain in $\mathbb{R}^N$ whose boundary ∂D is Liapunov and contains a portion Σ of finite area over which Cauchy data is specified. It is assumed that $\overline{\Sigma}$ is a C^1-surface.

Let $\rho: \mathbb{R} \to \mathbb{R}$ be a nonlinear continuously differentiable function which satisfies the inequalities

$$\rho(s) > 0, \ s > 0, \tag{2.1}$$

$$\int_0^s \rho(t)dt \leq s\rho(s), \tag{2.2}$$

and let $u: D \to \mathbb{R}^N$ be twice continuously differentiable with components u_i, $i = 1,\ldots,N$. We adopt the summation and comma conventions and consider the nonlinear system of equations

$$(a_{rk}\rho(\sigma)u_{i,r})_{,k} = 0, \ x \in D \tag{2.3}$$

in which

$$\sigma = a_{rk}u_{i,r}u_{i,k}, \tag{2.4}$$

and the a_{ij} are symmetric functions with uniformly bounded first derivatives. We suppose also that the a_{ij} are positive-definite in D;

$$a_{ij}\xi_i\xi_j \geq a_0\xi_i\xi_i, \ x \in D \tag{2.5}$$

for some positive constant a_0.

Let n_i be the components of the unit outward normal on Σ, and define the conormal derivative on Σ by

$$\frac{\partial u_i}{\partial \nu} = \rho(\sigma)a_{rk}n_k u_{i,r} \tag{2.6}$$

The Cauchy data associated with system (2.3) is then given by

$$u_i = g_i, \ x \in \Sigma, \tag{2.7}$$

$$\frac{\partial u_i}{\partial \nu} = h_i, \ x \in \Sigma. \tag{2.8}$$

where g_i, h_i are specified. It may be easily shown (see e.g., Payne and Weinberger [7] p. 553) that

$$\rho^2(\sigma)\sigma = n^{-1}[\frac{\partial u_i}{\partial \nu}\frac{\partial u_i}{\partial \nu} + \rho^2(\sigma)\frac{\partial u_i}{\partial t}\frac{\partial u_i}{\partial t}],\ x \in \Sigma, \tag{2.9}$$

in which $\partial u_i/\partial t$ designates tangential derivativesin Σ and $n = a_{rs}n_r n_s$. Thus, (2.9) may be solved implicitly for σ on Σ, and hence it may be assumed that $\rho(\sigma)$ and therefore $\sigma\rho(\sigma)$ are known on Σ in terms of the Cauchy data (2.7), (2.8).

We wish to investigate solutions of (2.3) subject to (2.1) and (2.2) together with the Cauchy data (2.7) and (2.8). We shall show that the null solution to this problem is unique and that perturbations of the null solution (in an appropriate class) depend continuously in the sense of Hölder on the Cauchy data (2.7), (2.8) as well as the value of $\sigma\rho(\sigma)$ on Σ. We do not consider the question of existence, but shall assume existence of a classical solution (in a fixed subdomain of D to be defined later) corresponding to the given data.

It should be remarked in passing that conditions (2.2) and (2.5) imply the ellipticityof the system of equations (2.3), thus the Cauchy problem posed in this paper will always be an improperly posed problem.

Before passing to the proof of the basic differential inequality from which the uniqueness and continuous dependence results follow, we indicate briefly some special cases of (2.3) and its relation to nonlinear elasticity.

When $\rho \equiv 1$, condition (2.2) is satisfied identically and the system (2.3) reduces to a linear elliptic system. When additionally $u = (0,0,\ldots,u)$ the system becomes the single linear elliptic equation, the Cauchy problem for which has been treated by Payne [5]. Cauchy problems for such equations are clearly of physical importance. Another example in continuum physics of the occurence of (2.3) is in nonlinear elasticity provided the first Piola-Kirchloff stress tensor has the particular form

$$T_{ki} = a_{rk}\rho(\sigma)u_{i,r}, \tag{2.10}$$

then (2.3) and (2.6) represent the associated equilibrium equations and surface traction, respectively. Alternatively, we may regard (2.10) as constitutive equations for a hyperelastic material in the special case in which the strain energy function $W(u_{i,j})$ is given by

$$W(u_{i,j}) = W(\sigma) \tag{2.11}$$

and

$$\rho = dW/d\sigma. \tag{2.12}$$

Then, inequality (2.2) becomes

$$2(W(\sigma) - W(0)) \leq u_{i,j}\ \partial W/\partial u_{i,j}, \tag{2.13}$$

which is known to imply infinitesimal stability of the corresponding displacement and traction boundary value problems. When the inequality in (2.13) is strict, the condition implies uniqueness of the solution in the same problems. (See, for instance, Aron [1]).

3. The fundamental inequality.

By means of a procedure used by Payne [5] for the single linear elliptic equation, a second order differential inequality is now derived.

First, let us introduce a family of not necessarily closed surfaces

$$f(x) = \text{const.},\ x \in \mathbb{R}^N, \tag{3.1}$$

chosen so that for $\alpha \in (0,1]$ the surface $f(x) = \alpha$ intersects D and forms a closed region D_α whose boundary consists only of points of Σ and the part of the surface $f(x) = \alpha$ in D. We also suppose that for $\alpha \in (0,1]$, the domains D_α has non-zero volume measure, but that D_0 has zero volume measure.

We make the following hypotheses on $f(x)$:

H1) f is twice continuous differentiable in D_1,

H2) $\beta \leq \gamma$ implies $D_\beta \subseteq D_\gamma$, $0 < \beta \leq \gamma \leq 1$,

H3) $\sup_{x\in D_1} \left|\left|\frac{\rho'(\sigma)}{\rho(\sigma)} a_{rk}u_{i,r}u_{i,kj}\right|\right| \leq M_j$, for some positive constant M_j, $x \in D_1$,

H4) $(a_{rk}\rho(\sigma)f_{,k})_{,r} \leq 0$, $x \in D_1$.

An example of a surface satisfying these hypotheses is supplied by

$$a_{ij} = \delta_{ij},\ f(x) = \frac{e^{-\lambda r_o^2} - e^{-\lambda r^2}}{e^{-\lambda r_o^2} - e^{-\lambda R^2}},\ r^2 = x_i x_i,\ r_0 < r < R_\sigma \tag{3.2}$$

To verify that H3) is satisfied, we have

$$\rho(\sigma)\Delta f + 2\rho'(\sigma)u_{i,j}u_{i,jk}f_{,k} \leq - \frac{\lambda\rho(\sigma)}{e^{-\lambda r_0^2} - e^{-\lambda R_2}}[4\lambda r^2 - 2N - 2Mr],\ M^2 = M_i M_i \tag{3.3}$$

which can clearly always be made negative for $r \in (r_0,R]$ by taking λ positive and sufficiently large.

Let Σ_α be the intersection of Σ with D_α and S_α be the intersection of the surface $f(x) = \alpha$ with D. Set

$$\Phi(\alpha) = \int_0^\alpha \int_D a_{k\ell}\rho(\sigma)u_{i,k}u_{i,\ell}\,dx\,d\eta + Q, \tag{3.4}$$

where $$Q = k_1 Q_1 + k_2 Q_2 + k_3 Q_3 \quad \text{and}$$

$$Q_1 = \int_\Sigma u_i u_i\, dS,\ Q_2 = \int_\Sigma \frac{\partial u_i}{\partial \nu}\frac{\partial u_i}{\partial \nu} dS,\ Q_3 = \int_\Sigma \sigma\rho(\sigma)dS \tag{3.5}$$

are given in terms of the Cauchy data. The constants $k_1 k_2, k_3$ are positive and will be determined later. The integral term in the definition (3.4) of the function Φ is positive-definite (the integrand is merely $\sigma\rho(\sigma)$) and because of our basic assumptions, Φ itself is twice continuous differentiable in D_1. Indeed, the continuous dependence results will be proved using Φ as the measure of the solution and to this end we now show that Φ satisfies a second order differential inequality. We have the following theorem.

<u>Theorem 1</u>. Suppose the surfaces $f(x) = \alpha$, $\alpha \in (0,1]$ satisfy hypotheses H1) - H4) and that a classical solution to the Cauchy problem (2.3), (2.7) and (2.8) exists in D_1. Then the function $\Phi(\alpha)$, defined by (3.4), satisfies the second order differential inequality

$$\Phi\Phi_{,\alpha\alpha} - (\Phi_{,\alpha})^2 \geq -c_1\Phi\Phi_{,\alpha} - c_2\Phi^2,\ x \in D_1, \tag{3.6}$$

where c_1, c_2 are positive constants computable in terms of the Cauchy data (2.7), (2.8).

To prove the theorem, we start by performing an integration by parts to alternatively express Φ as

$$\Phi(\alpha) = \int_0^\alpha \{\int_{S_\eta} \frac{a_{rk}\rho(\sigma)u_{i,r}u_i f_{,k}dS}{|\nabla f|}\}d\eta + \int_0^\alpha\{\int_{\Sigma_\eta} a_{rk}\rho(\sigma)u_{i,r}u_{i\ k}dS\}\ d\eta + Q, \qquad (3.7)$$

since the unit outward normal on S_α is $\nabla f/|\nabla f|$, where ∇ is the spatial vector gradient operator. The first term on the right in (3.7) may be written

$$\int_0^\alpha \{\int_{S_\eta} \frac{a_{rk}\rho(\sigma)u_{i,r}u_i f_{,k}dS}{|\nabla f|}\}\ d\eta = \int_{D_\alpha} a_{rk}\rho(\sigma)u_{i,r}u_i f_{,k}dx$$

$$= \frac{1}{2}\int_{S_\alpha} \frac{a_{rk}\rho(\sigma)u_i u_i f_{,k}f_{,k}\ dS}{|\nabla f|} + \frac{1}{2}\ a_{rk}\rho(\sigma)u_i u_i f_{,k}n_r dS$$

$$- \frac{1}{2}\int_{D_\alpha}(a_{rk}\rho(\sigma)f_{,k})_{,r}u_i u_i dx \qquad (3.8)$$

On inserting (3.8) into (3.7), using H4), and selecting k_1, k_2, k_3 to be sufficiently large, we then obtain

$$\Phi(\alpha) \geq \frac{1}{2}\int_{S_\alpha} a_{rk}\rho(\sigma)u_i u_i f_{,k}\ \frac{f_{,r}}{|\nabla f|}\ dS + \frac{1}{2}\ Q. \qquad (3.9)$$

Next, it follows immediately from (3.4) that

$$\Phi_{,\alpha}(\alpha) = \int_{D_\alpha} \sigma\rho(\sigma)dx \qquad (3.10)$$

$$= \int_0^\alpha \int_{S_\eta} \frac{\sigma\rho(\sigma)}{|\nabla f|}\ dSd\eta\ , \qquad (3.11)$$

and thus

$$\Phi_{,\alpha\alpha}(\alpha) = \int_{S_\alpha} \frac{\sigma\rho(\sigma)}{|\nabla f|}\ dS \qquad (3.12)$$

On the other hand, we have from (3.7), an equivalent expression for $\Phi_{,\alpha}(\alpha)$ namely,

$$\Phi_{,\alpha}(\alpha) = \int_{S_\alpha} a_{rk}\rho(\sigma)u_{i,r}u_i\ \frac{f_{,k}}{|\nabla f|}\ dS + \int_{\Sigma_\alpha} a_{rk}\rho(\sigma)u_{i,r}u_i n_k dS \qquad (3.13)$$

and therefore

$$(\Phi,_\alpha)^2 = (\int_{S_\alpha} a_{rk}\rho(\sigma)u_{i,r}u_i \frac{f,_k}{|\nabla f|} dS)^2 + 2\Phi,_\alpha \int_{\Sigma_\alpha} a_{rk}\rho(\sigma)u_{i,r}u_i n_k dS$$

$$- (\int_{\Sigma_\alpha} a_{rk}\rho(\sigma)u_{i,r}u_i n_k dS)^2 \tag{3.14}$$

We may then use (3.12) and (3.14) together with the Cauchy-Schwarz inequality, to easily derive the inequality

$$\Phi\Phi,_{\alpha\alpha}-(\Phi,_\alpha)^2 \geq (\frac{1}{2}\int_{S_\alpha} \frac{a_{rk}\rho(\sigma)u_i u_i f,_r f,_k}{|\nabla f|} dS)$$

$$[\int_{S_\alpha} \frac{\sigma\rho(\sigma)}{|\nabla f|} dS - 2\int_{S_\alpha} \frac{\rho(\sigma)a_{rk}u_{i,r}f,_k a_{jm}u_{i,j}f,_m}{|\nabla f| a_{pq}f,_p f,_q}$$

$$- 2\Phi,_\alpha \int_{\Sigma_\alpha} a_{rk}\rho(\sigma)u_{i,r}u_i n_k dS. \tag{3.15}$$

In order to bound the term in square brackets on the right of (3.15), we need the form of the Rellich identity provided in the following lemma.

Lemma. With smooth arbitrary functions g_i, the following identity holds for the class of solutions of system (2.3);

$$\int_{S_\alpha \cup \Sigma_\alpha} g_p u_{i,p}\rho(\sigma)a_{rk}u_{i,r}n_k dS - \int_{D_\alpha} g_{p,k}u_{i,p}\rho(\sigma)a_{rk}u_{i,r}dx$$

$$- \frac{1}{2}\int_{S_\alpha \cup \Sigma_\alpha} g_p n_p (\int_0^\sigma \rho(t)dt)dS$$

$$+ \frac{1}{2}\int_{D_\alpha} g_{p,p}\int_0^\sigma \rho(t)dt\, dx + \frac{1}{2}\int_{D_\alpha} g_p\rho(\sigma)a_{rk,p}u_{i,r}u_{i,k}dx = 0, \tag{3.16}$$

where in the first and third integrands on the left there is minor abuse of rotation.

The proof of the lemma follows immediately from the identity

$$\int_{D_\alpha} g_p u_{i,p}[\rho(\sigma)a_{rk}u_{i,r}],_k \, dx = 0, \tag{3.17}$$

(obtained from (2.3)) together with integration by parts.

We now set

$$g_p = a_{pk}f,_k/(a_{rk}f,_r f,_k) \tag{3.18}$$

and thus derive from (3.15) the identity

$$\frac{1}{2}\int_{S_\alpha}\frac{\int_0^\sigma\rho(t)dt}{|\nabla f|}dS-\int_{S_\alpha}\frac{\rho(\sigma)a_{mn}f,_n u_{i,m}a_{rk}u_{i,r}f,_k}{|\nabla f|a_{pq}f,_p f,_q}$$

$$=\int_{\Sigma_\alpha}\frac{\rho(\sigma)a_{mn}f,_n u_{i,m}a_{rk}u_{i,r}n_k}{a_{pq}f,_p f,_q}dS-\frac{1}{2}\int_{\Sigma_\alpha}\frac{1}{|\nabla f|}(\int_0^\sigma\rho(t)dt)dS-\int_{D_\alpha}g_{\ell,k}u_{i,\ell}a_{rk}u_{i,r}\rho(\sigma)dx$$

$$+\frac{1}{2}\int_{\Sigma_\alpha}g_{p,p}(\int_0^\sigma\rho(t)dt)dx+\frac{1}{2}\int_{D_\alpha}\rho(\sigma)g_p a_{mn,p}u_{i,m}u_{i,n}dx \tag{3.19}$$

$$\geq-\bar{c}_1\int_{\Sigma_\alpha}\sigma\rho(\sigma)dS-\bar{c}_2\int_{D_\alpha}\sigma\rho(\sigma)dx, \tag{3.20}$$

where we have used the positive-definiteness of a_{ij}, the boundedness of the derivatives of a_{ij} and of g_p, and assumption (2.2). The constants $\bar{c}_1,\bar{c}_2$ are positive and are determined from bounds for f, a_{ij} and their derivatives. Use of (2.2) again in the left of inequality (3.20) finally leads to a lower bound for the expression in square brackets in (3.15). In this way, we are led to

$$\Phi\Phi,_{\alpha\alpha}-(\Phi,_\alpha)^2\geq\frac{1}{2}\int_S\frac{\rho(\sigma)a_{rk}f,_r f,_k u_i u_i}{|\nabla f|}dS[-\bar{c}_1\int_{\Sigma_\alpha}\sigma\rho(\sigma)dS-\bar{c}_2\int_{D_\alpha}\sigma\rho(\sigma)dx]$$

$$-2\Phi,_\alpha\int_{\Sigma_\alpha}\rho(\sigma)a_{rk}u_{i,r}u_i n_k dS. \tag{3.21}$$

We may now use the arithmetic-geometric mean inequality together with (3.9), (3.10) and (3.4) to recover (3.6).

4. Uniqueness and continuous dependence.

The inequality (3.6) may be employed to establish uniqueness of the solution to (2.3) under null Cauchy data, and also continuous dependence of the solution on the Cauchy data in the sense of Hölder. These conclusions are derived in a standard way (see, for instance Payne [10]), and hence need only be briefly described.

As regards uniqueness, Levine [3] has shown that when a solution to (3.6) vanishes at any value of α in the interval (0,1], then it vanishes identically. Under null Cauchy data (2.7), (2.8), we have $\Phi(0) = 0$, and thus there is uniqueness of the null solution to (2.3).

To derive continuous dependence of the solution, we introduce a new variable through the relation

$$\beta = e^{-c_1\alpha} \tag{4.1}$$

and rewrite (3.6) as

$$(\ln\{\phi\beta^{-c_2/c_1^2}\})_{,\beta\beta} \geq 0 . \tag{4.2}$$

Jensen's inequality then yields

$$\phi(\alpha)\beta^{-c_2/c_1^2} \leq [\phi(1)\beta_1^{-c_2/c_1^2}]^{(\frac{1-\beta}{1-\beta_1})} [\phi(0)]^{(\frac{\beta-\beta_1}{1-\beta_1})} , \tag{4.3}$$

where

$$\beta_1 = e^{-c_1} \tag{4.4}$$

We note that $\phi(0) = Q$, an expression involving only data terms.

Let us now assume that the solution lies in the class κ defined by

$$\kappa = \{v: \phi(1) < K^2\} \tag{4.5}$$

for some perscribed constant K. The continuous dependence conclusion is then immediate from the next theorem.

Theorem 2. Suppose a classical solution u exists in the class κ, and the hypotheses of Theorem 1 are satisfied. Then

$$\int_0^\alpha \{\int_{D_\eta} \sigma\rho(\sigma)dx\}d\eta \leq C\, K^{2\delta(\alpha)}Q^{1-\delta(\alpha)}, \tag{4.6}$$

for computable constant C, and $\delta(\alpha) \in [0,1]$.

It would be of interest to examine the consequences of different assumptions on the form of the function $\rho(s)$. If the requirement (2.2) were replaced by a suitable alternate inequality would it for instance be possible to obtain global nonexistence results similar to those obtained by Knops, Levine and Payne [22]? It would also be useful to have unique continuation formulas for systems which are of mixed type in D. It will hopefully be possible to extend the results of this paper in several directions.

References

[1] Aron, M. On the uniqueness of solutions in finite elasticity, Rend. Inst. Lombardo, A109 (1975), pp. 424-432.

[2] Knops, R.J., H.A. Levine and L.E. Payne, Non existence, instability and growth theorems for solutions of a class of abstract nonlinear equations with applications to nonlinear elastodynamics, Arch. Rat. Mech. Anal. 55 (1974), pp. 52-72.

[3] Levine, H.A., Logarithmic Convexity and the Cauchy problem for some abstract differential inequailities, J. Diff. Eqtns. 8 (1970), pp. 34-55.

[4] Payne, L.E., Bounds in the Cauchy problem for elliptic equations, Arch. Rat. Mech. Anal. 5 (1960), pp. 35-45.

[5] Payne, L.E., On a priori bounds in the Cauchy problem for elliptic equations, S.I.A.M. J. Math. Anal. 1 (1970), pp. 82-89.

[6] Payne, L.E. Improperly posed problems in partial differential equations, Regional Conference Series in Applied Math. #22, S.I.A.M., Philadelphis (1975).

[7] Payne, L.E. and H.F. Weinberger, New bounds for solutions of second order elliptic partial differential equations, Pac. J. Math. 8 (1958), pp. 551-573.

[8] Schaefer, P.W., On the Cauchy Problem for an elliptic system, Arch. Rat. Mech. Anal. 20 (1965), pp. 391-412.

[9] Schaefer, P.W., On uniqueness, stability and pointwise estimates in the Cauchy problem for coupled elliptic systems. Quart. Appl. Math. 31 (1973), pp. 321-328.

[10] Trytten, G.N., Pointwise bounds for solutions of the Cauchy problem for elliptic equations, Arch. Rat. Mech. Anal. 13 (1963), pp. 222-244.

BEMERKUNG ZU EINER KLASSISCHEN ARBEIT VON J. BLUMENFELD UND W. MAYER

REINER KÜHNAU*)

Inhalt: Das Problem der expliziten Bestimmung des Spektrums der Fredholmschen Eigenwerte zum klassischen Neumannschen Kern ist in der Literatur nur in ganz wenigen Fällen vollständig behandelt. Diese Aufgabe wird hier mit funktionentheoretischen Methoden gelöst für den Fall einer Cassinischen Kurve. Diese Frage ist schon einmal in einer klassischen Arbeit von J. BLUMENFELD und W. MAYER untersucht worden, allerdings - wie sich hier zeigt - fehlerhaft und unvollständig.

I. Die Lösung inverser Probleme setzt meistens die Lösung des direkten Problems voraus. Bekanntlich spielt mit dem Neumannschen Kerne K(s,t) zu gegebenen Kurvensystemen in der Ebene die Integralgleichung

$$(1) \qquad \theta(s) = \lambda \int K(s,t)\theta(t)dt + g(s)$$

mit ihren nichttrivialen Eigenwerten λ_i, $|\lambda_2| \leq |\lambda_3| \leq \ldots$, eine wichtige Rolle bei direkten Problemen der Potentialtheorie und in der Theorie der konformen Abbildung (vgl. z. B. [2], Seite 27 ff.). Die Frage nach den Eigenwerten läßt sich auch funktionentheoretisch formulieren [6], [9] und in Zusammenhang mit der Theorie der quasikonformen Abbildungen bringen [7]. Die Eigenwerte und -funktionen sind explizit nur in ganz wenigen Fällen bekannt. In [1] wurden diese Eigenwerte für einen Typ Cassinischer Kurven notiert, jedoch falsch, wie schon in [7] bemerkt. Dies soll hier berichtigt und vervollständigt werden unter Zugrundelegung von Satz 5 in [7]. Wir bestimmen dazu die dort w(z) genannten Funktionen als "komplexe Eigenfunktionen" mit den zugehörigen Eigenwerten $-q$, $q > 1$. Dabei kann die Methode in [4], [5] zur Konstruktion dieser komplexen Eigenfunktion verwendet werden.

II. Die Cassinischen Kurven seien durch $|z^2 + \mathfrak{z}_1| = 1$ mit $\mathfrak{z}_1 > 0$ in der komplexen z-Ebene gegeben. Diese Lage ist wegen der Invarianz der Fredholmschen Eigenwerte bei linearer Transformation keine Einschränkung der Allgemeinheit. Zunächst sieht man, daß es keine geraden Eigenfunktionen w(z) geben kann, da diese vermöge

*) Doz. Dr. Reiner Kühnau, Sektion Mathematik der Martin-Luther-Universität Halle - Wittenberg, DDR-401 Halle (Saale), Universitätsplatz 8/9

(2) $$\mathfrak{z} = z^2 + \mathfrak{z}_1$$

Anlaß gäben zu nichttrivialen Eigenfunktionen $w(z(\mathfrak{z}))$ zur Einheitskreislinie. Darüber hinaus müssen alle Eigenfunktionen $w(z)$ ungerade sein, da andernfalls $w(z) + w(-z) \not\equiv 0$ eine gerade Eigenfunktion wäre.

III. Wir betrachten nun zunächst den Fall $\mathfrak{z}_1 > 1$, der auf eine Cassinische Kurve führt, die aus zwei Zügen besteht. Unter der Annahme, $w(z)$ sei eine (also ungerade) Eigenfunktion mit Eigenwert $-q < -1$ entsprechend Satz 5 in [7], erhalten wir über (2) eine zweiwertige Funktion $w(z(\mathfrak{z})) = F(\mathfrak{z})$ mit Verzweigungsverhalten in $\mathfrak{z} = \mathfrak{z}_1$. Dann gehen wir analog der Prozedur in [4], [5] über zu $f(\mathfrak{z}) = F(\mathfrak{z}) - q\overline{F(\mathfrak{z})}$, $G(\mathfrak{z}) = \overline{f(1/\bar{\mathfrak{z}})}$. F und G sind für $|\mathfrak{z}| \geq 1$ analytisch und $h = (1 - q^2) \cdot F^2 + G^2$ ist sogar auch eindeutig. Wie in [4], [5] erkennt man, daß h auf $|\mathfrak{z}| = 1$ reell ist, woraus wegen der Regularität auch in $\mathfrak{z} = \infty$ $h \equiv C$ (reelle Konstante) folgt. Setzt man noch $k(\mathfrak{z}) = FG' - GF'$, dann erhält man wieder aus dem Verhalten in $\mathfrak{z} = \mathfrak{z}_1$ und $\mathfrak{z} = \infty$ und der Tatsache, daß diese Funktion auf $|\mathfrak{z}| = 1$ offensichtlich reell ist, daß

(3) $$\mathfrak{z}^2 k(\mathfrak{z}) \, [(\mathfrak{z} - \mathfrak{z}_1)(1 - \mathfrak{z}_1\mathfrak{z})/\mathfrak{z}] = c^2$$

mit einer reellen Konstanten c gilt. Nachdem so h und k bekannt, rechnet man wie in [4], [5] zurück auf

(4) $$F(\mathfrak{z}) = \frac{\sqrt{C}}{\sqrt{q^2 - 1}} \, \mathfrak{Sin} \left\{ \sqrt{q^2 - 1} \, \frac{c}{C} \int^{\mathfrak{z}} \frac{d\mathfrak{z}}{\sqrt{\mathfrak{z}(\mathfrak{z} - \mathfrak{z}_1)(1 - \mathfrak{z}_1\mathfrak{z})}} \right\},$$

$$G(\mathfrak{z}) = \sqrt{C} \, \mathfrak{Cof} \left\{ \sqrt{q^2 - 1} \, \frac{c}{C} \int^{\mathfrak{z}} \right\}.$$

Die untere Integrationsgrenze kann hierbei $= \mathfrak{z}_1$ gesetzt werden wegen $F(\mathfrak{z}_1) = 0$. Die Eindeutigkeit von $F(\mathfrak{z})$ auf $|\mathfrak{z}| = 1$ bringt die Bestimmungsgleichung

(5) $$\sqrt{q^2 - 1} \, \frac{c}{C} \oint_{|\mathfrak{z}|=1} = 2N\pi i, \quad N \text{ ganz},$$

die Übereinstimmung der Funktionen $F(\mathfrak{z})$ und $[\overline{G(1/\bar{\mathfrak{z}})} + qG(1/\bar{\mathfrak{z}})]/(1-q^2)$ auf $|\mathfrak{z}| = 1$ wie in [4] unter (17) und in [6] unter (12)

(6) $$q = \mathfrak{Cof} \left\{ 2 \sqrt{q^2 - 1} \, \frac{c}{C} \int_{\mathfrak{z}_1}^{1} \right\}$$

mit jeweils gleichem Integranden wie in (4). Damit haben wir, nachdem wir aus (5) $\sqrt{q^2 - 1}\, c/C$ errechnet und in (6) eingesetzt und schließlich die elliptischen Integrale auf Normalform transformiert haben, die Eigenwerte $-q = -q_N$ mit

$$(7) \qquad q_N = \mathfrak{Cof}\left\{N\pi\,\frac{K'(1/\mathfrak{z}_1)}{K\,(1/\mathfrak{z}_1)}\right\}, \quad N = 1,\ 2,\ 3,\ \ldots$$

Die zugehörigen Eigenfunktionen errechnen sich noch dazu über (4).

IV. Nun sei <u>$0 < \mathfrak{z}_1 < 1$, so daß eine aus einem einzigen Zuge bestehende Cassinische Kurve vorliegt.</u> Genau wie oben bei $\mathfrak{z}_1 > 1$ werden aus einer komplexen Eigenfunktion w(z) mit Eigenwert $-q < -1$ Funktionen F , G usw. mit leicht zu modifizierendem Mehrdeutigkeitsverhalten gebildet. Wieder folgt $h \equiv C$ (= reelle Konstante). In (3) ist $\mathfrak{z}_1$ durch $1/\mathfrak{z}_1$ zu ersetzen, entsprechend in (4), in (5) zusätzlich noch 2N durch 2N − 1, N ganz. Zur Bestimmung der Integrationskonstanten in (4) nutzen wir $G(1/\mathfrak{z}_1) = 0$ aus entsprechend der Tatsache, daß G in z eine ungerade Funktion ist. Dem wird durch

$$\sqrt{q^2 - 1}\,\frac{c}{C}\int^{\mathfrak{z}} = \sqrt{q^2 - 1}\,\frac{c}{C}\int_{1/\mathfrak{z}_1}^{\mathfrak{z}} + \left(\frac{\pi}{2} + n\pi\right) i\ , \quad n \text{ ganz},$$

Rechnung getragen. So erhalten wir dann

$$(8) \qquad q = \mathfrak{Cof}\left\{2\sqrt{q^2 - 1}\,\frac{c}{C}\int_{1/\mathfrak{z}_1}^{1}\right\}.$$

Nach entsprechender Prozedur wie oben haben wir bei

$$(9) \qquad q_N = \mathfrak{Cof}\left\{\frac{2N - 1}{2}\,\pi\,\frac{K'(\mathfrak{z}_1)}{K(\mathfrak{z}_1)}\right\}, \quad N = 1,\ 2,\ 3,\ \ldots,$$

die Eigenwerte $-q = -q_N$ der aus einem Zuge bestehenden Cassinischen Kurve für diesen Fall $0 < \mathfrak{z}_1 < 1$.

V. Im Gegensatz zu (9) werden für diesen Fall $0 < \mathfrak{z}_1 < 1$ in [1] Werte angegeben, die sich zu

$$(10) \quad \lambda_N = \frac{2(1 + \mathfrak{z}_1)}{\pi}\,K(\mathfrak{z}_1)\,\mathfrak{Cof}\left\{\frac{2N - 1}{2}\,\pi\,\frac{K'(\mathfrak{z}_1)}{K(\mathfrak{z}_1)}\right\}, \quad N = 1,\ 2,\ 3,\ \ldots,$$

umschreiben lassen. (Der Fall $\mathfrak{z}_1 > 1$ wird in [1] nicht behandelt). Und zwar ergibt sich (10) aus [1] (Seiten 2043/44), weil dort die Summe über die ungeraden m = Differenz der Summe über alle m und der Summe über die geraden m. Erstere läßt sich nach [3], 16.23.3 auf Seite 575 (dort u = 0 gesetzt), zusammenfassen, letztere ebenso nach vorheriger Landenscher Transformation des Ausdruckes K'/K (vgl. [10], Seiten 237/8). Daß (10) tatsächlich nicht die Eigenwerte darstellen kann, erkennt man übrigens auch schon aus $\lambda_N \to \infty$ für $z_1 \to 1$ (bei festem N). Dies geht nämlich nicht an, da λ_N dabei nach [8], Seite 1209, monoton fallend sein muß.

Schrifttum

[1] BLUMENFELD, J. und MAYER, W.: Über Poincaré'sche Fundamentalfunktionen Sitzungsber. d. Kaiserl. Akad. d. Wiss., Math.-Nat. Kl., 123, Abt. IIa (1914), 2011-2047.

[2] GAIER, D.: Konstruktive Methoden der konformen Abbildung. Berlin-Göttingen-Heidelberg 1964.

[3] Handbook of mathematical functions with formulas, graphs, and mathematical tables. Edited by M. ABRAMOWITZ and I. A. STEGUN New York 1965.

[4] KÜHNAU, R.: Eine funktionentheoretische Randwertaufgabe in der Theorie der quasikonformen Abbildungen. Indiana Univ. Math. J. 21 (1971), 1-10.

[5] ---, Zur analytischen Darstellung gewisser Extremalfunktionen der quasikonformen Abbildung. Math. Nachr. 60 (1974), 53-62.

[6] ---, Eine Verschärfung des Koebeschen Viertelsatzes für quasikonform fortsetzbare Abbildungen. Ann. Acad. Sci. Fenn., A. I. Math., 1 (1975), 77-83.

[7] ---, Eine Integralgleichung in der Theorie der quasikonformen Abbildungen. Math. Nachr. 76 (1977), 139-152.

[8] SCHIFFER, M.: The Fredholm eigen values of plane domains. Pac. J. Math. 7 (1957), 1187-1225.

[9] ---, Fredholm eigen values of multiply-connected domains. Pac. J. Math. 9 (1959), 211-269.

[10] TRICOMI, F.: Elliptische Funktionen. Übersetzt und bearbeitet von M. KRAFFT. Leipzig 1948.

SOME PROBLEMS OF ANALYTIC CONTINUATION

M.M.LAVRENTJEV

1. A large class of an analytic continuation problems for the functions of one variable can be formulated in the following way.

Let $f(z)$ be analytic function of complex variable in the domain $\mathcal{D}$ of complex plane. Some additional information on function $f(z)$ can be given. For example, $f(z)$ is continuous in $\overline{\mathcal{D}}$ (closure of $\mathcal{D}$) and bounded in $\overline{\mathcal{D}}$ with given constant:

$$|f(z)| \leq C, \quad z \in \overline{\mathcal{D}}$$

Let A , B be some subsets of $\overline{\mathcal{D}}$,

$$A \subset B \subset \overline{\mathcal{D}}$$

and values of $f(z)$ on the set A are known. The problem is to find the values of $f(z)$ on the set B .

2. Analytic continuation problem are classical problems in analytic functions theory and in the theory of complex variable functions; some uniqueness theorems for this problems were proved in the last century and are presented in most of monographs.

If we consider the analytic functions in $\mathcal{D}$ with no additional restrictions, then analytic continuation problems are linear. Thus the uniqueness of solution in the analytic continuation problem is equivalent to the statement:

Let $f(z)$ be analytic function in the domain $\mathcal{D}$,

$$f(z) = 0, \quad z \in A$$

Then

$$f(z) = 0, \quad z \in B$$

Note, that in uniqueness theorem for analytic continuation problems set B includes the domain of analyticity $\mathcal{D}$:

$$B \supset \mathcal{D}$$

3. We present now the classical results on uniqueness of a solution in analytic continuation problems (see 2, 3, 5).

<u>Theorem 1.</u> Let $f(z)$ be analytic function in the domain $\mathcal{D}$, and $\mathcal{D}_1$ is a subset of $\mathcal{D}$. Then, if

$$f(z) = 0, \qquad z \in \mathcal{D}_1,$$

we have

$$f(z) = 0, \qquad z \in \mathcal{D}$$

<u>Theorem 2.</u> Let $f(z)$ be analytic function in $\mathcal{D}$, and set with limiting point from $\mathcal{D}$. Then, if

$$f(z) = 0, \qquad z \in A,$$

we have

$$f(z) = 0, \qquad z \in \mathcal{D}.$$

<u>Theorem 3.</u> Let $f(z)$ be analytic function in the domain $\mathcal{D}$ with boundary Γ , and $\Gamma_1 \subset \Gamma$ be some straightened curve. Then, if

$$f(z) = 0, \qquad z \in \Gamma_1,$$

we have

$$f(z) = 0, \qquad z \in \mathcal{D}.$$

4. A notion of riemannian manifold is connected with analytic continuation. We give here some corresponding definitions (see 5).

<u>Definition 1.</u> We call "an analytic element" the totality $\{\mathcal{D}, f\}$ of a domain $\mathcal{D}$ in complex plane and analytic function $f(z)$ in $\mathcal{D}$.

<u>Definition 2.</u> Two analytic elements $\{\mathcal{D}_1, f_1\}$, $\{\mathcal{D}_2, f_2\}$ are called the direct analytic continuation of each other through the domain $\tilde{\mathcal{D}}$, $\tilde{\mathcal{D}} \subset \mathcal{D}_1 \cap \mathcal{D}_2$, if

$$f_1(z) = f_2(z), \qquad z \in \tilde{\mathcal{D}}.$$

<u>Definition 3.</u> Analytic elements $\{\mathcal{D}_0, f_0\}$ and $\{\mathcal{D}_n, f_n\}$ are called an analytic continuation of each other, if there exists such analytic elements $\{\mathcal{D}_k, f_k\}$ and domain $\tilde{\mathcal{D}}_k$, that analytic elements $\{\mathcal{D}_k, f_k\}$, $\{\mathcal{D}_{k+1}, f_{k+1}\}$ are direct analytic continuation of each other through $\tilde{\mathcal{D}}_k$, $k = 0, \ldots, n-1$.

<u>Definition 4.</u> We call an anlytic function the totality of analytic elements which are analytic continuations of some analytic

element. Functions $f_a(z)$ belonging to one analytic function are called branches of this function.

Consider the totality of pairs:

$$\{a,\ f_a(z)\},$$

where a - is a point of complex plane, $f_a(z)$ is analytic in some circle U_a with origin a :

$$f_a(z) = \sum_0^\infty c_k (z-a)^k$$

Now introduce the topology in the set of pairs $\{a, f_a(z)\}$: We call an ε - neighbourhood of element $\{a, f_a(z)\}$ the totality of elements $\{b, f_b(z)\}$, such that

1) $|a-b| < \varepsilon$

2) an analytic element $\{U_b, f_b\}$ is a direct analytic continuation of the element $\{U_a, f_a\}$.

It can be shown, that topological space of pairs $\{a, f_a\}$ is Hausdorff space, i.e. the separation axiom is valid for its neighbourhoods.

<u>Definition 5.</u> The Hausdorff space of pairs $\{a, f_a(z)\}$, where is a point of complex plane and $f_a(z)$ is analytic function in a circle U_a with above introduced topology, is called Riemannian manifold.

Mapping (projection) of Riemannian manifold onto complex plane:

$$\{a,\ f_a(z)\} \rightarrow a$$

is local homeomorfism (i.e. some neighbourhoods of pair $\{a, f_a(z)\}$ and of point a this mapping is one-to-one and continuous).

<u>Definition 6.</u> The part of Riemannian manifold, such that all functions belonging to the points of this part are branches of some analytic function, is called Riemannian surface of this function.

5. In connection with this considerations the following formulations of analytic continuation problems arise:

Let A be subset of the Riemannian surface of the analytic function $f(z)$, and values $f(z)$ are given in the set A .

1) To determine values $f(z)$ in some set $B \supset A$.

2) To determine some characteristics (say, topological) of the Riemannian surface of the function $f(z)$.

3) One needs to determine the singular points of the function $f(z)$ and behaviour of the function in the neighbourhoods of this points, in particular, types of the singular points, and behaviour of the function in the neighbourhoods of the singular points of the Riemannien surface.

The following theorem is valid (see 3, 5).

Theorem 4. Let the set A includes limiting point inside the Riemannian surface of analytic function $f(z)$. Then Riemannian surface of function $f(z)$ is uniquely determined by values of $f(z)$ on the set A

6. We present now the formulation of analytic continuation problem for the function of several variables, an analogue of formulation in one-dimensional case.

Let $f(z_1, z_2, \dots, z_n)$ be analytic function of n complex variables in the domain $\mathcal{D}$ of n -dimensional complex space, A and B be some subsets of $\bar{\mathcal{D}}$:

$$A \subset B \subset \bar{\mathcal{D}}$$

The values of a function $f(z_1, z_2, \dots, z_n)$ are known in the set A. To determine values $f(z_1, \dots, z_n)$ in the set B.

Problems of analytic continuation theory for the functions of several variables were treated from the begining of our century. Analytic continuation playes an important role in the monographs on the theory of analytic functions of several variables (see 1, 4, 5).

The space of n complex variables $(z_1, z_2, \dots, z_n)$ will be denoted by C^n.

Definition 7. A domain $G \subset C^n$, strictly including $\mathcal{D}$, is called holomorphic expansion of $\mathcal{D}$, if any in $\mathcal{D}$ analytic function $f(z_1, \dots, z_n)$ has an analytic continuation onto G.

It is known, that the domains in complex plane have no holomorphic expansion, but in the spaces of two and more variables such domains do exist. Next theorem establishes the existence of one class of such domains:

Theorem 5. (Hartogs). Let $\mathcal{D}$ be a domain in C^n, $\mathcal{D}_1$ be a domain in the complex plane, Γ_1 be a boundary of $\mathcal{D}_1$, M be a set

$$M = [\bar{\mathcal{D}} \times \Gamma_1] \cup [(z_1^0, \ldots, z_n^0) \times \bar{\mathcal{D}}_1]$$

where $(z_1^0, \ldots, z_n^0) \in \mathcal{D}$

Then every function analytic in some neighbourhood of set M (in the space C^{n+1}) can be analytically continuated to the domain:

$$\tilde{\mathcal{D}} = \mathcal{D} \times \mathcal{D}_1$$

7. The solutions of some analytic contipuation problems continuously depend on data. In case of one variable it is the case, when set A is equal to Γ, boundary of the domain $\mathcal{D}$. For several variables it is, for example, an analytic continuation into holomorphic expansion, which was mentioned in the theorem 5.

For the function of one variable the analytic continuation problems from the inner set of the regularity domain and from the part of the boundary of the domain are uncorrect in a classical sense,and character of the instability is the same, as instability in the uncorrect Cauchy problems for Laplace and heat conduction equations.

8. As known, the solutions of some differential equations are the analytic functions in the regularity domains. In particular, the solutions of elliptic equations with analytic coefficients have this property.

In this connection the following problems arise (also call them analytic continuation problems).

Let $\mathcal{L}$ be a differential operator in the space $(x_1, \ldots, x_n)$ with the following property:

Every solution of the equation

$$\mathcal{L} u = 0 \tag{1}$$

is an analytic function of valiables $(x_1, \ldots, x_n)$ in the regularity domain.

It is known, that equation (1) has regular solution in the domain $\mathcal{D}$, $u(x_1, \ldots, x_n)$

Some additional information on the function $u(x_1, \ldots, x_n)$ can be given, for example,

$$|u| \leq C, \qquad (x_1, \ldots, x_n) \in \bar{\mathcal{D}}$$

Let A, B be some subsets of $\bar{\mathcal{D}}$. The values of the solution (1) are known in the set A, and we need to determine the values of the solution in the set B.

References

1. Vladimirov V.S. Methods in the theory of the functions in several complex variables. Moscow, Nauka, 1964 (Russian).
2. Lavrentjev M.A., Shabat B.V. Method in the theory of the functions of complex variable. Moscow, Fizmatgiz, 1958 (Russian).
3. Markushevich A.I. Theory of analytic functions. Moscow-Leningrad, Gostehizdat, 1950 (Russian).
4. Fux B.A. Introduction to the theory of analytic functions in several complex variable. Moscow, Fizmatgiz, 1962 (Russian).
5. Shabat B.V. Introduction to the complex analysis. Moscow, Nauka, 1969 (Russian).

Prof. Dr. M. M. Lavrentjev, USSR Academy of Sciences, Sibirian Branch, Computing Centre, Novosibirsk 630090

ABOUT INVERSE PROBLEMS IN CONTINUUM MECHANICS AND SOLID STATE PHYSICS

BERND M I C H E L

Summary:

Applying linear continuum mechanics of solids containing structural defects, the author deals with different physical problems leading to inverse and improperly posed problems of potential theory. Higher-order elliptic equations, such as biharmonic and modified biharmonic equations, are derived for special cases of crystal symmetry. It is shown that essential advances in the determination both of the sources of field quantities and related material parameters can be achieved, if the solutions of inverse problems as presented in this paper are available.

1. Introduction

The present paper is not concerned with the derivation of solutions for special cases of inverse problems, but an attempt is made to formulate certain inverse problems corresponding to elliptic differential equations. For simplification the author restricts his investigations within the field of linear continuum mechanics of solids. Continuum mechanics of solids has become more and more important in recent years, since it is necessary for the progress of many branches of solid state physics and modern material sciences, in general, e.g. fracture mechenics /21/ and soil mechanics /4/,/22/.

Practically all of modern physics deals with fields: potential fields, probability fields, electromagnetic fields, tensor fields, and spinor fields. Classical theory of elasticity as a typical field theory is strongly related to potential theory. The famous works of PAPKOVITCH /19/ and NEUBER /20/ are concerned with this subject. Clas-

---

Dr. B. Michel, Sektion Fertigungstechnik, Technische Hochschule, DDR 901 Karl-Marx-Stadt sical theory of elasticity of solid bodies deals with their behaviour under the action excerted by external loads. Most modern branches of linear continuum mechanics of the solid state, on the other hand, are concerned with a medium containing structural defects(dislocation theory etc. /8/). The properties of these defects are essential for the physical (not only the mechanical!)behaviour both on a microscopic and on a macroscopic scale. From the geometrical point of view one can divide the defects into several classes;

1. zero-dimensional or point defects (impurity atoms, such as interstitials,vacancies etc.),
2. one-dimensional or line defects (e.g. dislocations),
3. two-dimensional defects or plane imperfections (stacking faults and grain boundaries), and
4. three-dimensional defects or space imperfections (volume defects such as inclusions, precipitations, interface layers, regions of decomposition etc.).

The theoretical study of the deformation and stress fields in the defected region presents extremely difficult problems, since both the exact distribution function of the defects and the local material parameters usually are not available. Continuum mechanics describes the deformed solid by suitable field functions such as stress and strain tensors /10/. The sources of the field are represented by so-called "source functions" (usually density functions). The latter are strongly related both to the defect properties and to the material parameters in the defected region /13/-/15/. For the physisist the following two basic problems result:

1. To get information about the source functions (without being able to obtain experimental values immediately within the defected region).
2. To get information about the material properties in the defected region(e.g. this problem is related to questions about the exact values of elastic constants,so-called local elastic constants).

It is impossible, in general, to separate the two problems from each other, because the "strength" of the source function, as a rule, is not independent of the real values of the material parameters immediately at the defects. That is why it is an approximation only to apply, for instance, YOUNG's modulus, obtained by a compact tension test, for a workpiece after machining in order to calculate the deformations remaining in the strongly distorted inhomogeneous surface region of the same solid. An exact calculation should,however, take into account the changed material constants in the sub-surface layer of the solid /17/.

Local material parameters usually are not available, but their values may be very different frome those, which were obtained by conventional experimental methods for the whole solid. This problem has not yet been solved up to now. So the values of elastic parameters outside the defects are different from those inside them. This kind of questions has become typical for many branches of modern physics and material sciences /15/,atomic physics /25/ and soil mechanics as well /4/. It always leads to inverse and improperly posed problems.

2. Inverse problems for solids with "quasiplastic" deformation due to structural defects

2.1. Distortions in an anisotropic medium

In the framework of linear continuum mechanics it is possible to carry out separate investigations for stress and strain fields connected with structural defects on the one hand and for stress and strain fields due to external loads on the other. This fact is expressed by the theorem of COLONETTI, which states that the elastic energy of static interaction between a source of internal stress and a field of external stress vanishes. The structural defects, described above, are the sources of additional stress and strain fields /8/,/11/-/13/,/16/. The elastic medium (matrix), with possible exceptions of some singular points, lines, or planes, can be described by a regular displacement field $\vec{u}$, the tensor of total distortion β being given by the relation

$$\beta = \vec{u}\,\nabla \qquad , \qquad (1)$$

where β is composed of two terms

$$\beta = \beta_{el} + \beta_{pl} \qquad . \qquad (2)$$

It is assumed that the elastic part β_{el} is related to the stress tensor σ by generalized HOOKE's law

$$\sigma = C : \beta_{el} \qquad , \qquad (3)$$

C being a material tensor of rank four, containing 21 independent elastic parameters in the general case of anisotropy. The part β_{pl} of the total distortion tensor β is called tensor of "quasiplastic" distortions. Taking into account the equation of mechanical equilibrium

$$\sigma \cdot \nabla = \rho \ddot{\vec{u}} \qquad , \qquad (4)$$

when body forces are absent /10/, we arrive at the fundamental displacement equation /9/

$$C : \vec{u}\,\nabla \cdot \nabla - \rho \ddot{\vec{u}} = C : \beta_{pl} \cdot \nabla \qquad , \qquad (5)$$

in which ρ is the mass density function. We do not further regard the general case of equation (5) here (see /9/). In the steady state case (5) leads to elliptic equations. It is necessary to mention the appearence both of the material tensor C and the source density function β_{pl} of quasiplastic distortions on the right-hand side of equation (5). For a steady state of plane strain it is possible to replace (5) by the following set of scalar equations /9/:

$$L^{(4)}U = D \,, \quad L^{(4)}V = S \,, \quad L^{(4)}W = E \,, \qquad (6)$$

where the differential operator $L^{(4)}$ is given by the expression /5/

$$L^{(4)} = s_{11}\frac{\partial^4}{\partial x^4} + (2s_{12}+s_{44})\frac{\partial^4}{\partial x^2 \partial y^2} + s_{11}\frac{\partial^4}{\partial y^4} \,. \qquad (7)$$

The coefficients s_{11}, s_{12}, and s_{44} are VOIGT's elastic constants. The source functions D, S, and E are related to the cartesian components $\beta_{pl}^{(i,j)}$ of β_{pl} by

$$D = \frac{1}{2}\left(\beta_{pl}^{(1,1)} + \beta_{pl}^{(2,2)}\right) - \frac{s_{12}}{s_{11} - s_{12}}\,\beta_{pl}^{(3,3)}$$
$$S = \beta_{pl}^{(1,2)} + \beta_{pl}^{(2,1)} \qquad (8)$$
$$E = \frac{1}{2}\left(\beta_{pl}^{(2,2)} - \beta_{pl}^{(1,1)}\right) \,.$$

The components σ_{ij} of σ can be derived from the potentials U, V, and W /5/, /9/. It should be pointed out here that equation (6) describes the behaviour of continuous distributions of dislocations in crystals with cubic and hexagonal symmetry /5/,/9/. The most interesting physical problem, however, does not consist in solving the direct boundary value problems for (6), but in getting information about β_{pl}, that means about the quantities D,S, and E within the defected regions, where experimental results are not or very difficult to obtain. This leads to inverse problems, since both the right-hand side of equation (6) and the generalized potential functions U,V,W (more precisely their second partial derivations) are known outside the sources only. That is why the physisist, as a rule, introduces certain simplifications of the model, i.e. he prescribes more or less certain restrictions for the source functions β_{pl} and for the whole right- hand sides of (6). Very often piecewise constant and linear dependent functions are assumed (/7/,/12/,/13/). These rough approximations do neither take into consideration any internal structure of the defected regions, nor do they base upon exact local material constants, but they may already lead to very good results for special cases /16/.

2.2. Isotropic media

In the case of an isotropic, homogeneous medium containing regions with an additional quasiplastic volume dilatation η (local fractional change of volume $\eta = \Delta V/V$ /12/) for a plane state of strain the following biharmonic equation can be derived /6/:

$$\Delta\Delta F + \frac{2G(1+\nu)\Delta\eta}{3(1-\nu)} = 0 \qquad , (9)$$

where G denotes the shear modulus and ν POISSON's ratio, respectively, The cartesian components of the stress tensor σ are derived from the AIRY function F by $\sigma_{xx} = \frac{\partial^2 F}{\partial y^2}$, $\sigma_{yy} = \frac{\partial^2 F}{\partial x^2}$, $\sigma_{xy} = -\frac{\partial^2 F}{\partial x \partial y}$. (10)

If the solid is not acted upon by external forces, the following boundary conditions are given at the "free" surface of the solid

$$F = 0 \quad , \quad \frac{\partial F}{\partial n} = 0 \qquad , (11)$$

where $\frac{\partial F}{\partial n}$ denotes the normal derivative of F. A corresponding inverse problem consists in the calculation of the source function

η and the elastic constants G and ν in the region, where η depends on position (G and ν are assumed to be piecewise constant here) The solution of this mathematical problem is very important, as it provides new possibilities for a real check of the assumtions of the basic model conceptions used today /14/. Additional informations about the possible set of source distribution functions η , which result from physics, have to be considered too to make the problem unique. To find those restrictions usually is already a very difficult problem /15/. For plane thermoelastic deformations equation

$$\Delta\Delta F + \frac{E\alpha\Delta T}{1-\nu} = 0 \qquad (12)$$

holds /6/,/7/. It results from setting $\eta = 3\alpha(T-T_0)$, T_0 = const., and applying $E = 2G(1+\nu)$. α denotes the coefficient of linear thermal expansion and E Young's modulus, respectively.

An extension of the problem for composite materials leads to equations of the kind $\Delta\Delta F_i = \gamma_i \Delta \eta_i$, (13)
with different coefficients γ_i. Additional coupling conditions at the boundaries (to find these boundaries leads to inverse problems too) of the homogeneous subdomains of the solid have to be taken into account /10/. A class of inverse problems consists in the determination of the material parameters γ_i and the source functions η_i again. The problems can be further generalized by regarding γ_i piecewise continuous. But this leads to other differential equations /10/.

2.3. Spinodal decomposition in solids

The process of spinodal decomposition in isotropic elastic solids (e.g. glasses) can be described by the following differential equation (or similar kinds of equations) for the concentration c/2/

$$f(c) = \Delta\Delta c + b\Delta c \quad , \qquad (14)$$

where f(c) is an analytic function of concentration. f is proportional to so-called interdiffusion coefficient, which has to be determined by an inverse problem, because it is not exactly known in the regions of decomposition. It is especially difficult to get information about f and M in regions of the solid, where grad c is very high. Equation (14) w_as derived by taking into account the elastic constituent of GIBBS free energy of the system. The problem described here is very important for investigations both of the mechanical and the physico-mechanical and physico-chemical behavior of several kinds of solid materials (e.g. strength of glasses etc.).

2.4. Coupling between drift-diffusion and deformation around moving defects

In 1970 SEEGER /23/ derived the following differential equation

$$\Delta \psi + U_1(E)\psi = 0 \qquad (15)$$

in order to describe the phenomenon of stress-assisted diffusion of point defects in an elastic solid. This equation has now become the subject of many investigations /24/. (15) is a special case of a stationary SCHROEDINGER equation. The "wave function" ψ both depends on the concentration c of the defects, on the temperature T and, above all, on the elastic drift potential energy E:

$$\psi = c \exp\left(\frac{E}{2kT}\right) \qquad . (16)$$

The quantity $U_1(E)$ is given by the expression

$$U_1(E) = \frac{2kT\Delta E - (\operatorname{grad} E)^2}{4k^2T^2} \qquad (17)$$

where k is the BOLTZMANN constant. The corresponding inverse problem consists in the determination of the drift potential E, especially in the immediate vicinity of drifting defects. In recent works of the author a way is shown to find an approach to understand drift diffusion processes near solid surfaces and interfaces /11/,/18/. From the drift potential one can deduce new assertions about the governing elementary processes on microscopic scale and about interaction phenomena between the defects /11/. The newly developed methods and results known from the inverse quantum mechanical scattering problem should be applied to solve this kind of inverse problems too /25/.

2.5. Elastic coupling between temperature field and deformations

The well known basic equations for the displacement field $\vec{u}$ of a thermoelastic solid

$$\operatorname{div}\operatorname{grad}\vec{u}-\frac{1}{1-2\nu}\operatorname{grad}\operatorname{div}\vec{u}-\frac{2(1+\nu)}{1-2\nu}\operatorname{grad}(\alpha T)+\frac{\rho}{G}\vec{K}=\frac{\rho}{G}\ddot{\vec{u}} \qquad (18)$$

is coupled with a generalized heat conduction equation of the kind

$$k_1\,\Delta T + a_1\,\operatorname{div}\dot{\vec{u}} + b_1 = c_1\dot{T} \quad . \qquad (19)$$

K denotes mass forces, whereas k_1, a_1, b_1, and c_1 are material parameters. We have to mention the importance of the term $a_1 \operatorname{div}\dot{\vec{u}}$ in (19), which is responsible for the coupling between heat production and elastic deformation. It represents the change of elastic dilatation $\varepsilon_I = \operatorname{div}\vec{u}$ with time. Usually this term is not taken into consideration. (18) and (19) can be uncoupled /3/. This leads to a temperature equation /3/,/4/. For the case of a quickly changing periodic temperature field (of constant frequancy ω) one can derive equations of the following type /14/,/15/

$$\sum_{i=1}^{5} c_i\,\Delta^{(i)}T = f \qquad ,(20)$$

where $\Delta^{(i)}$ is the polyharmonic operator of the order i. f is known. outside the sources. It is a single -valued function of the heat sources and of several thermomechanical parameters. The inverse problem is the determination of f and T inside the source region, T and f being given outside the sources only. The progress in this field is connected with real advances in several branches of material sciences, which are of technological importance, since the sources of heat production play an important role in mechanical engineering sciences such as metal cutting and metal forming /15/,/17/.

3. Concluding remarks

The kind of inverse problems, which have been outlined within this paper, has become typical for coupled problems in many branches of solid state physics and continuum mechanics as well. For this reason one likes to speak about the continuum thermodynamics of solids today. For simplification we were forced to restrict our investigations to linear models here. The more general conception of both nonlinear constitutive laws (general constitutive theory, which also includes plastic and viscous flow) and geometrically non-linear relations(such as large deformations) provide the basis for new and more complicated types of equations and corresponding inverse problems. Including higher-order terms of the stresses,distortions and further field quantities (e.g. magnetoelastic or electro-mechanical ones) is possible, but sensible only after achieving real advances in the study of the analysis of inverse problems for the linear case.

Acknoledgments

The author is grateful to Professor G.Anger and Professor F.Loges of the Martin Luther University Halle-Wittenberg and to Professor H.Weber of the Technische Hochschule Karl-Marx-Stadt for useful discussions.

Bibliography

/1/ Anger,G.,Die Rolle der modernen Potentialtheorie in der inversen Aufgabenstellung,Gerlands Beitr.z.Geophys. 85(1976),1-20.

/2/ Cahn,J.W.,On spinodal decomposition,Acta Metallurg. 9(1961),795.

/3/ Carlson,D.E.,Thermoelasticity,in: Handbuch d.Physik,Bd.VIa/2, Springer-Verlag,Berlin,Heidelberg,New York, 1972,328.

/4/ Chamecky,S.,Soil structure interactions in the analysis of raft formulations,2o.PolishSolid Mech. Conf.,Porabka-Kozubnik,1978.

/5/ Deuretzbacher,G.,Das Eigenspannungsfeld eines Gleitbandes in einem kubischen Kristall,Diss.Univ.Halle-Wittenberg,1973.

/6/ Hieke,M.,Eine indirekte Bestimmung der Airyschen Fläche bei unstetigen Wärmespannungen,ZAMM 35,8(1955),285-294.

/7/ Hieke,M.,Die Anwendung der inhomogenen Differentialgleichung 4. Ordnung und die Festkörperphysik,Beitr.z.Analysis,4(1972),61-67.

/8/ Kovács,I.,Zsoldos,L.,Dislocations and plastic deformation, Budapest,Akademiai Kiado,1973.

/9/ Maschke,H.,Frischbier,R., Zur Integration der Grundgleichungen für Eigenspannungsquellen in kubischen Kristallen, Kristall u. Technik 12,1o(1977),1001-1oo6.

/1o/ Michel,B.,Loges,F.,Probleme der Kontinuumsmechanik inhomogener Werkstoffe,Wiss.Z. Univ.Halle 25M,5(1976),5-15.

/11/ Michel,B.,Influence of solid surfaces on the energy of point defects and inclusions,2o.Polish Solid Mech.Conf.,Porabka-Kozubnik, 1978.

/12/ Michel,B.,Beitrag zur Ermittlung der elastischen Energie von Punktdefekten, Einschlüssen und Entmischungsbezirken, Acta Phys. Hung. 44,2(1978),147-156.

/13/ Michel,B.,Diss.z.Prom.B., Univ.Halle-Wittenberg,1979.

/14/ Michel,B.,Spannungen und Verformungen in Festkörpern und inverse Aufgaben der Potentialtheorie,Sommerschule über inverse Probleme der Geophysik, Freiberg,June 1978, u. Veröff. d. Nationalkom. f. Geodäsie u. Geophys., Potsdam (in print).

/15/ Michel,B.,Über das inverse Problem in der Kontinuumsmechanik fester Körper, Vortrag,Math.Koll.Sektion Math. Univ.Halle-Wittenberg, 12.10.1978.

/16/ Michel,B., A note on the interface interaction of point defects and volume defects, phys.stat.sol.(b) 89(1978),K179-K181.

/17/ Michel,B., A study on calculations of self-stresses in machined surfaces,Research Report, Techn. Hochschule Karl-Marx-Stadt, Mart 1977.

/18/ Michel,B., phys.stat.sol.(a) 1979(in print).

/19/ Papkovitch,P.F.,On the general solution of thermal stress (in Russian),Prikl.Math.Mech.1,1(1937).

/2o/ Neuber,H.,ZAMM 14(1934),203.

/21/ Rice,J.R.,The elastic-plastic mechanics of crack growth,2o.Polish Solid Mech. Conf.,Porabka-Kozubnik,1978.

/22/ Sawczuk,A.,Stutz,P.,On the formulation of stress-strain relationships for soils at failure,ZAMP 19(1968),770-778.

/23/ Seeger,A.,A separability of the diffusion equation with drift term, phys.stat.sol.(b) 41(197o),509.

/24/ Yoo,H., Butler,W.H., Steady-state diffusion of point defects in the interaction force field,phys.stat.sol.(b) 77(1976),181.

/25/ Chadan,K.,Sabatier,R.C., Inverse problems in quantum scattering theory,Texts and Monographs in Physics,Springer,Berlin,Heidelberg, New York, 1977.

SH-WAVE PROPAGATION IN QUASIANISOTROPIC ROCKS: PROBLEM OF PARTIAL DIFFERENTIAL EQUATION WITH RAPIDLY OSCILLATING COEFFICIENTS

JIŘI NEDOMA +)

Summary: In this paper we propose a solution of partial differential equations with rapidly oscillating periodic coefficients and its applications to the study of problems in SH-wave propagation in a quasianisotropic rock.

1. Introduction

In studies of the propagation of elastic waves through the earth it is usually assumed that the medium composing the earth is isotropic. But any solid material which forms the core of the earth and the upper mantle is usually equiped by a crystallic structure. Its crystallic structure is characterized by a regular form and unequal properties at the other directions - by anisotropy. On the strength of prevailing theories regarding the composition and the mechanical behaviour of the upper mantle these results suggest an upper mantle which is transversely isotropic to seismic wave propagation.
Backus [1] shows that in the long wavelength limit a sequence of isotropic layers acts as a transversely isotropic medium (quasianisotropic medium) which is described by five independent elastic constants. Elastic constants are related to the velocities of elastic waves in a medium through the equations of small motion, i. e.

$$\frac{\partial}{\partial x_j} \left(c_{ijkl} \frac{\partial u_k}{\partial x_l} \right) + \varrho f_i = \varrho \frac{\partial^2 u_i}{\partial t^2} ,$$

where u_i is the particle displacement vector in Cartesian coordinates x_i, ϱ is the density, f_i are the body forces per unit masses, and

$$c_{1111} = c_{2222} = c_{11}, \quad c_{1133} = c_{2233} = c_{13},$$

$$c_{1131} = c_{1123} = c_{1112} = c_{2223} = c_{2231} = c_{2212} = c_{3323} =$$

$$c_{3331} = c_{3312} = c_{2331} = c_{2312} = c_{3112} = 0,$$

$$c_{2323} = c_{3131} = c_{44},$$

$$c_{1122} = c_{1111} - 2\,c_{1212}, \text{ i. e. } c_{12} = c_{11} - 2\,c_{66},$$

$$c_{3333} = c_{33}.$$

As $c_{11} = c_{22}$, $c_{66} = \frac{1}{2}(c_{11} - c_{12})$ hold in the hexagonal system, the hexagonal materials are transversely isotropic (quasianisotropic) and all directions in a plane (x_1,x_3) are equivalent.

In the present contribution we shall study the SH-wave propagation in quasianisotropic materials, approximated by a medium in which N isotropic materials are alternating periodically and forming a sequence of N thin layers. By a thin layer we understand a layer, the thickness of which is considerably smaller than the wavelength. In our mathematical studies we assume that the thickness of layers tends to zero. Let us assume that N is even.

2. Mathematical Description of our Problem and its Solution

The model of a part of the earth in which we shall investigate a propagation of SH-waves in a quasianisotropic medium is approximated by a bounded region consisting of thin layers created by the planes parallel to the (x_1,x_3) plane. Let us assume a source to be a line harmonic source parallel to x_2 axis. Then the problem can be studied in the bounded region

$$G = \bigcup_{k=1}^{r} G_k^h \doteq G_s \subset R^2 , \quad k = (k_1,\dots,k_N)$$

defined as follows:

$$\begin{aligned}
&G = \{(x_1,x_3) \mid |x_1| \leq R,\ 0 \leq x_3 \leq H\}, \\
&G_s = \{(x_1,x_3) \mid x_1^2 + x_3^2 \leq r^2,\ r>0\}, \\
(1) \qquad &\partial G = \bigcup_{i=0}^{4} \Gamma_i ,\ \Gamma_0 \equiv \partial G_s = \{(x_1,x_3) \mid x_1^2 + x_3^2 = r^2,\ r>0\}, \\
&\Gamma_1 = \{(x_1,x_3) \mid |x_1| \leq R,\ x_3 = 0\},\ \Gamma_2 = \{(x_1,x_3) \mid x_1 = R,\ 0 \leq x_3 \leq H\}, \\
&\Gamma_3 = \{(x_1,x_3) \mid |x_1| \leq R,\ x_3 = H\},\ \Gamma_4 = \{(x_1,x_3) \mid x_1 = -R,\ 0 \leq x_3 \leq H\}.
\end{aligned}$$

We shall be interested in the boundary value problem for Helmholtz type equation describing propagation of SH-waves in a quasianisotropic medium on a domain G

$$(2) \qquad \begin{aligned} L^h u^h \equiv \frac{\partial}{\partial x_1}\left(\mu^h(x_3)\frac{\partial u^h}{\partial x_1}\right) + \frac{\partial}{\partial x_3}\left(\mu^h(x_3)\frac{\partial u^h}{\partial x_3}\right) + \\ + g^h(x_1,x_3)u^h = f, \end{aligned}$$

with the boundary conditions

$$\text{(3a-d)} \quad \left.\frac{\partial u^h}{\partial n}\right|_{\Gamma_1} = 0, \quad \left.u^h\right|_{\Gamma_2\cup\Gamma_3\cup\Gamma_4\cup\Gamma_0} = 0,$$

$$u^h_{k_i} = u^h_{k_i+1}, \quad \mu^h_{k_i}\frac{\partial u_{k_i}}{\partial x_3} = \frac{\partial u^h_{k_i+1}}{\partial x_3}\mu^h_{k_i+1},$$

where f, g^h, $\mu^h(x_3)$ are known functions (real or complex) on G. Let N be an integer, and let N be even. Let $h = \sum_{i=1}^{N} h_i = 1$. Then we define

$$G^h_{k_i} = G \cap \left\{(x_1,x_3) \mid |x_1| \leq R,\ k_i h_i \leq x_3 \leq (k_i+1)h_i\right\},$$

$$i = 1,\dots,N;\ k = 1,\dots,r.$$

Let

$$\mu^h(x_3) = \begin{cases} \mu_1 & \text{for } x_3 \in G^h_{k_1}, \\ \mu_2 & x_3 \in G^h_{k_2}, \\ \vdots & \\ \mu_N & x_3 \in G^h_{k_N}. \end{cases}$$

For $u,v \in H^1(G) \equiv \left\{v \mid v, \frac{\partial v}{\partial x_i} \in L_2(G)\right\}$ we define the bilinear form

$$b^h(u^h,v) = -\int_G \sum_{i,j=1,3} a^h_{ij}\frac{\partial u^h}{\partial x_j}\frac{\partial \bar{v}}{\partial x_i}\,dx + \int_G g^h(x_1,x_3)\,\bar{v}\,dx$$

$$\text{(5)} \qquad = b^h_0(u^h,v) + \int_G g^h\,\bar{v}\,dx, \quad b^h_1(u^h,v) = \int_G g^h\bar{v}\,dx,$$

$$b^h(u^h,v) = b^h_0(u^h,v) + b^h_1(u^h,v).$$

Then u^h is the unique solution of

$$\text{(6)} \qquad b^h(u^h,v) = (f,v) \quad \forall v \in H^1_0(G) \equiv \left\{v \mid v \in H^1(G),\ v|_{\Gamma_0\cup\Gamma_2\cup\Gamma_3\cup\Gamma_4} = 0\right\}$$

where $(f,v) = \int_G f\,\bar{v}\,dx$, $a^h_{11} = a^h_{22} = \mu^h(x_3)$, $a^h_{12} = a^h_{21} = 0$.

Homogenization ([2,3]) tends to solve the problem

(7) $\quad b(U,V) = (f,v) \qquad \forall\ U,V \in H^1(G),\ b = b_0 + b_1$,

where b(U,V) is a bilinear form associated with homogenized equation

(8) $$\sum_{i,j=1,3} \frac{\partial}{\partial x_i} M_{ij} \frac{\partial U}{\partial x_j} + g\,U = f ,$$

where

$$M_{ij} = M_{ji} ,\ \sum_{i,j=1,3} M_{ij}\, \xi_i\ \xi_j > M_0 \sum_{i=1,3} \xi_i^{\,2} ,\ M_0 > 0.$$

The coefficients M_{lk} are defined as follows ([2,3])

(9) $$M_{lk} = b_0(w_k,v_l) = \int_{G_0} \Big(\sum_{i,j=1,3} a_{ij} \Big(\delta_{ik} - \frac{\partial \chi_k}{\partial x_j} \Big) \frac{\partial v_l}{\partial x_j} \Big)\, dx,$$

(10) $$w_k = x_k - \chi_k ,\ v_l = x_l - \chi_l ,\quad l,k = 1,3 ,$$

δ_{ik} - Kronecker's delta, $G_0 = \{ (x_1,x_3) \,|\, |x_1| \leq \frac{1}{2} ,\ |x_3| \leq \frac{1}{2} \}$
and where the functions χ_k satisfy the integral relation

(11) $$\int_{G_0} \Big(\sum_{i,j=1,3} a_{ij} \frac{\partial \chi_l}{\partial x_i} \frac{\partial v_l}{\partial x_j} \Big)\, dx = \int_{G_0} \sum_{j=1,3} a_{lj} \frac{\partial v_l}{\partial x_j}\, dx$$

and

(12) $$g = \int_{G_0} g^h(x_1,x_3)\, dx.$$

Since (9) and (12) are constant coefficients it is evident that the homogenized problem has a unique solution. According to the results of [2] and [3] it can be proved that $\| U - u^h \| \leq ch$, where c is a constant independent of h.

In this special case of quasianisotropic materials for coefficients M_{lk} and g we obtain

(13) $$M_{11} = \sum_{i=1}^{N} h_i \mu_i ,\ M_{12} = 0,$$

$$(14) \qquad M_{21} = 0 \ , \ M_{22} = \left(\sum_{i=1}^{N} h_i \frac{1}{\mu_i} \right)^{-1}.$$

For the numerical solution the finite element technique was used. We have the following results:

There exists exactly one approximate solution $U_{0k} \in S_k$, S_k is the space created by basis functions $\{\varphi_i^k\}_{i=1}^M$, of our problem with the bilinear form (7) of our homogenized problem for k sufficiently small, k is the largest side of all triangles of the given triangulation. The proof is based on Babuška's theorem [4].

The approximate solution $U_{0k} \in S_k$ can be found in the form

$$U_{0k} = \sum_{s=1}^{M} a_s^k \varphi_s^k \ ,$$

where $\{\varphi_i^k\}_{i=1}^M$ is the basis of the space S_k. Then we obtain the system of linear algebraic equations for unknown coefficients a_s^k. This system was solved by Gauss-Seidel or SOR methods. The matrix $b_0(\varphi_i^k, \varphi_j^k)$ $i,j = 1,\dots,M$, is positive definite and

$$(15) \qquad \frac{b_1(\varphi_i^k, \varphi_j^k)}{b_0(\varphi_i^k, \varphi_j^k)} \longrightarrow 0 \quad \text{if } k \longrightarrow 0.$$

Let us put

$$e_{1ij} = b_0(\varphi_i^k, \varphi_j^k) \ , \ i < j,$$

$$e_{2ij} = b_1(\varphi_i^k, \varphi_j^k) \ , \ i < j,$$

$$d_{1j} = b_0(\varphi_j^k, \varphi_j^k) \ ,$$

$$d_{2j} = b_1(\varphi_j^k, \varphi_j^k),$$

and let us denote the corresponding matrices by symbols E_k, F_k, D_k, D_{1k}. From the positive definiteness of the matrix $b_0(\varphi_i^k, \varphi_j^k)$ and from (15) follows that

$$(D_k - E_k)^{-1} (D_{1k} + F_k + F_k^T) \longrightarrow 0 \quad \text{for } k \longrightarrow 0,$$

where F_k^T is the transpose of the matrix F_k. This condition guarantees the convergence of Gauss-Seidel method of the system of linear algebraic equations for every $k \to 0$. The proof is analogous to the proof demonstrated in [5,6]. It can also be shown that for the guarantee of the convergence of this iterative process we have to reckon with a certain minimum number of mesh points per wavelength, analogous to the estimation given in [8].

3. Discussion

At first we shall study the relations between elastic constants c_{ij} and strains τ_{ij} in the elementary layers. As $M_{11} = c_{66} > 0$, $M_{12} = M_{21} = 0$, $M_{22} = c_{44} > 0$ then according to $\tau_{23} = c_{44}\,\varepsilon_{23}$, $\tau_{12} = c_{66}\,\varepsilon_{12}$ we have

$$\tau_{12} = c_{66}\,\varepsilon_{12} = \sum_{i=1}^{N} h_i\,\mu_i\,\varepsilon_{12} = \sum_{i=1}^{N} h_i\,\tau_{12}^{(i)}\ ,\ \tau_{12}^{(i)} = \mu_i\,\varepsilon_{12}\ ,$$

$$\varepsilon_{12} = \varepsilon_{12}^{(i)}\ ,\ i = 1,\dots,N\ ,\ \text{and}\ \ \varepsilon_{12} = \Big(\sum_{i=1}^{N} h_i\,\mu_i\Big)^{-1}\tau_{12} \neq 0\ ,$$

$$\varepsilon_{11} = \varepsilon_{22} = \varepsilon_{33} = \varepsilon_{23} = \varepsilon_{31} = 0.$$

Analogically we have

$$\tau_{23} = c_{44}\,\varepsilon_{23} = M_{22}\,\varepsilon_{23} = \Big(\sum_{i=1}^{N} h_i\,\frac{1}{\mu_i}\Big)^{-1}\varepsilon_{23}\ .$$

Then

$$\varepsilon_{23} = M_{22}^{-1}\,\tau_{23} = \Big(\sum_{i=1}^{N} h_i\,\frac{1}{\mu_i}\Big)\tau_{23} = \sum_{i=1}^{N} h_i\,\varepsilon_{23}^{(i)} \neq 0.$$

So that $\tau_{23} = \mu_i\,\varepsilon_{23}^{(i)}$, $i = 1,\dots,N$, and then

$$\tau_{23}^{(1)} = \tau_{23}^{(2)} = \dots = \tau_{23}^{(N)} = \tau_{23}$$

and

$$\varepsilon_{11} = \varepsilon_{22} = \varepsilon_{33} = \varepsilon_{32} = \varepsilon_{12} = 0.$$

As $\mu_i > 0$, $i = 1,\dots,N$ then $M_{11} > 0$, $M_{22} > 0$, and then $M_{11} \geq M_{22}$, i.e. $c_{66} \geq c_{44}$. It is true as $M_{11} = c_{66}$ is an arithmetic mean and $M_{22} = c_{44}$ is a geometric (harmonic) mean of μ_i , $i = 1,\dots,N$.

The assumption that the wavelength is sufficiently large compared to the thickness of the system of layers is essential as only in these cases we can take these materials as homogeneous quasianisotropic media. As in our investigations of quasianisotropic media $h \to 0$, the assumption discussed above is fulfilled. Now we shall study the limit of velocity of SH-waves for $\lambda \to 0$ if waves propagate perpendicular to the layer. The velocity in the direction perpendicular to the system of layers is like the marginal value the average velocity $\bar{c}_{SH}$ defined as $\bar{c}_{SH}^{-1} = (\sum_{i=1}^{N} h_i c_{SH}^{(i)})^{-1}$, where $c_{SH}^{(i)} = (\frac{\mu_i}{\varrho_i})^{1/2}$. We compare this value with the marginal value of velocity for $\lambda \to \infty$ (λ be a wavelength), i.e. $c_{SH}^{2} = \frac{c_{44}}{\varrho}$. As

$$g = \int_{G_0} g^h(x_1,x_3)\, dx = g_1 + i\, g_2$$

$$g_1 = \varrho(x_3)\omega^2,\ g_2 = g_2(x_1,x_3),$$

we have

$$\bar{\varrho} = \bar{g}_1(x_3)\omega^{-2} = \int_{G_0} \varrho(x_3)\, dx = \sum_{i=1}^{N} h_i \varrho_i .$$

Then

$$c_{SH}^{-2} = \frac{\varrho}{c_{44}},\ \bar{c}_{SH}^{-2} = (\sum_{i=1}^{N} h_i \frac{\varrho_i}{\mu_i})^2,\ c_{SH}^{-2} - \bar{c}_{SH}^{-2} \geqslant 0.$$

Especially for N = 2 we have

$$c_{SH}^{-2} = \frac{1}{2}(\varrho_1 + \varrho_2)\frac{1}{2}(\frac{1}{\mu_1} + \frac{1}{\mu_2}),$$

$$\bar{c}_{SH}^{-2} = \frac{1}{4}(\frac{1}{c_1} + \frac{1}{c_2})^2 = \frac{1}{4}(\frac{\varrho_1}{\mu_1} + \frac{\varrho_2}{\mu_2} + \frac{2}{c_1 c_2}).$$

From there

$$c_{SH}^{-2} - \bar{c}_{SH}^{-2} = \frac{1}{4}(\sqrt{\frac{\varrho_2}{\mu_1}} - \sqrt{\frac{\varrho_1}{\mu_2}})^2 \geqslant 0.$$

Then in the direction perpendicular to the system of layers we have

$$\lim_{\lambda \to 0} c_{SH} \geqslant \lim_{\lambda \to \infty} c_{SH} .$$

This relation also holds in the case if the SH-waves propagate parallel to the system of layers. This result follows from the fact that c_{44} and c_{66} as the mean values are smaller than the greatest of the μ_i, $i = 1,\dots,N$.

References

[1] Backus, G. E.: Long-wave Elastic Anisotropy Produced by Horizontal Layering. J. Geophys. Res. 67(1962), 4427-4441.

[2] Babuška, I.: Solution of Interface Problems by Homogenization. I. Tech. Note BN-782, January 1974, II. Tech. Note BN-787, March 1974, SIAM J. Math. Anal. 7(1976), No. 5, 603-645.

[3] Lions, J. L.: Asyptotic Behaviour of Solution of Variational Inequalities with Highly Oscillating Coefficients. In Applications of Methods of Functional Analysis to Problems in Mechanics, Lecture Notes in Mathematics, No. 503, Berlin-Heidelberg-New York 1975.

[4] Babuška, I.: Error Bounds for Finite Element Method. Numer. Math. 16(1971), 322-333.

[5] Marek, I. and Nedoma, J.: Using Finite Element Procedure in the SH-wave Propagation (manuscript).

[6] Handlová, J., Nedoma, J. and Šmid, J.: Issledovanie razprostranenija uprugich SH-voln v sloistoj srede s lokalnoj neodnorodnost'ju metodami vyčislitelnovo analiza. Numerische Methoden in der Geophysik, Geoph. Inst. Czech. Acad. Sci., Prague 1973.

[7] Nedoma, J.: On the Application of Homogenization to Study of the Ultrabasic Rocks. In IV. Conference on Basic Problems of Numerical Analysis. Plzeň, September 4-8, 1978.

[8] Nedoma, J.: Investigation of Linear Harmonic Field of SH-waves in a Stratified Inhomogeneous Medium Using the Finite Difference Method. Zeitschrift f. Geophys. 38(1972), 529-542.

+) J. Nedoma, Geofysikálni Ústav ČSAV, Bočni II,
14131 Praha - Spořilov (ČSSR)

ON THE DETERMINATION OF FUNCTIONAL PARAMETERS IN NONLINEAR EVOLUTION EQUATIONS OF THE NAVIER-STOKES TYPE

Reiner Nürnberg *)

In this paper we deal with the problem of determination of viscosity properties of an incompressible fluid, by means of optimality criteria using measurement data of a corresponding flow. The viscous behavior of a non-newtonian fluid are characterized by a nonlinear function which gives a relation between the stresses and the velocity of deformations. The basic equations describing the flow of a viscous fluid are the Navier-Stokes equations. The corresponding boundary and initial value problems of these partial differential equations lead to evolution equations in function spaces.

In a first part we give an answer to the question of existence of an "optimal" functional parameter for evolution equations in a general case. To be able to take Bingham effects into considerations we regard evolution equations with multivalued mappings. In a second part we apply these results to Navier-Stokes equations and to the problem of determination of the viscosity of a fluid.

1.

Let X_0 be a real reflexive Banach space imbedded in a Hilbert space H_0, where the injection of X_0 into H_0 is continuous and X_0 is dense in H_0. Identifying H_0 with H_0^*(the dual of H_0) and H_0^* with a linear subspace of X_0^*(the dual of X_0) we have

$$X_0 \subset H_0 \subset X_0^* ,$$

where the injections are continuous and each space is dense in the following one. We denote by $(.,.)_0$ the scalar product in H_0 as well as between X_0 and X_0^*.

For $0 < T < +\infty$ and $1 < p \leq q < +\infty$ let X be a reflexive Banach space with

$$L^q(0,T;X_0) \subset X \subset L^p(0,T;X_0) ,$$

which is imbedded in the Hilbert space $H = L^2(0,T;H_0)$.

---

*) ZIMM der AdW , DDR 108 Berlin , Mohrenstraße 39

Identifying H with H^* and H^* with a linear subspace of X^* we obtain

$$X \subset H \subset X^* ,$$

where the injections are continuous and dense. By $(.,.) = \int_0^T (.,.)_0 dt$ we denote the scalar product in H as well as between X and X^*, and we have

$$L^{p'}(0,T;X_0^*) \subset X^* \subset L^{q'}(0,T;X_0^*)$$

with $\frac{1}{p} + \frac{1}{p'} = 1$ and $\frac{1}{q} + \frac{1}{q'} = 1$.
More precisely we make the following assumption to the space X.
Let F_{0i} $(i=1,\dots,m)$ be reflexive Banach spaces and $F_i = L^{p_i}(0,T;F_{0i})$ with $p \leq p_i \leq q$. Then let $\delta_0 = [\delta_{01},\dots,\delta_{0m}] : X_0 \longrightarrow F_0 = F_{01} \times \dots \times F_{0m}$ and $\delta = [\delta_1,\dots,\delta_m] : X \longrightarrow F = F_1 \times \dots \times F_m$ be continuous linear mappings with $(\delta .)(t) = \delta_0(.(t))$ f.a.a. $t \in (0,T)$, $\|\delta_0 .\|_{F_0} = \|.\|_{X_0}$ and $\|\delta .\|_F = \|.\|_X$. Further let $\pi_0 : F_0 \longrightarrow H_0$ and $\pi : F \longrightarrow H$ be continuous linear mappings with $(\pi .)(t) = \pi_0(.(t))$ f.a.a. $t \in (0,T)$, where $\pi_0 \circ \delta_0 = I_0 : H_0 \longrightarrow H_0$ and $\pi \circ \delta = I : H \longrightarrow H$ are the identities.

Next we introduce the reflexive Banach space

$$Y = \left\{ x \in X : \frac{dx}{dt} \in X^* \right\},$$

where $\frac{d}{dt}$ means the derivative in the sense of distributions.

<u>Remark 1.</u> For x, $y \in X$ with $\frac{dx}{dt}$, $\frac{dy}{dt} \in X^*$ we have x, $y \in C(0,T;H_0)$ and

$$\int_0^T \left(\frac{dx}{dt} , y \right)_0 dt + \int_0^T \left(x , \frac{dy}{dt} \right)_0 dt = (x(T), y(T))_0 - (x(0), y(0))_0 .$$

Finally let U be a nonempty bounded closed convex subset of a reflexive Banach space V. Then we consider the following minimum problem:

Find a pair $[x, u] \in G(S)$ such that

$$\Phi(x, u) = \min\left\{ \Phi(y, v) : [y, v] \in G(S) \right\} , \tag{1}$$

where G(S) is the graph of the solution operator $S : U \longrightarrow 2^Y$ of the parametric evolution equation

$$\left.\begin{aligned} \frac{dx}{dt} + A(x, u) &\ni f , \\ x(0) &= x_0 , \end{aligned}\right\} \tag{2}$$

here $\Phi : Y \times U \longrightarrow R^1$ is a functional, $A : X \times U \longrightarrow 2^{X^*}$ is a multivalued mapping, $f \in X^*$ and $x_0 \in H_0$.

Now we introduce some notions.

Definition 1. The mapping $A(.,u) : X \longrightarrow 2^{X^*}$ is called monotone if for all $[x, f]$, $[y, g] \in G(A(., u))$ the relation

$$(f - g , x - y) \geq 0$$

holds; maximal monotone if it is monotone and for each $[x, f] \in X \times X^*$ the inequality

$$(f - g , x - y) \geq 0 \quad \text{for all } [y, g] \in G(A(., u))$$

implies $[x, f] \in G(A(., u))$.

Definition 2. The mapping $A(x, .) : U \longrightarrow 2^{X^*}$ is called increasing lower semicontinuous if for each sequence $u_i \in U$ with $u_i \longrightarrow u$ in V and each f $A(x, u)$ there exists a sequence $f_i \in A(x, u_i)$ with $f_i \longrightarrow f$ in X^*.

Definition 3. The mapping $A : X \times U \longrightarrow 2^{X^*}$ is called pseudomonotone if
(1) the set $A(x, u)$ is closed and convex for all $[x, u] \in D(A)$, and
(2) for each sequence $[x_i, u_i, f_i] \in G(A)$,
$[x_i, u_i, f_i] \longrightarrow [x, u, f]$ in $X \times V \times X^*$ and $\overline{\lim} (f_i, x_i) \leq (f, x)$ imply $[x, u, f] \in G(A)$ and $\lim (f_i, x_i) = (f, x)$.

Lemma 1. The mapping $A_*: X \times U \longrightarrow 2^{X^*}$ satisfies the following conditions:
(1) $A(., u) : X \longrightarrow 2^{X^*}$ is maximal monotone for all $u \in U$, and
(2) $A(x, .) : U \longrightarrow 2^{X^*}$ is increasing lower semicontinuous for all $x \in X$.
Then the mapping A is pseudomonotone.

The proof is given in KLUGE [5], NÜRNBERG [9].

We are going to study the properties of the solution operator S.

Theorem 1. The mapping A has the properties:
(1) $A : X \times U \longrightarrow 2^{X^*}$ is bounded with $D(A) = X \times U$ and there exist a constant c_0 and a function $c : R^+ \longrightarrow R^+$ with $c(r) \longrightarrow +\infty$ for $r \longrightarrow +\infty$ such that

$$\min_{t \in [0,T]} \int_0^t (g, x)_0 ds \geq - c_0 \cdot \|x\| \quad \text{and}$$

$$\|x\|^2_{C(0,T;H_0)} + (g, x) \geq c(\|x\|) \cdot \|x\| \quad \text{for all } [x, u, g] \in G(A) ,$$

(2) $A : Y \times U \longrightarrow 2^{Y^*}$ is pseudomonotone.
Then the solution operator $S : U \longrightarrow 2^Y$ is defined on the whole set U and S is a weakly closed and bounded mapping.

Proof. A complete proof is given in NÜRNBERG [9], here we give a scetch only.

Firstly we define a monotone and closed linear mapping L from $X \times H_0$ into $X^* \times H_0$ by

$$L(x, x_0) = \left[\frac{dx}{dt}, x_0 \right] \text{ for each } x \in X \text{ with } \frac{dx}{dt} \in X^* \text{ and } x(0) = x_0,$$

where D(L) is dense in $X \times H_0$. If we take L and A canonically as mappings from $X \times U \times H_0$ into $X^* \times H_0$ the parametric evolution equation (2) is equivalent to

$$(L + A)[x, u, x(0)] \ni [f, x_0] .$$

The monotone linear mapping L^* from $X \times H_0$ into $X^* \times H_0$ defined by

$$L^*(y, y_0) = \left[- \frac{dy}{dt}, y_0 - y(0) \right] \text{ for each } [y, y_0] \in X \times H_0$$

with $\frac{dy}{dt} \in X^*$ and $y(T) = 0$, where $D(L^*)$ is dense in $X \times H_0$, is the adjoint mapping to L. Therefore the mapping L is maximal monotone (see LIONS [8]).

So we obtain the fact that the sum $L + A : Y \times U \times H_0 \longrightarrow 2^{Y^* \times H_0}$ is pseudomonotone (see also BROWDER-HESS [2]). That means the graph G(S) is weakly closed in $Y \times V$.

Using the coercivity assumtion to A it is easy to show that the mapping $(L + A)^{-1} : X^* \times H_0 \longrightarrow 2^{Y \times U \times H_0}$ is bounded.

Finally we prove the existence of a solution $x \in Su$ for a fixed $u \in U$ with the help of the method of elliptic regularisation (see also LIONS [8]).

<u>Theorem 2.</u> Let the functional $\Phi : Y \times U \longrightarrow R^1$ be a weakly lower semi-continuous one and assume the conditions of Theorem 1 to be fulfilled. Then the minimum problem (1) has a solution.

The proof is based on the Weierstrass' theorem in its extension which was investigated in VAINBERG [11] .

2.

Let Ω be a bounded domain of R^n ($n \leq 3$) with a sufficiently regular boundary Γ, i.e. let Γ be locally Lipschitzian and Ω be locally located on one side of Γ. Further let Γ_1 and Γ_2 be two parts of Γ with $\Gamma = \Gamma_1 \cup \Gamma_2$ and $\Gamma_1 \cap \Gamma_2 = \emptyset$. By ν we denote the exterior unit vector normal to the boundary. For a finite interval $[0,T]$ we seek a vector field $x : [0,T] \times \Omega \longrightarrow R^n$, a scalar field $\bar{p} : [0,T] \times \Omega \longrightarrow R^1$ and a tensor field $\sigma : [0,T] \times \Omega \longrightarrow R^{n \cdot n}$ representing the velocities, the pressure and the stresses, respectively, which satisfy the following equations and conditions:

$$\left.\begin{array}{lll} \frac{\partial x}{\partial t} + (x\cdot\mathrm{grad})\cdot x = \mathrm{div}\,\sigma + \bar{f} - \mathrm{grad}\,\bar{p} & \text{in } Q=(0,T)\times\Omega, & \\ \text{where} \quad |\sigma| \leq u(|\mathrm{grad}\, x|) & & \\ \text{and} \quad \sigma = u(|\mathrm{grad}\, x|)\cdot \frac{\mathrm{grad}\, x}{|\mathrm{grad}\, x|} & \text{if } \mathrm{grad}\, x \neq 0, & \\ \mathrm{div}\, x = 0 & \text{in } Q, & \\ \sigma\cdot\nu + \frac{1}{2}|x|^2\cdot\nu = \bar{p}\cdot\nu + \bar{g} & \text{on } \Sigma_2=(0,T)\times\Gamma_2, & \\ x = 0 & \text{on } \Sigma_1=(0,T)\times\Gamma_1, & \\ x(0,.) = x_0 & \text{in } \Omega, & \end{array}\right\} \quad (3)$$

where $u : R^+ \longrightarrow R^+$ is a material function which satisfies

$$0 < c_1\cdot(1 + r^{p-2}) \leq u'(r) \leq c_2\cdot(1 + r^{p-2}) \quad \text{and}$$
$$0 \leq u(0) \leq c_0$$

for positive constants c_0, c_1, c_2 and $p \geq \frac{n+2}{2}$, moreover let $\bar{f} \in L^2(Q)^n$, $\bar{g} \in L^2(\Sigma_2)^n$ and $x_0 \in L^2(\Omega)^n$ be given functions.

The variational formulation of this problem leads to a generalized problem in function spaces (see also LIONS [8] , TEMAM [10]). Therefore we introduce the following spaces:

$$X_0 = \left\{ x \in (W^{1,p}(\Omega)\cap L^q(\Omega))^n : \mathrm{div}\, x = 0 \text{ in } \Omega,\ x = 0 \text{ on } \Gamma_1 \right\},\quad q = \frac{2\cdot p}{p-1},$$
$$H_0 = \left\{ x \in (L^2(\Omega))^n : \mathrm{div}\, x = 0 \text{ in } \Omega,\ x\cdot\nu = 0 \text{ on } \Gamma_1 \right\},$$
$$X = \left\{ x \in (L^p(0,T;W^{1,p}(\Omega))\cap L^q(0,T;L^q(\Omega)))^n : \mathrm{div}\, x = 0 \text{ in } Q,\ x = 0 \text{ on } \Sigma_1 \right\},$$
$$H = \left\{ x \in (L^2(0,T;L^2(\Omega)))^n : \mathrm{div}\, x = 0 \text{ in } Q,\ x\cdot\nu = 0 \text{ on } \Sigma_1 \right\}.$$

Furthermore we define mappings $A_1(.,\ u) : X \longrightarrow 2^{X^*}$ and $A_2 : X \longrightarrow X^*$ by

$$A_1(x,\ u) = \Big\{ g \in X^* : (g,\ y) = \int_0^T\!\!\int_\Omega u(|\mathrm{grad}\, x|)\cdot\xi\cdot\mathrm{grad}\, y \,ds\,dt, \text{ f.a. } y \in X, \text{ where } \xi \in L^\infty(Q)^{n\cdot n} \text{ with } |\xi| \leq 1 \text{ and } \xi = \frac{\mathrm{grad}\, x}{|\mathrm{grad}\, x|} \text{ if } \mathrm{grad}\, x \neq 0 \Big\},$$

$$(A_2\, x,\ y) = -\int_0^T\!\!\int_\Omega (x \times \mathrm{rot}\, x)\cdot y \,ds\,dt \text{ for all } y \in X,$$

where $x \in X$.
Then for $f \in X^*$ with $(f,\ y) = \int_0^T\!\!\int_\Omega \bar{f}\cdot y \,ds\,dt + \int_0^T\!\!\int_\Gamma \bar{g}\cdot y \,d\Gamma\,dt$ and for the mapping $A(.,\ u) = A_1(.,\ u) + A_2$ the evolution equation (2) can be interpreted as a weak formulation of the problem (3).

In what follows the function u is supposed to be unknown and has to be determined from observed data of the state x. In connection with (2) we consider the minimum problem (1) where $V = W^{1,2}(R^+;e^{-r})$,

$$U = \left\{ u \in V : c_1\cdot(1 + r^{p-2}) \leq u'(r) \leq c_2\cdot(1 + r^{p-2}),\ 0 \leq u(0) \leq c_0 \right\}$$

and $\Phi(x,\ u) = \Phi(x)$ is a continuous convex functional (see also KLUGE-LANGMACH [6] , LANGMACH [7]).

We are going to prove properties of the mapping A which are important for the application of the results of part 1.

Lemma 2. The mapping $A_1(.,u) : X \longrightarrow 2^{X^*}$ is maximal monotone for each $u \in U$.

Proof. A bounded monotone mapping $A_1(.,u)$ which has the property that the set $A_1(x,u)$ is nonempty and convex for all $x \in X$, is maximal monotone if it is demiclosed, i.e. the graph $G(A_1(.,u))$ is (strongly, weakly)- closed in the space $X \times X^*$ (see BROWDER [1]).

1. The set $A_1(x,u)$ is nonempty and convex for all x X.

Let $x \in X$ be arbitrary. Then by setting $\xi = \frac{\text{grad } x}{|\text{grad } x|}$ if grad $x \neq 0$ and $\xi = 0$ if grad $x = 0$ we define a tensor function $\xi \in L^\infty(Q)^{n \cdot n}$. On the other hand we have $u(|\text{grad } x|) \in L^{p'}(Q)$ since the function u is continuous and satisfies the growth condition $|u(r)| \leq (c_0 + c_2)(1 + r^{p-1})$. Hence there exists a linear functional $g \in X^*$ with

$$(g, y) = \int_0^T \int_\Omega u(|\text{grad } x|) \cdot \xi \cdot \text{grad } y \, ds dt \quad \text{for all } y \in X$$

that means $g \in A_1(x,u)$.

Finally it follows directly from the definition that the set $A_1(x,u)$ is convex.

2. The mapping $A_1(.,u)$ is monotone.

Let $x \in X$ and $y \in X$. Then we have

$$\left(u(|\text{grad } x|) \cdot \frac{\text{grad } x}{|\text{grad } x|} - u(|\text{grad } y|) \cdot \frac{\text{grad } y}{|\text{grad } y|}\right)(\text{grad } x - \text{grad } y) \geq$$

$$(u(|\text{grad } x|) - u(|\text{grad } y|))(|\text{grad } x| - |\text{grad } y|) \geq 0$$

$$\text{if grad } x \neq 0 \text{ and grad } y \neq 0,$$

$$\left(u(|\text{grad } x|) \cdot \frac{\text{grad } x}{|\text{grad } x|} - u(0) \cdot \xi\right) \text{grad } x \geq$$

$$(u(|\text{grad } x|) - u(0))|\text{grad } x| \geq 0 \quad \text{if grad } x \neq 0 \text{ and grad } y = 0,$$

where $\xi \in L^\infty(Q)^{n \cdot n}$ with $|\xi| \leq 1$.

Taking into account the definition of the mapping A_1 we obtain from the last inequalities by integration

$$(f - g, x - y) \geq 0 \quad \text{for all } f \in A_1(x,u) \text{ and } g \in A_1(y,u),$$

that means, $A_1(.,u)$ is monotone.

3. The mapping $A_1(.,u)$ is demiclosed.

Let $x_i \in X$ be a sequence with $x_i \longrightarrow x$ in X and let $f_i \in A_1(x_i, u)$ be a sequence with $f_i \longrightarrow f$ in X^*. By definition of the mapping A_1 there exists a sequence $\xi_i \in L^\infty(Q)^{n \cdot n}$ such that

$$(f_i, y) = \int_0^T \int_\Omega u(|\text{grad } x_i|) \cdot \xi_i \cdot \text{grad } y \, ds dt \quad \text{for all } y \in X,$$

where $|\xi_i| \leq 1$ and $\xi_i = \frac{\text{grad } x_i}{|\text{grad } x_i|}$ if grad $x_i \neq 0$.

Using the relation

$$|u(|\text{grad } x_i|) - u(|\text{grad } x|)| \leq c_2 \cdot (1 + |\text{grad } x_i|^{p-2} + |\text{grad } x|^{p-2}) \cdot (|\text{grad } x_i| - |\text{grad } x|)$$

we obtain $u(|\text{grad } x_i|) \longrightarrow u(|\text{grad } x|)$ in $L^{p'}(Q)$. On the other hand there exists a subsequence $\xi_{i'} \xrightarrow{*} \xi$ in $L^\infty(Q)^{n\cdot n}$. We have $|\xi| \le 1$, hence the unit ball in the space $L^\infty(Q)^{n\cdot n}$ is weakly-*-closed. Further, for almost all $(t,s) \in Q$ with grad $x(t,s) \neq 0$ there exists a subsequence grad $x_{i'}(t,s) \neq 0$ such that grad $x_{i'}(t,s) \longrightarrow$ grad $x(t,s)$. So we obtain $\xi_{i'}(t,s) = \frac{\text{grad } x_{i'}(t,s)}{|\text{grad } x_{i'}(t,s)|} \longrightarrow \xi(t,s) = \frac{\text{grad } x(t,s)}{|\text{grad } x(t,s)|}$.
Finally it follows that

$$(f, y) = \int_0^T \int_\Omega u(|\text{grad } x|)\cdot\xi\cdot\text{grad } y \, ds dt \quad \text{for all } y \in X.$$

that means $f \in A_1(x, u)$.

Obviously the mapping $A_1(., u)$ is bounded. This completes the proof.

<u>Lemma 3.</u> The mapping $A_1(x, .) : U \longrightarrow 2^{X^*}$ is increasing lower semi-continuous for all $x \in X$.

<u>Proof.</u> Let $u_i \in U$ be a sequence with $u_i \longrightarrow u$ in V. Then we have $u_i \in C(R^+)$ and $u_i(r) \longrightarrow u(r)$ for all $r \in R^+$. Hence we obtain
$u_i(|\text{grad } x|) \longrightarrow u(|\text{grad } x|)$ a.e. in Q for an arbitrary $x \in X$. Because of the growth condition $|u_i(|\text{grad } x|)| \le (c_0+c_2)(1+|\text{grad } x|^{p-1})$ a.e. in Q we can apply the theorem of Lebesgue. So it follows
$u_i(|\text{grad } x|) \longrightarrow u(|\text{grad } x|)$ in $L^{p'}(Q)$.
Now let $f \in A_1(x, u)$, i.e. there is a tensor function $\xi \in L^\infty(Q)^{n\cdot n}$ with $|\xi| \le 1$ and $\xi = \frac{\text{grad } x}{|\text{grad } x|}$ if grad $x \neq 0$, which satisfies the equation

$$(f, y) = \int_0^T \int_\Omega u(|\text{grad } x|)\cdot\xi\cdot\text{grad } y \, ds dt \quad \text{for all } y \in X .$$

Then we define a sequence $f_i \in A_1(x, u_i)$ by the equations

$$(f_i, y) = \int_0^T \int_\Omega u_i(|\text{grad } x|)\cdot\xi\cdot\text{grad } y \, ds dt \quad \text{for all } y \in X.$$

Using the convergence of the sequence $u_i(|\text{grad } x|)$ in the space $L^{p'}(Q)$ we obtain $f_i \longrightarrow f$ in X^*, and the lemma is proved.

<u>Lemma 4.</u> The mapping $A_2 : Y \longrightarrow Y^*$ is pseudomonotone.

<u>Proof.</u> The choice of the space X guaranties that the mapping A_2 is bounded and continuous from X into X^*. Using the compact imbedding of the space Y into H we confirm that the mapping A_2 is weakly continuous from Y into $L^1(Q)^n$, hence A_2 is weakly continuous from Y into X^*, too (see also LIONS [8], TEMAM [10]).
Finally we remark that A_2 has the property $(A_2 x, x) = 0$ for all $x \in X$. From this it follows directly that the mapping $A_2 : Y \longrightarrow Y^*$ is pseudomonotone.

<u>Corollary 1.</u> The mapping $A : Y \times U \longrightarrow 2^{Y^*}$ is pseudomonotone.

<u>Proof.</u> It follows from Lemma 1, 2 and 3 that the mapping $A_1 : X \times U \longrightarrow 2^{X^*}$ is pseudomonotone. Furthermore the sum of two bounded

pseudomonotone mappings is pseudomonotone, too (see BROWDER-HESS [2]).

Remark 2. Using the continuous imbedding

$$L^p(0,T;W^{1,p}(\Omega)) \cap C(0,T;L^2(\Omega)) \subset L^q(0,T;L^q(\Omega))$$

it is easy to show that the mapping A fulfils the coercivity assumption of Theorem 1.

Based on the properties of A given here, we are able to apply the results of the part 1 on the problem of the existence of an optimal parameter formulated in this part.

Remark 3. Under the assumption above it can be shown that the solution of the evolution equation (2) is unique for each $u \in U$.

References

[1] Browder, F. E., Problèmes non linéaires. Les Presses de l'Université de Montréal 1966.

[2] Browder, F. E., Hess, P., Nonlinear Mappings of Monotone Type in Banach Spaces. J. Funct. Anal. 11 (1972), 251-294.

[3] Kluge, R., Optimal Control with Minimum Problems and Variational Inequalities. Lecture Notes in Computer Science 27, Springer-Verlag, Berlin 1975, 377-382.

[4] Kluge, R., Über eine Klasse von Minimumproblemen mit Nebenbedingungen. Math. Nachr. 78 (1977), 37-64.

[5] Kluge, R., On some Parameter Determination Problems and Quasi-variational Inequalities. Abhandlungen Akad. Wiss. DDR N6, Akademie-Verlag, Berlin 1978, 129-139.

[6] Kluge, R., Langmach, H., On some Problems of Determination of Functional Parameters in Partial Differential Equations. Lecture Notes in Control and Information Sciences 1, Springer-Verlag, Berlin 1978, 298-309.

[7] Langmach, H., On the Determination of Functional Parameters in some Parabolic Differential Equation. Abhandlungen Akad. Wiss. DDR N6, Akademie-Verlag, Berlin 1978, 175-184.

[8] Lions, J. L., Quelques méthodes de résolution des problèmes aux limites non linéaires. Dunod, Gauthier-Villars, Paris 1969.

[9] Nürnberg, R., Evolutionsungleichungen und Parameterbestimmungs-probleme. Diss. A, Berlin 1979.

[10] Temam, R., Navier-Stokes Equations, Theory and Numerical Analysis. Studies in Mathematics and its Applications 2, North-Holland 1977.

[11] Vainberg, M. M., Variational Methods in the Theory of Nonlinear Operators (Russ.). Moscow 1956.

INVERSE PROBLEMS IN GROWTH DYNAMICS[1)]

Jürgen PEIL[2)]

Summary

The meaning of the term "inverse problem" in a general biocybernetical context will be explained. The mathematical tasks of general inverse problems as well as special inverse problems are demonstrated at the analysis of growth dynamics. To this end at first the main features of a phenomenologic-mathematical kind of modelling of body length growth processes of man are sketched to give the concrete meaning of general inverse problem within this framework, and to show the embedding of the special inverse problem on different levels of the general inverse problem. Then the special inverse problem will be specified as the task of parameter estimation for growth differential equations and for the corresponding functional expressions on the basis of measured growth courses. The numerical results of solving these special inverse problems are presented.

1. The meaning of Inverse Problem in Biocybernetics

In Biocybernetics as Inverse Problem the task is named to find a mathematical expression for a connection between variables which is experimentally given by a series of measured values. Such a collection of data may be a time-series as the result of examinations on the dynamical behaviour of the system in question, or it is the picture of a functional relationship between variables of the system. In every case the measured values are the starting-point in the search for a theoretical description of the observed phenomenon.

In respect to what end the mathematical expression is used one can meet two extreme situations. On the one hand the expression is intended only for a quantitative description of the observed process. It transforms the information which is diffused in the measured values to a comprehended form, and in this way it is suitable to serve as an objective

---

1 This work was supported by the Martin-Luther-Universität Halle on the basis of the research theme "Biomathematische Modellierung"

2 Dr.rer.nat.habil. Jürgen Peil, Anatomisches Institut des Bereichs Medizin der MLU Halle, DDR-402 Halle, Große Steinstr. 52

basis for comparisons. In most cases there is to settle a class of empirical functions and the numerical task remains of determining that expression out of this class which gives the best fitted quantitative description of the measured course. This special inverse problem is called "parameter estimation problem".

On the other hand, the general inverse problem means the construction of mathematical models, i. e. the deriving or defining an appropriate mathematical expression from the data. Such a mathematical model has not only to give a quantitative description of a measured connection between variables, but it should express the main dynamical features of the system, its structural or functional properties, and moreover, the underlying general principles by which the system is governed. But this cannot be done by mathematics alone. As a rule, there is no unequivocal way from experimental data to the mathematical expression, and in the multi-step process of model building and model proving one has to consider some empirical facts and hints from the background of the field from which the problem arises. It should be noticed that in this sense a mathematical model carries also prospective and designing features.

Sometimes general inverse problems also are called as system identification or system specification. Among the mathematical tasks of solving a general inverse problem parameter estimations (as special inverse problems) play an important role. But before an estimation can be performed the general inverse problem must be specified: One has to decide what kind of mathematical description is appropriate for reflecting the structural, dynamical and so on features of the system, and one has to settle the class of mathematical expressions out of which the "best" expression is to determine by parameter estimation.

In the following we use differential equations and analytical functions for a quantitative description or for modelling of growth processes. (This settlement is qualified because we are not intended to consider aspects of time-delays which, from a control-theoretical point of view, may be important in certain growth processes of populations, see e. g. CUSHING /1977/. Our kind of modelling, see 2., has also a control-theoretical background, indeed, but the details of the control mechanisms are not yet well understood. So we don't take into consideration any time-delays). As differential equations several growth models come into question. The solutions of these growth differential equations constitute the mentioned classes of analytical functions.

2. General inverse problem of growth dynamics

Fig. 1 shows measured courses of the body length growth of man. The course at the top in fig. 1 is a mean value course reaching from conception until the time of adolescens (the values were taken from anthropometric literature). There are a lot of single measurements (cross-sectional data) which enter the respective mean value at every time point.

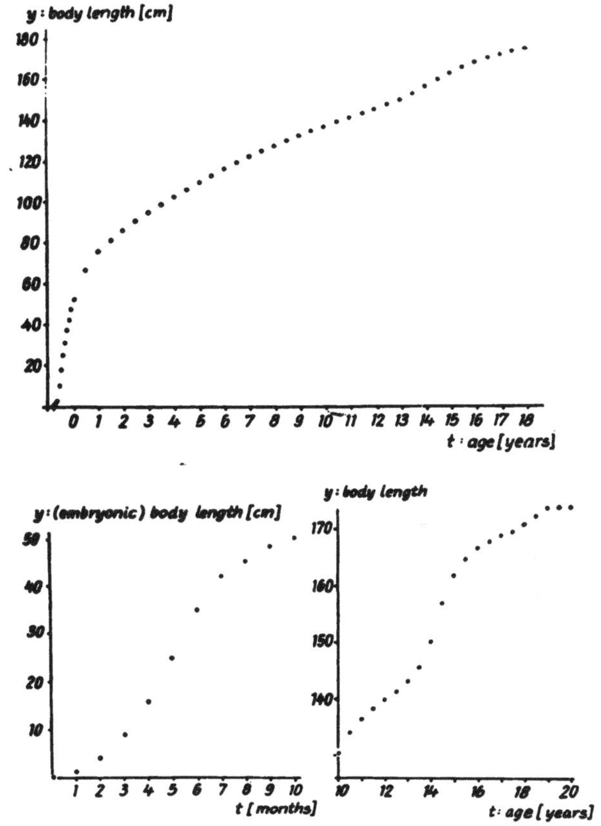

Fig. 1:
Examples of measured growth courses (body length growth of man)

At the left on the bottom in fig. 1 the embryonic part of body length increase is drawn. These values are also mean values of cross-sectional data. The right part shows a longitudinal data series of the body length of an individual (one of twin-brothers), see also fig. 2. There may be several occasions, besides the purpose of modelling the growth dynamics, to give for a measured course a quantitative description, e. g. to provide an objective basis for comparisons of growth processes of different populations in examining acceleration phenomena. The ways of choosing appropriate classes of functions and solving the special inverse problems of parameter estimation are discussed in detail in PEIL /1979/.

For modelling growth processes at first differential equations are to prefer as mathematical expressions because for constructing a differential equation one has to know the underlying physical, physico-chemical, or biochemical etc. principles and processes which generate the observed phenomenon. (Attempts to construct such differential equations were undertaken by SCHARF /1974/ on the basis of laws of mechanics). The differential equations which will be used in 3. rest on simple and obvious assumptions. An analytical expression as solution of a differential equation of course carries the same aspects in modelling as the differential equation does it.

The general inverse problem of growth dynamics consists in seeking for a mathematical expression which is able to reflect the quantitative and qualitative features of growth dynamics represented by measured courses like those shown in fig. 1. It would be without chance of success to seek for an expression in form of a differential equation which is valid for the whole time interval of body length growth of man (fig. 1 at the top). Our attempts to get a quantitative description for an as large as possible time interval of growth (PEIL /1979/) showed the generalized logistic function to be proper for meeting this task to a certain extent. But its generating differential equation contains the time explicitely (see PEIL /1978/).

Starting-point for our kind of modelling of the growth process is the well known empirical fact that growth of human beings takes place in form of more or less clearly marked growth spurts succeeding one another. Taking this into account we try to separate the whole growth into growth spurts mathematically. Following the principle of simplicity we assume spurts to have a symmetric-sigmoidal shape (in its time course). Furthermore it seems to be reasonable that the type of functions used for describing the single spurt must have such asymptotical properties which don't limit the validity of the corresponding mathematical term only to that time interval in which this spurt happens. (Otherwise one would be obliged to cut off the improper parts). The superposition of the single spurts has to ensure the smooth course of the growth character y as it is in reality (see fig. 1).

Fig. 2 shows an example of analyzing two longitudinal data series of twin-brothers according to the just mentioned manner. The data cover the time interval of the last part of the praepuberal development as well as the time interval in which the puberal and the postpuberal growth spurts take place. After setting up a 3-parametric hyperbolic tangent - the solution (4a) of the differential equation (2) can be written in this form - for describing a single spurt the solution of the parameter estimation problem is also given in fig. 2.

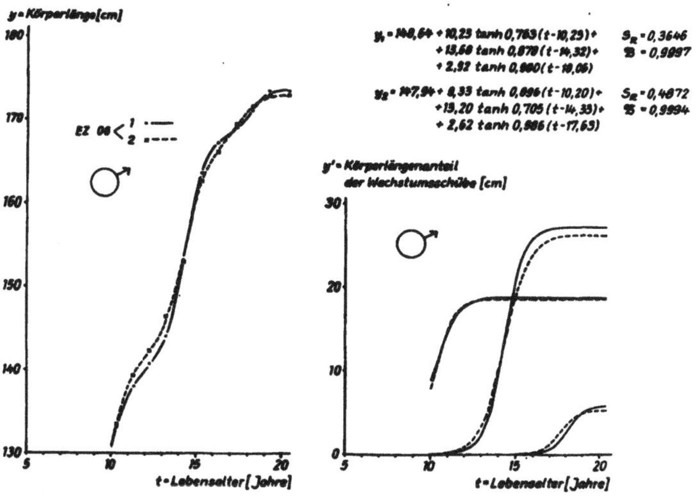

Fig. 2:

Example of longitudinal data series of body length development of twins (to the left) and phenomenologic-mathematical separation of growth spurts (to the right) with their quantitative description by means of terms of a 3-parametric hyperbolic tangent function (from HELWIN et al. /1979/). (ϑ: coefficient of determination, S_R: residual standard deviation). As the graphs of the estimated expressions written down in the right-top corner of the figure the 2 curves in the left part are best fitted to the measured courses.

The mathematical separation of growth spurts is only one step within the process of model construction. The biological relevance may be proved by considering the social and the biological circumstances in which the growth process happens (see HELWIN and PEIL /1978/, HELWIN et al. /1979/).

Details of the suggested phenomenologic-mathematical model are discussed in former papers (HELWIN and PEIL 1977, PEIL and HELWIN 1979). There are quoted also examples of performed separations. For the present theme it was the intention to show that special as well as general inverse problems may occur on different hierarchical levels within the process of model building and that they are mutually connected.

3. Mathematical models of growth spurts

3.1. Differential equations

In the phenomenologic-mathematical modelling of growth processes as mentioned before an essential step is to set up the class of functions for a quantitative description of the single growth spurts. Because such a functional relation between time t and body length y should also have a meaning as a mathematical model we go back to the traditional growth differential equations.

The general structure of a growth differential equation (of first order) is

$$\dot{y} = a_w \cdot f_1(y) - a_d \cdot f_2(y) \qquad (a_w, a_d > 0) \qquad (1)$$

The growth velocity $\dot{y}$ is a function only of the growth variable y itself in form of the balance between the growth term $f_1(y)$ (with intensity a_w) and the damping term $f_2(y)$ (with intensity a_d). The functions f_1 and f_2 must be chosen in such a way that the solution of (1) can reflect the sigmoidal shape of real growth processes[1]. Besides this more formal argument the choice of the functions f_1 and f_2 is an essential point in the model construction. They reflect the ideas about the (dynamical) properties of anabolism and catabolism the balance of which determines the growth of the organism.

The most simple setting up of f_1 and f_2 already was done by VERHULST in 1838. The equation

$$\dot{y} = a_w y - a_d y^2 \qquad (2)$$

is the so-called logistic law of growth. (For a summarizing and detailed discussion of the logistic law of growth and its generalizations in respect to the differential equation forms as well as to their solutions see PEIL /1978/). The meaning of (2) is a quadratically damped exponential growth[2].

To get more general expressions there are 3 possibilities to parameterize VERHULST's law of growth. Parameterization of the exponent of the growth term alone yields

$$\dot{y} = a_w y^n - a_d y^2 \qquad (0 < n < 2) \qquad (3a)$$

An other version of this is the partly parameterized form of BERTALANFFY's growth equation

$$\dot{y} = a_w y^n - a_d y \qquad (2/3 \leq n < 1) \qquad (3b)$$

---

1 For instance, the choice $f_1(y)=y$ and $f_2(y)=y \cdot \ln y$ leads to the well known law of growth which was suggested by GOMPERTZ.

2 The growth term $a_w y$ alone would describe unlimited exponential growth.

(As to the setting up of the interval in which n may vary and concerning the linear form of the damping term see BERTALANFFY /1960/).

Parameterization of the damping term gives

$$\dot{y} = a_w y - a_d y^m \qquad (1 < m < \infty) \qquad (3c)$$

The differential equation

$$\dot{y} = a_w y^n - a_d y^m \qquad (m,n > 0;\ m > n) \qquad (3d)$$

is the completely parameterized version of the former logistic law of growth[1] being able to describe sigmoidal growth courses[2].

At this stage one of the general inverse problems of growth dynamics was to fix the structure of the growth differential equation as a first order differential equation being composed of growth term and damping term, and the choice of the power form for f_1 and f_2.

Now it is of interest to estimate the parameters in (3), the intensities of the growth term and the damping term as well as the exponential parameters n resp. m proceeding from measured values of y. To a certain extent this special inverse problem - related to growth differential equations - appears on the level of growth functions as a general inverse problem. The differential equation (3d) has, if at all, different classes of functions as solutions dependent on the interval in which the values of the exponential parameters vary (e. g. see SCHARF /1975/, SAGER /1978/).

3.2. Growth functions

Here we consider only those analytical expressions as growth functions which are integrals of the above mentioned differential equations.

As solution of (2) we have the expression

$$y = a_1/[1+\exp(a_2-a_3 t)] \qquad (4a)$$[3]

with $a_1=a_w/a_d$, $a_2=\ln[a_w/(a_d \cdot y_o)-1]$, $a_3=a_w$ and $y_o=y_{|t=0}$ as the initial value of the growth variable.

The solution of (3c) is, written with the parameters of the differen-

---

1 This version sometimes is called BERTALANFFY's (general) growth equation. For growth processes of the mass he sets the catabolism to be proportional to weight, that means m=1 in (3d), see (3b).

2 Growth velocity becomes zero for y=0 and for $y=(a_d/a_w)^{1/(m-n)}$ according to formulae (3).

3 This functional expression of logistic growth is the starting-point for another generalization which has a polynom of t as exponent in the e-function (for a discussion also of this generalization see PEIL /1978/).

tial equation:

$$y=(a_w/a_d)^{1/(m-1)}[1+(a_w y_o^{(1-m)}/a_d-1)\exp(-(m-1)a_w t)]^{1/(1-m)} \quad (4b)$$

or, without the structure of the parameters:

$$y=a_1[1+\exp(a_2-a_3t)]^{-a_4} \qquad (a_3,a_4 > 0) \quad (4b')$$

As solution of (3b) one can get

$$y=[(a_w/a_d)-\{(a_w/a_d)-y_o^{(1-n)}\}\exp(-(1-n)a_w t)]^{1/(1-n)} \quad (4c)$$

also written in the simpler form (i. e. neglecting the parameter structures influenced by the parameters and coefficients of the differential equation):

$$y=b_1[1-\exp(b_2-b_3t)]^{b_4} \qquad (b_3 > 0,\ b_4 \geq 3) \quad (4c')$$

4. Nonlinear regression for solving the inverse problems

For solving the special inverse problems of growth dynamics we estimate the values of the parameters in the differential equations (3b,c,d) as well as separately those in the analytical expressions (4b',c') and (4a). The appropriate principle for the estimation procedure is the GAUSS' principle of least squares because in the measured data series always a random component is contained.

As measured data series we choose the values of the body length of the embryonic development of man. The mean values are shown to the left at the bottom of fig. 1, each of them represents a large amount of single measurements. The reason for this particular choice is the idea that the embryonic body length growth seems to be a more "uniform" growth period than the following periods because it may be determined, first at all, genetically. Therefore we are allowed to expect that our main concern now - the examination of numerical aspects in solving the special inverse problems - will be disturbed not too much by a possible superposition of growth spurts as one can see it in the other parts of fig. 1.

The inverse problem related to the differential equations will be solved by means of internal regression which is in the present cases a nonlinear[1] approximation procedure because of the exponential parame-

1 Usually, internal regressions will be performed to get initial values for parameters which are nonlinearly contained in an analytical expression and which can be calculated from linear parameters (coefficients) in the differential equation the solution of which is the analytical expression in question (e. g. see HARTLEY /1948/).

ters. To this end the divided central differences DDy_i will be calculated from the measured values $y(t_{i-1})$, $y(t_{i+1})$ ($i=2,\ldots,N-1$, N: number of measured values). The DDy_i are plotted as dots in fig. 3. These first order divided differences serve as approximations for the values of the differential quotient $\dot{y}$ at the corresponding time-points t_i. Then the course of DDy in its dependence of y will be least-squares approximated according to the respective growth differential equation.

As to the estimation problem of the parameters of the growth functions two remarks must be stated: First, according to the parameters a_w, a_d and m or n in (3b), (3c) there are in the expressions (4b'), (4c') only 3 independent parameters. The fourth parameter (a_2 resp. b_2) is formed by the two linear parameters of the differential equation and by the initial value y_o which usually is to settle. But in (4b'), (4c') this parameter is treated also as an adjustable one. (For the numerical integration of (3b) this would mean to optimize the initial value y_o, too). Secondly, in completion the approximation tasks posed by (4b'), (4c') we furnish these expressions with a time-constant term consisting of a parameter a_o or b_o respectively. (This would allow to describe also growth processes which don't start from the zero-level. In the present connection of the embryonic growth period such a parameter may be of advantage if there are small "inconsistencies" between the value y_o and the course of measured values for $t > 0$[1].) Such a "level-parameter were to be taken into account also in the differential equations by replacing y by $(y-a_o)$ allowing the growth velocity to become zero for values of y unequal zero.

The nonlinear least-squares estimation of the parameter values exactly is a constrained optimization because the parameters have to vary in the above specified intervals. In the integrals of the differential equations (3b), (3c) these constraints are partly reflected by the structure of the analytical expression. The common structure of (4b) and (4c) could be written

$$y=A_1[1+A_2\exp(-A_3t)]^{A_4} \qquad (5)$$

Because of $A_2=(y_o/A_1)^{1/A_4}-1$ and $y_o < A_1$ (A_1 is the asymptotically attained largest value of the growth variable) for $A_4 < 0$ follows $A_2 > 0$ (and the differential equation (3c)), for $A_4 > 0$ follows $A_2 < 0$ (and the differential equation (3b)).

---

1 The time $t=0$ is a singular point in respect to the embryonic growth dynamics and our considerations concerning the time-variations of the growth variable should be restricted to the time interval $t > 0$.

The parameter estimations were performed by means of a computer program system for nonlinear approximation (PEIL /1975/). In this program the GAUSS-NEWTON iterative algorithm is completed by several damped versions of this algorithm which show different properties in relation to reliability and velocity of convergence. Since the values of linear parameters can be determined without any difficulties by a linear regression step an expression like (4c') is preferable to (4c).

Two other computer programs for nonlinear regression were used to verify the numerical results because certain numerical effects (see 5.) have been happened. It should be excluded that these phenomena are artefacts caused by the used special procedure.

In one of these two other programs (KNAUFF and HENNIG /1978/) there was the possibility to regard the constraints in a direct manner in the optimization algorithm, and parameter estimations were performed with constraints as well as without constraints[1]. In the third program (NAGEL and WOLFF /1974/) a GAUSS-NEWTON step can be followed by a "dog-leg" search if the first one did not reduce the residual sum of squares (SQ).

For numerical integration of the differential equation (3d) after parameter estimation the BURLISCH-STOER procedure was used (ORZ MLU /1972/).

5. Numerical results and discussion

Starting values for the nonlinear parameters m, n in (3d) were chosen according to BERTALANFFY's approach: $m^{(o)}=1$, $n^{(o)}= 0.67$. The iterative approximation converged rapidly from $SQ^{(o)}= 13.8$ down to $SQ^{(4)}= 1.68$ already in the fourth step. Only after the first and after the second GAUSS-NEWTON step dog-leg-searches were in need of. Without such intermediate searches the GAUSS-NEWTON iteration did not converge starting with the above specified values for m and n. The estimated parameter values written in fig. 3 are the result obtained in the 8th step of the iteration. The curve I in fig. 4 was calculated by numerical integration of the differential equation I of fig. 3 assuming $y_{|t=o}=0.1$ as initial value.

As measure of goodness of approximation we take the sum of squared differences between measured and calculated values (SQ), and the sum of (linear) differences (SD) may possibly indicate systematical deviations between measured and calculated courses. It should be mentioned that

---

1 I am indebted to H. KNAUFF, ORZ MLU Halle, for performing these estimations

2 For these verifying calculations I am grateful to Dr. R. SCHMIDT, VEB Chemische Werke Buna

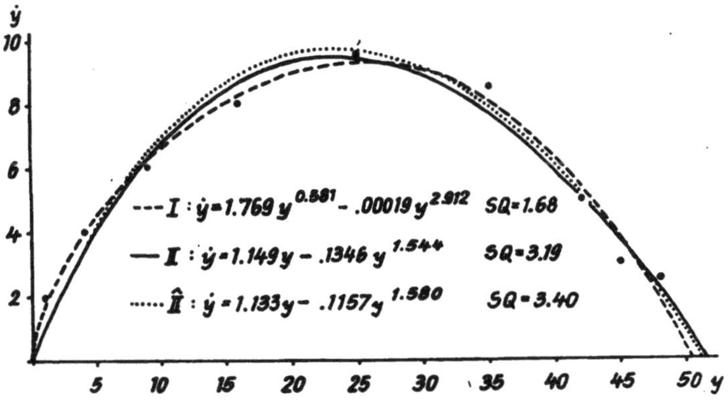

Fig. 3:
Course of divided central differences (dots) calculated from the measured values $y(t_i)$ of (embryonic) body length. Differential equations fitted by internal least squares regression. Parameter values of case $\hat{II}$ are calculated from those of the growth function (4b') after its approximation to the measured course of y_i (see case II in fig. 6).

in some cases there may be a flat minimum of the SQ-surface in the parameter space[1]. For discriminating between various sets of parameter values one has to take into account the amount of the "natural variability" and any inconsistencies[2] (relative to the applied model) which could be contained in the measured course. From this point of view hardly we can decide whether the curve I or II in fig. 4 better reflects the measured course. (This statement will not be influenced by the fact that the initial value - $y_{|t=0}=0.1$ for numerical integration in case I, $y_{|t=10}=50$ for the backward numerical integration in case II - is preassigned and therefore not optimized in the former sense).

Notice that in case II of fig. 4 the parameter values are quite different to those in case I which yield the attainable minimum value of SQ, see also fig. 3. They were obtained by a damped GAUSS-NEWTON procedure starting from the above mentioned parameter values. This set of parameter values is of interest as far as it lies in the "range" of the partly parameterized model (3c), the least squares fit and the estimated parameter values of which is shown by case II in fig. 3. Thus the goodness of fit of curve II, fig. 4, is nearly the same (SQ=3.37) as that

1 see for instance fig. 5

2 Such an inconsistency in respect of the here applied growth differential equations and growth functions could revealed by the measured values of body length in the neighbourhood of t=8 months (corresponding to y=45 in fig. 3). In a former paper we showed by separation of growth spurts that at this time a growth spurt begins which is responsible for a smooth course of the body length increase also over the period of birth.

of the 3-parametric differential equation (3c) in fig. 3. We can state that the fitting of the differential equation (3d) to the course of divided differences shows clearly the favourable status of the least-squares-estimated set of parameter values. This could not be demonstrated by means of the integral curves (of fig. 4).

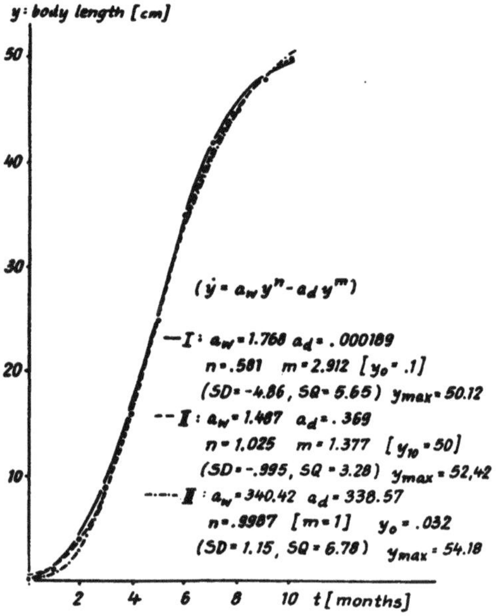

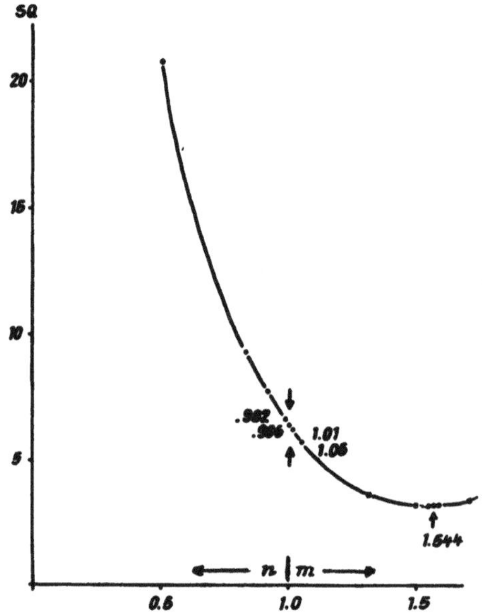

Fig. 4:

Integral curves of the differential equation (3d) (curves I and II) obtained by numerical integration. Noted parameter values are estimated by fitting (3d) to the course of divided differences (see fig. 3), the values of III are calculated from the estimated one of the integral expression (4c'), see text and fig. 7 (curve I). With the exception of case III the SQ-values are not minimized totally since y_0 is not adjusted.

Fig. 5:

SQ-"surface" in the space of the parameter n of (3b) resp. m of (3c), linear parameters eliminated. Obtained values, in turns n resp. m, a_w, a_d, SQ:

.982, 2.54, 2.36, 6.63
.996, 104.1, 102.3, 6.43
1.01, 43.67, 41.96, 6.22
1.05, 9.04, 7.40, 5.72
1.544, 1.149, .1346, 3.19, see fig. 3
(see also fig. 4, case III)

The results of parameter estimation relative to the differential equation (3c) as well as to (3b) without constraints are also shown in fig. 3 (curve II). Constrained optimizations of BERTALANFFY's partly parameterized growth equation (3b) were not feasible. The reason for this explains the SQ-curve of fig. 5: Within that part of the parameter space (here denoted by n) where the special growth model of BERTALANFFY is defined the SQ-function of this model hasn't got a minimum. In view of our approach to solve the special inverse problem by least squares estimation the partly parameterized version (3b) (with its integral (4c)) is an improperly posed inverse problem. In runs of unconstrained optimization starting with various parameter values of $n < 1$ spontaneous type changes happened from (3b) to (3c) and the noted SQ-minimum of 3.19 was achieved. In these cases the iteration skipped over the singular point n=m=1[1] without difficulties because far from the minimum point the parameter increments were large enough. Using more slowly convergent damped versions of the GAUSS-NEWTON iteration n approached the point n=1 to a certain extent with monotonically decreasing SQ-values but without to skip to the m-range.

In fig. 5 SQ(n) resp. SQ(m) is drawn. The dots mark some SQ-values which were obtained in the course of the mentioned iteration runs. The numerical values of n resp. m are written beside the dots belonging to. The linear parameters a_w and a_d were calculated by a linear regression step fitted to the respective value of the nonlinear parameter. (Therefore the SQ-values in the one-dimensional parameter space are minimum values in relation to a_w and a_d). The graph in fig. 5 shows a smooth transition of SQ-values from $n < 1$ to $m > 1$. The parameter values of a_w and a_d increase for $n,m \to 1$ (limited by the accuracy with which the calculations can be performed by the computer), see legend to fig. 5.

Naturally, the just discussed effect of type changing could also be seen in adjusting the growth functions (4b'), (4c') to the measured courses of the body length. There didn't occur any obstacles in parameter estimation for expressions (4a), (4b'). The results are shown in fig. 6. (Notice that one can get a very reduced SD-value if a_o is an adjustable parameter). The same results may be obtained by using (5) and starting with parameter values which lie in the allowed region of BERTALANFFY's function (4c') ($A_2 < 0$, $A_4 > 3$). But if we operate with (4c') a type change to (4b') can't happen and numerical effects can be seen analogous to those just described: The approach of $n \to 1$ means for the value of parameter b_4 in (4c') to go to infinity. The minimum of the SQ-surface in the space of the parameters of (5) has the numerical value $SQ_{min}=2.52$ ($SQ_{min}=1.88$ if an adjustable level-parameter is added, see fig. 6). This SQ-minimum is localized in that part of the parameter space which corresponds with the space spanned by the parame-

1 Disregarding the numerical aspects of parameter estimation n=1 in (3b) (m=1 in (3c)) would mean an unlimited exponential growth process whereas exponential damped growth would be described by (3b) for n=0.

ters of (4b).

Fig. 6:

Results of fitting the integrals of the partly parameterized versions of the logistic law of growth to the measured course. I and II: Unconstrained SQ-minimization as for (5) and constrained SQ-minimization as for (4b'), see text. III: Result of parameter estimation for the logistic function (4a) with adjustable a_o.

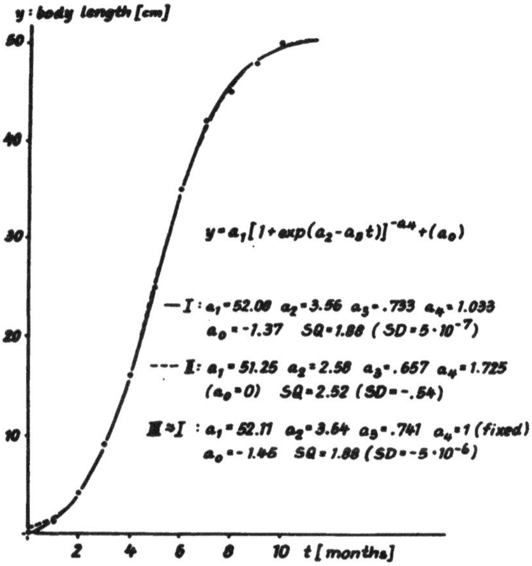

To give an impression to what extent the BERTALANFFY growth function (4c') is able to approximate the measured course though its SQ-function hasn't a feasible minimum in fig. 7 sets of parameter values are noted which were calculated for (4c') in different intermediate stages of the iterative process of parameter estimation. The value a_4=732.2 corresponds to n=.9987, and a_4=6433.7 corresponds to n=.9998. The curve I of fig. 7 also is drawn as curve III in fig. 4 (there in competition to integral curves obtained by numerical integration). The numerical values of a_w, a_d, and n as well as that of y_o were calculated from set I of parameter values in fig. 7. They illustrate numerically the above discussed statements.

To asses the goodness of fit achieved by the competing expressions here we also should notice of the numerical differences in the least squares estimated parameter values as the case may be of adjusting the differential equation or its integral. The specified parameters for curve II in fig. 3 were obtained by converting those yielded by approximation of (4b') to the measured course (see case II in fig. 6). The calculated SQ for this is greater than that for the approximation of (3c) to the course of divided differences (curve II in fig. 3).

As mentioned before, such differences in the parameter values must be taken into account in relation to that of the SQ-values if one has to criticize several proposed models for a further specification the general inverse problem of growth dynamics.

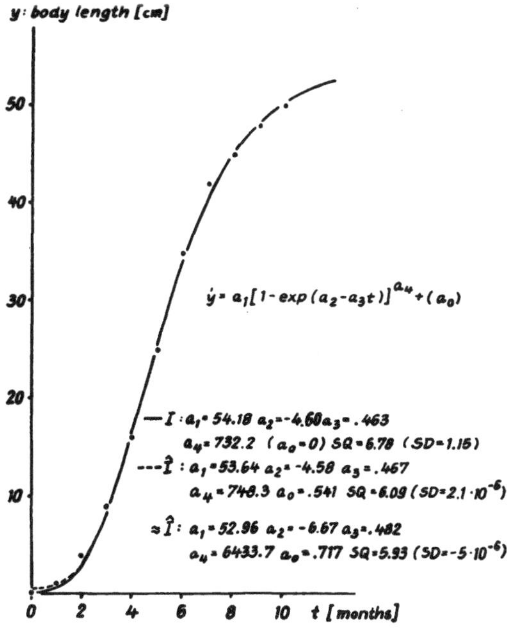

Fig. 7:
Intermediate results of fitting the BERTALANFFY growth function (4c') to the measured course. Constrained SQ-minimization without a feasible minimum, see text.

Summarizing the described results of solving the special inverse problem we can state:

1. From a numerical point of view the appealing inverse problem which was formulated in terms of the BERTALANFFY growth function (4c') or in form of its differential equation (3b) turns out to be improperly posed. This fact may contribute to model discrimination by specifying the general inverse problem of growth dynamics.
2. The special inverse problems formulated in terms of the other growth equations resp. growth functions could satisfactorily be solved by least squares parameter estimation. But because there are relative flat minima of the respective SQ-surfaces in the parameter spaces the sets of parameter values which achieve a reasonable good fit are not too well determined by the measured course.
3. The least squares estimates for the exponential parameters of the completely parameterized logistic law of growth (3d) are 0.58 for the growth term and 2.91 for the damping term.

4. The best fit of the integral expressions of the partly parameterized versions of logistic law of growth was achieved for an exponential parameter value a_4=-1.03 (corresponding to m=1.97) (adjustable a_0=-1.37) numerical closely neighbouring the values of the logistic function (a_4=1, see (4a), m=2, see (2)). (This fact would support the use of hyperbolic tangent terms for describing growth spurts as mentioned before).
5. Further work must be done with regard to including of a level-parameter and of the initial value y_0 in the estimation procedure.

Zusammenfassung

Die Bedeutung des Begriffs "Inverses Problem" im biokybernetischen Rahmen wird erläutert. Die bei der Modellierung biologischer Systeme zu gebende mathematische Beschreibung stellt das sog. allgemeine inverse Problem dar, in das eingebettet die speziellen inversen Probleme der Parameterschätzung sind. Die (zumeist numerische) Lösung letzterer kann auf die Formulierung und Spezifizierung des ersteren zurückwirken. Nach einer kurzen Schilderung der Hauptzüge einer phänomenologisch-mathematischen Modellierung von Wachstumsprozessen werden am Beispiel der embryonalen Körperlängenentwicklung des Menschen die konkreten Aufgabenstellungen bei der Formulierung der inversen Probleme der Wachstumsdynamik herausgearbeitet. Sie beziehen sich sowohl auf Differentialgleichungsmodelle des Wachstums, die durch Parametrisierungen des Logistischen Wachstumsgesetzes erhalten werden, als auch auf Wachstumsfunktionen, soweit sie Integralklassen dieser Differentialgleichungen darstellen. Im letzten Teil der Arbeit werden die numerischen Lösungen der speziellen inversen Probleme der Wachstumsdynamik mitgeteilt, die Bedeutung dieser Ergebnisse diskutiert und kritisch eingeschätzt und gezeigt, in welcher Weise sie auf die Lösung des allgemeinen inversen Problems Einfluß haben. Das auf die spezielle BERTALANFFYsche Wachstums-Differentialgleichung bzw. auf ihre Integralklasse bezogene spezielle inverse Problem erweist sich vom numerischen Standpunkt der Parameterschätzung aus als ein ungeeignet formuliertes inverses Problem.

References

BERTALANFFY, L. v.: Principles and Theory of Growth. In: W. W. NOWINSKI (ed.): Fundamental Aspects of Normal and Malignant Growth. Amsterdam: Elsevier 1960, 137-259

CUSHING, J. M.: Integrodifferential Equations and Delay Models in Population Dynamics. Berlin/Heidelberg/New York: Springer 1977

HARTLEY, H. O.: The Estimation of Non-linear Parameters by Internal Least Squares. Biometrika 35 (1948) 32-45

HELWIN, H., and J. PEIL: Phänomenologisch-mathematische Modellierung des Körperhöhenwachstums des Menschen durch Zerlegung in Wachstumsschübe und deren quantitativ-analytische Erfassung. Gegenbaurs morph. Jahrb., Leipzig 123 (1977) 641-655

HELWIN, H., and J. PEIL: Quantifizierung milieubedingter Entwicklungsunterschiede von Kleinkindern aufgrund eines phänomenologisch-mathematischen Modells. Gegenbaurs morph. Jahrb., Leipzig 124 (1978) 191-206

HELWIN, H., J. PEIL, and P. BERGMAN: Quantitativ-mathematische Auswertung von Wachstumsuntersuchungen unter dem Gesichtspunkt genetisch bzw. exogen beeinflußter Variabilität. Gegenbaurs morph. Jahrb., Leipzig 124 (1979) (in press)

KNAUFF, L., and Th. HENNIG: Bestimmung von Parametern in nichtlinearen Modellen nach der Methode der kleinsten Quadrate. Symposion "Chronobiologie und Chronomedizin", Halle, July 10th-15th, 1978. Proceedings of the symposion, Berlin: Akademie-Verlag 1979

NAGEL, G., and W. WOLFF: Ein Verfahren zur Minimierung einer Quadratsumme nichtlinearer Funktionen. Biometr. Z. 16 (1974) 431-439

ORZ MLU Halle: ALGOL-Prozedur DIFFSYS, Programm-Nr. 35 67 00; 1972

PEIL, J.: Ein ALGOL-Programm zur nichtlinearen Regression - Programminformation über ein biomathematisches Verfahren. Depot-paper 505/75 of Wiss. Z. Univ. Halle; short communication in: Wiss. Z. Univ. Halle, math. nat. R. 24 (1975)3, 77-78

PEIL, J.: Das Logistische Wachstumsgesetz und seine Erweiterungen. Gegenbaurs morph. Jahrb., Leipzig 124 (1978) 546-558

PEIL, J.: Quantitative Description and Mathematical Modelling of Biological Systems by Means of Nonlinear Approximations - Demonstrated at the Example of the Body Length Growth of Man. Biom. J. 21 (1979) 57-70

PEIL, J., and H. HELWIN: Ein phänomenologisch-mathematisches Modell der Wachstumsdynamik. 5. Biometr. Kolloquium, Eisenach, March 20th-23rd, 1979

SAGER, G.: Zuwachsfunktionen vom Typ $dW/dt=kW^m(E-W)^n$ und ihre Integrale. Anat. Anz. 144 (1978) 235-245

SCHARF, J. H.: Über Wachstumsmechanik. Biometr. Z. 16 (1974) 283-299

SCHARF, J. H.: Was ist Wachstum? Nova Acta Leopold., NF 40 (1975), No. 214

INVERSE PROBLEMS AND ENERGY INEQUALITIES

V.G.ROMANOV

The important role of energy integrals in studying of the correctivity questions for hyperbolic equations and systems is well-known. In the theory of inverse problems the method of energy estimates for the first time was used by R.G.Muhometov, who obtained the stability estimates for the inverse kinematic problem on a plane, and for similar integral geometry problem on the plane curves [1]. Later the energy estimates were used in the same way by G.Ya.Beilkin, I.N.Bernstein, M.L.Gerver, R.G.Muhometov and V.G.Romanov (see [2 - 5]). In this paper we give a method for studing of the correctivity of the coefficient's determination problems for linear hyperbolic and parabolic equations. The general description of this method can be found in [6].

As known, the most interesting applied inverse problems are uncorrect (see [7 , 8]). Therefore the most important in studying of the correctivity are the questions, connected with the uniqueness of a solution and with stability estimates. Both these questions will be considered in this paper.

We start with differential equation

$$u_{tt} - L_q u = f(x,t), \tag{1}$$

where L_q is uniformly elliptic operator:

$$\left.\begin{aligned} L_q u &= \sum_{i,j=1}^{n} a_{ij}(x)\, u_{x_i x_j} + \sum_{i=1}^{n} b_i(x)\, u_{x_i} + c(x)\, u, \\ &\sum_{i,j=1}^{n} a_{ij}(x)\, \alpha_i \alpha_j \geq \mu \sum_{i=1}^{n} \alpha_i^2, \quad \mu > 0, \quad a_{ij} = a_{ji} \end{aligned}\right\} \tag{2}$$

Here q is an ordered totality of the coefficients:

$$q = (a_{ij},\ 1 \le i \le j \le n;\ b_i,\ 1 \le i \le n;\ c).$$

Now consider the following problem of determination of the coefficients q for operator L_q in the equation (1): let $\mathcal{D}$ be some bounded domain of a space $X = (x_1, \dots, x_n)$ with smooth boundary S, Ω be its inner subdomain, and let operator L_q be known outside of Ω. The problem is to find the operator in Ω if we know, that the solution of a Cauchy problem

$$U_{tt} - L_q U = F(x,t),\quad U|_{t=0} = U_0(x),\quad U_t|_{t=0} = U_1(x) \tag{3}$$

takes in the set $\mathfrak{M} = S \times [0,T]$, $0<T<\infty$, given values

$$U|_{\mathfrak{M}} = U_2(x,t). \tag{4}$$

Here F, U, U_K, $K = 0,1,2$, are square matricesof order $N \times N$, where N is the number of components of the vector q.

Denote by Q a set, composed by the functions $q(x)$ The study of uniqueness and stability for the inverse problem (3)-(4) in the set Q can be reduced, under certain conditions, to the study of the same questions for some linear problem.

Let us introduce the square matrix $B(x)$ by formula

$$L_q U_0 \equiv q B. \tag{5}$$

<u>Lemma 1.</u> Let $\det B(x) \neq 0$, $x \in \bar{\Omega}$. Then studing of the uniqueness and stability (with respect to data (4)) for the inverse problem (3)-(4) in the set Q reduces to the same questions for the problem of determination of a function $\varphi(x)$, $\varphi = (\varphi_1, \dots, \varphi_N)$, $\operatorname{supp} \varphi \subset \bar{\Omega}$, from

$$V_{tt} - L_q V = \varphi(x) R(x,t), \tag{6}$$

$$V|_{t=0} = 0,\quad V_t|_{t=0} = \varphi(x),\quad V|_{\mathfrak{M}} = g(x,t), \tag{7}$$

where L_q is given operator, $q \in Q$; $R(x,t)$, $g(x,t)$ are given functions, $R(x,t)$ is N by N matrix, $V = (V_1, \dots, V_N)$.

Indeed, if (q, U) and $(\tilde{q}, \tilde{U})$ are the two solutions of a problem (3)-(4), corresponding to the same initial data, but different data (4), then, denoting

$$V = U - \tilde{U}\ , \quad \varphi = (q - \tilde{q}) B\ , \quad \frac{\partial}{\partial t}(U_2 - \tilde{U}_2) = g\ , \quad L_{\varphi B^{-1}} \tilde{U}_t = \varphi R,$$

we can easily obtain (6), (7). The norm estimate for the difference $q - \tilde{q}$ can be found, if we will know the corresponding estimate for the function φ , uniform with respect to the set Q

Note, that finding the function $\varphi(x)$ from (6), (7) is some special integral geometry problem. It will appear, if we represent the solution of the equation (6) with Cauchy data (7) with the help of fundamental solution $G_q(x, \xi, t-\tau)$:

$$\int_{\Omega} \varphi(\xi) \left[G_q(x,\xi,t) E + \int_0^t G_q(x,\xi,t-\tau) R(\xi,\tau) d\tau \right] d\xi = g(x,t), \quad (x,t) \in \mathfrak{M}, \tag{8}$$

Here E is a unit matrix of dimension $N \times N$.

Let us continue with the problem (6), (7). Let Q be the set of functions $q(x) \in C^{\nu}$, where $\nu = [\frac{n}{2}] + 4$ with the norm in C^{ν} , which is less then given constant M , and fixed constant $\mu > 0$ in (2). Now assume, that for $q \in Q$ the family of characteristic conoids is regular in the sence, that bicharacteristics, arising in one point, nowhere intersect each other.

Denote by Φ the set of functions $\varphi(x)$ with support in Ω , which $[\frac{n}{2}]$ -th derivative is bounded by given constant K . Assume that $R(x,t)$ is continuous and bounded together with its partial derivatives with respect to x up to the order $[\frac{n}{2}]$ in the domain $G = \bar{\mathcal{D}} \times [0, T]$. Let

$$\|\varphi\|^2_{L_2(\Omega)} = \int_{\Omega} |\varphi(x)|^2 dx \ , \qquad \|g\|^2_{L_2(\mathfrak{M})} = \int_0^T dt \int_S |g(x,t)|^2 dS,$$

$$\|R(x,t)\|_{C(G)} = \max_{1 \le i,j \le n} \|R_{ij}(x,t)\|_{C(G)} .$$

Here dS is an area element of the surface S , and $R_{ij}(x,t)$ are the components of matrix $R(x,t)$.

The two following stability theorems are valid

<u>Theorem 1.</u> Let

$$T > \frac{1}{\mu} \operatorname{diam} \mathcal{D}, \tag{9}$$

Then there exist such $\delta > 0$, $\delta = \delta(T, \mu, M, K, R)$, that, under conditions

$$\operatorname{diam} \Omega < \delta \quad , \quad \| R(x,t) \|_{C(G)} < \delta , \tag{10}$$

the solution of a problem (6), (7) for $\varphi \in \Phi$ satisfy the estimate

$$\| \varphi \|_{L_2(\Omega)} \leq C \sqrt{\| g \|_{L_2(\mathfrak{M})}} , \quad C > 0 , \tag{11}$$

uniformly with respect to set Q .

<u>Theorem 2.</u> There exists such $\delta > 0$, that under the conditions

$$\operatorname{diam} \mathcal{D} < \delta \quad , \quad \operatorname{diam} \mathcal{D} < \mu T ,$$

the estimate (11) is valid for the function $\varphi \in \Phi$, uniformly with respect the set Q .

Proofs of both theorems are similar. Assuming for simplicity that $c(x) \leq 0$, we introduce an energy integral

$$\mathcal{J}(t) = \sum_{k=1}^{N} \int_{t=const} \Big[(V_k)_t^2 + \sum_{i,j=1}^{n} a_{ij} (V_k)_{x_i} (V_k)_{x_j} - c(x) V^2 \Big] dx$$

The integration is performed along the intersection of the domain $G = \mathcal{D} \times [0,T]$ with the plane t = const. Using standard method of obtaining the energy relations, we find;

$$\mathcal{J}(t_2) - \mathcal{J}(t_1) = 2 \int_{t_1}^{t_2} d\tau \int_{S} \sum_{k=1}^{N} (V_k)_\tau \sum_{i,j=1}^{n} (V_k)_{x_i} \cos(n, x_j)\, dS +$$

$$+ 2 \int_{t_1}^{t_2} d\tau \int_{\mathcal{D}} \sum_{k=1}^{N} \Big[(\varphi R)_k (V_k)_\tau + (V_k)_\tau \sum_{i=1}^{n} \Big(b_i - \sum_{j=1}^{n} (a_{ij})_{x_j} \Big) (V_k)_{x_i} + c (V_k)_\tau^2 \Big] dx$$

By the condition (7), taking into account that the derivatives V_{k_t} in $\mathfrak{M}$ are bounded, we obtain

$$| \mathcal{J}'(t) | \leq c_1 \| g_t \|_{L_2(\mathfrak{M})} + c_2 \| \varphi \|^2_{L_2(\Omega)} + c_3 \mathcal{J}(t) , \tag{13}$$

where $C_k \geq 0$ and are independent of $t \in [0,T]$, and

$$C_2 = \| R(x,t) \|^2_{C(G)} .$$

From the inequality (13) we have

$$\|\varphi\|^2_{L_2(\Omega)} = \mathcal{I}(0) \leq [\mathcal{I}(T) + (C_1 \|g_t\|_{L_2(\mathfrak{M})} + C_2 \|\varphi\|^2_{L_2(\Omega)})T] e^{C_3 T},$$

or

$$\|\varphi\|^2_{L_2(\Omega)} \leq C_4 [\mathcal{I}(T) + \|g_t\|_{L_2(\mathfrak{M})}], \qquad (14)$$

if $C_2 T e^{C_3 T} < 1$. This inequality automatically holds if the condition (10) is satisfied, or if T (in theorem 2) is sufficiently small and aligned with δ, so that (12) is true. To estimate $\mathcal{I}(T)$ we represent the solution of the equation (6) with Cauchy data from (7) in the form $V = V_1 + V_2$, where V_1 is a solution of nonhomogeneous equation (6) with zero Cauchy data, and V_2 is a solution of homogeneous equation (6) with Cauchy data from (7). Let $\mathcal{I}_1(T)$ and $\mathcal{I}_2(T)$ be the integrals, corresponding to V with $V = V_1$ and $V = V_2$. Naturally, $\mathcal{I}(T) \leq 2[\mathcal{I}_1(T) + \mathcal{I}_2(T)]$

By energy estimates we have

$$\mathcal{I}_1(T) \leq C_2 T e^{C_3 T} \|\varphi\|^2_{L_2(\Omega)} \qquad (15)$$

At the same time

$$V_2(x,t) = \int_\Omega G_q(x,\xi,t)\varphi(\xi)\,d\xi$$

It is known, that all singularities of the function $G_q(x,\xi,t)$ are situated on the lateral part of dependency conoid with cone point (x,t). Under the given assumptions the function G_q is continuous in the conoid with its first derivatives $(G_q)_{x_i}$, $(G_q)_t$. But to estimate $\mathcal{I}_2$ we need the continuity of the function G_q and its first derivatives only at the point $t=T$, $x \in \Omega$, $\xi \in \Omega$. Under this restriction it is easy to see, that the values of the function $G_q(x,\xi,t)$ are used only strictly inside the dependency conoids, where G_q is a smooth function.

It is easy to find, that

$$\mathcal{J}_2(T) \leq \|\varphi\|^2_{L_2(\Omega)} \, c_5(\mu, M, T, \mathcal{D}, \Omega) , \qquad (16)$$

where

$$c_5 = \int_{\mathcal{D}} dx \left\{ \int_{\Omega} \left[\left(\tfrac{\partial}{\partial t} G_q\right)^2 - c(x)\, G_q^2 \right] d\xi + \sum_{i,j=1}^{n} a_{ij}(x) \left[\int_{\Omega} \left(\tfrac{\partial}{\partial x_i} G_q\right)^2 d\xi \right]^{1/2} \cdot \left[\int_{\Omega} \left(\tfrac{\partial}{\partial x_j} G_q\right)^2 d\xi \right]^{1/2} \right\} \leq c_6(\mu, M, T) \, (\mathrm{diam}\,\mathcal{D} \cdot \mathrm{diam}\,\Omega)^n$$

The estimates (15), (16) imply

$$\mathcal{J}(T) \leq c_7 \, \|\varphi\|^2_{L_2(\Omega)} ,$$

where $c_7 \to 0$, if δ in (10), (12) goes to zero. For small from the inequality (14) we obtain (11).

Now we will consider the inverse problem of the type (3), (4), but now for parabolic equation. We look for an operator L_q , assuming that the solution of the Cauchy problem

$$W_t - L_q W = F_1(x,t) \quad , \quad W|_{t=0} = W_0(x) \qquad (17)$$

is known in the set $\mathfrak{M} = S \times [0,\infty)$:

$$W|_{\mathfrak{M}} = H(x,t) \qquad (18)$$

As above, we assume here that the operator L_q is known out of S , and we have to find it in S , using the data (18). Here W , F_1 , W_0 , H are the matrices of dimension $N \times N$.

The study of the problem (17), (18) can be done by its reduction to the problem of the type (3)-(4).

<u>Lemma 2.</u> The problem (17)-(18) is equivalent to the problem (3)-(4) provided $U_0(x) = W_0(x)$, $U_1(x) = 0$ and F , U_2 are connected with F_1 , H by

$$\left. \begin{aligned} F_1(x,t) &= \frac{1}{2\sqrt{\pi t^3}} \int_0^\infty F(x,\tau)\, \tau \exp\left(-\frac{\tau^2}{4t}\right) d\tau , \\ H(x,t) &= \frac{1}{2\sqrt{\pi t^3}} \int_0^\infty U_2(x,\tau)\, \tau \exp\left(-\frac{\tau^2}{4t}\right) d\tau \end{aligned} \right\} \qquad (19)$$

Indeed, if we denote by $\widetilde{V}(x,p)$ the Laplace image of an arbitrary function $V(x,t)$ by variable t, then the Laplace images of the functions U, W satisfy the equations

$$p\widetilde{W} - L_q \widetilde{W} = W_0(x) + \widetilde{F}_1(x,p)\ , \quad \widetilde{W}|_{\mathfrak{M}} = \widetilde{H}(x,p)\ ,$$

$$p^2\widetilde{U} - L_q \widetilde{U} = U_0(x) + p\,U_1(x) + \widetilde{F}(x,p)\ , \quad \widetilde{U}|_{\mathfrak{M}} = \widetilde{U}_2(x,p)$$

Hence, if

$$U_0 \equiv W_0\ , \quad U_1(x) \equiv 0\ , \quad \widetilde{F}(x,\sqrt{p}) = \widetilde{F}_1(x,p)\ ,$$

then

$$\widetilde{U}(x,\sqrt{p}) = \widetilde{W}(x,p)\ , \quad \widetilde{U}_2(x,\sqrt{p}) = \widetilde{H}(x,p)$$

This connection between Laplace images leads to the relations (19). The relations (19) are invertible in $\mathfrak{M}$, so that one can uniquelly determine F, U_2 in $\mathfrak{M}$ by the functions F_1, H. This fact allows to study the correctness of the problem (17), (18), using the described scheme.

References

1. R.G.Muhometov, DAN SSSR, 232, N° 1, 32 (1977).
2. G.Ja.Beilkin, Proceedings of the LOMI seminars (1979) (to appear).
3. I.N.Bernstein, M.L.Gerver, DAN SSSR, 243, N° 2, 302 (1978).
4. V.G.Romanov, DAN SSSR, 241, N° 2, 290 (1978).
5. R.G.Muhometov, V.G.Romanov, Dan SSSR, 243, N° 1, 41 (1978).
6. V.G.Romanov, DAN SSSR, 242, N° 3, 541 (1978).
7. V.G.Romanov, Integral Geometry and Inverse Problems for Hyperbolic Equations. Springer Tracts in Natural Philosophy, vol. 26, Springer Verlag, Berlin, Heidelberg, New York, 1974.
8. V.G.Romanov, Inverse Problems for Differential Equations, Novosibirsk State University, Novosibirsk, 1973 (Russian).

Prof. Dr. V. G. Romanov, USSR Academy of Sciences, Sibirian Branch, Computing Centre, Novosibirsk 630090

ON EXTREMAL SOLUTIONS OF STURM LIOUVILLE INVERSE PROBLEMS

Pierre C. SABATIER*

Summary The problem of reconstructing the density $\rho(x)$ of a vibrating string given the N first eigenfrequencies of two vibrating configurations admits solutions that minimize certain weighted averages of the density. A new study of the expansions of functions on the squares of eigenfunctions enables the autor to give a simple set of necessary conditions on these weights. In particular, it is shown that the only weight functions $f(x)$ that can be consistent in all cases with the existence of an extremal density which is made up of a finite number of point masses are polynomials of degree two.

Introduction.

The problem of reconstructing the density $\rho(x)$ of a vibrating string given the first N eigenfrequencies for two vibrating configurations admits an infinite number of solutions. Among all such strings compatible with the truncated data set, it is possible to select solutions which achieve an extremum for a weighted average of the density :

$$M[\rho] = \int_0^1 f(x)\rho(x)dx \qquad (f(x) \geq 0) \tag{1}$$

These solutions have been studied recently by V. Barcilon[1], who gave a precise example where $M(\rho)$ is a true minimum, the extremal solution is made up by a finite number of concentrated point masses, and $f(\rho)$ is a quadratic function. Besides, Barcilon made plausible that, for large classes of weight functions $f(x)$, all extremal solutions are made up of separate point masses. Barcilon's approach of Sturm Liouville inverse problems is interesting, because information coming from experiments always is truncated, and it is sound to class the equivalent solutions. Extremal properties is a seducing way to do it, since it works very well for constrained linear inverse problems.

The aim of the present note is to write down necessary conditions for which $M(\rho)$ may be a local minimum in the class of strings made up of separate point masses, and to show that f(x) cannot be chosen arbitrarily. This result unfortunately reduces the interest of this kind of approach. However, the method which is described, and several tools that are introduced, in particular the "fundamental lemma" which is

* Laboratoire de Physique Mathématique, Equipe de Recherche associée au C.N.R.S., U.S.T.L. 34060 Montpellier Cedex.

proved, contribute to develop new points of view on inverse Sturm Liouville problems

The Problem

We study the eigenvalue problems

$$w_n'' + \nu_n \rho w_n = 0 \qquad n = 1,2,... \tag{2}$$

$$\left.\begin{aligned} w_n(0)\cos\alpha - w_n'(0)\sin\alpha = 0 \\ w_n(1)\cos\gamma + w_n'(1)\sin\gamma = 0 \end{aligned}\right\} \text{(even n)} \tag{3}$$

$$\left.\begin{aligned} w_n(0)\cos\beta - w_n'(0)\sin\beta = 0 \\ w_n(1)\cos\gamma + w_n'(1)\sin\gamma = 0 \end{aligned}\right\} \text{(odd n)} \tag{4}$$

where $0 \leq \alpha < \beta \leq \frac{\pi}{2}$ $\quad 0 \leq \gamma \leq \pi/2$, and $\rho(x)$ is made up by a finite number of point masses :

$$\rho(x) = \sum_{j=1}^{J} m_j \, \delta(x - x_j) \qquad (m_j \geq 0) \tag{5}$$

This problem has exactly $2J$ different eigenvalues, which can be organized in an increasing sequence $\nu_1, \nu_2, ... \nu_{2J}$. Let us be given an increasing sequence of $2N$ numbers $\nu_i^{(o)}$ $(i = 1,..2N)$, and let us call C_o the set of functions $\rho(x)$ which are of the form (5) and yield $\nu_i = \nu_i^{(o)}$ for $i = 1,2,...2N$. Since one can choose any $J \geq N$, and any value ν_n for $2N < n \leq J$, C_o is either void or contains an infinity of elements. It is convenient to define provisionnally a very large number J such that all densities we consider are of the form (5), with vanishing or non vanishing m_j's. Then one can define a distance between two densities, e.g. by $d(\rho,\rho') = \sup_{1\leq j\leq J} \{|x_j - x_j'|, |m_j - m_j'|\}$. In the metric space K_J of all functions like (5), the definition of C_o is achieved by imposing what we shall call the measurements constraint $\nu_i = \nu_i^{(o)}$ $(i = 1,2,...2N)$, and the positivity constraint $(m_j \geq 0)$.

We now seek conditions such that a given density $\rho^o \in C_o$ may yield a local minimum of the functional (1), ie such that there exists $R > 0$ with $M(\rho) \geq M(\rho^o)$ for any ρ that belongs to the intersection of C_o and the ball $B(\rho_o,R)$. In all the following, we assume for convenience $C_o \subset K_J$, $x_1^{(o)} \neq 0$, $x_J^{(o)} \neq 1$.

For a density of the form (5), the equation (2) is solved in such a way that a Dirac measure at x_i is included if the upper bound of the integral is x_i. This means that (2) is equivalent to the equations :

$$w(x) - w(0) - xw'(0) = -\nu \sum_i m_i w(x_i) \int_o^x \theta(t - x_i)dt \tag{6}$$

$$w'(x) - w'(0) = -\nu \sum_i m_i w(x_i) \theta(x - x_i) \tag{7}$$

where θ is the "Heaviside function". These equations are consistent with the continuity conditions that should be fixed for w and w' across a point mass, i.e. w

continuous and $w'(x_i^+) - w'(x_i^-) = -\nu\, m_i w(x_i)$.

Let us now write down the "wronskian" relation for two different densities, ρ^o and ρ, and the eigenvalues ν_o and ν (of same rank) that correspond ot them :

$$\int_o^1 (\nu\rho - \nu^o\rho^o)\, w^o w\, dx = 0 \tag{8}$$

The measurements constraint imposes the value of ν_n for $n = 1,\ldots 2N$. Therefore, all ρ's in C_o must satisfy $\int_o^1 (\rho - \rho^o) w_n^o w_n dx = 0$ for $n = 1,2,\ldots 2N$. Letting $d(\rho,\rho^o) \to 0$, and taking care of the continuity conditions, we obtain the first order "measurement constraints" on ρ :

$$\sum_{i=1}^{J} \{\delta m_i [w_n(x_i)]^2 + m_i \delta x_i w_n(x_i)[w_n'(x_i^-) + w_n'(x_i^+)]\} = 0 \text{ for } n = 1,2,\ldots 2N \tag{9}$$

where δm_i can have any sign if $m_i \neq 0$, and can only be positive if $m_i = 0$. The ℓ.h.s. of (9) define $2N$ linear forms on the space R_{2J} of vectors with components δm_i, δx_i. Equaling these forms to zero defines a manifold in R_{2J}. So as to characterize it, we prove the fundamental lemma.

<u>Fundamental lemma</u> Let us be given the problem (2-5), with $m_i \neq 0$, $i = 1,\ldots J$, and $2J$ real numbers, say, $f(x_i)$ and $f'(x_i)$ $(i = 1,2,\ldots J)$. There exists a unique function $F(x)$, which is a linear combination of the eigenfunctions squares, $\sum_{n=1}^{2J} F_n w_n^2(x)$, and such that, for any $i = 1,2,\ldots J$

$$F(x_i) = f(x_i) \qquad \tfrac{1}{2}[F'(x_i^-) + F'(x_i^+)] = f'(x_i) \tag{10}$$

<u>PROOF</u> 1° We first prove that if one knows that there exist numbers F_n such that (10) is satisfied, then $F(x) = \sum_1^{2J} F_n w_n^2(x)$ is completely determined, and is easy to construct.

Indeed, on each interval (x_i, x_{i+1}), $w_n(x)$ is a linear function, so that $F(x)$ is on this interval a 2nd degree polynomial.

$$F(x) = \frac{F'(x_i^+) - F'(x_{i+1}^-)}{2(x_{i+1} - x_i)}(x - x_i)(x_{i+1} - x) + F(x_i)\frac{x_{i+1}-x}{x_{i+1}-x_i} + F(x_{i+1})\frac{x-x_i}{x_{i+1}-x_i} \tag{11}$$

which obviously is determined when F and F' are known at x_i and x_{i+1}, and $F'(x_i^+)$ and $F'(x_{i+1}^-)$ are related by

$$F'(x_i^+) + F'(x_{i+1}^-) = 2(x_{i+1} - x_i)^{-1}[F(x_{i+1}) - F(x_i)] \tag{12}$$

On the interval $(x_N,1)$, it follows from (3) and (4) that $F(x)$ is equal to $[1 + (1 - x)\cot g\,\gamma]^2 F(1)$. Taking into account (10), we obtain

$$\delta_N = F'(x_N^-) - F'(x_N^+) = 2[f'(x_N) + \frac{2\cot g\, \gamma}{1+(1-x_N)\cot g\, \gamma} f(x_N)] \tag{13}$$

For any other value i, we obtain from (10) and (12) the recurrence relation $\delta_i = \Delta_i + \delta_{i+1}$, with

$$\Delta_i = 2\, f'(x_i) + 2f'(x_{i+1}) - 4\, \frac{f(x_{i+1}) - f(x_i)}{x_{i+1} - x_i} \tag{14}$$

Setting $\Delta_n = \delta_N$, we obtain

$$F'(x_i^-) - F'(x_i^+) = \delta_i = \Delta_i + \Delta_{i+1} + \dots \Delta_N \tag{15}$$

Thus, from (15), (10), and (11), $F(x)$ is fixed between x_1 and x_N. Between x_N and A, $F(x)$ is equal to $f(x_N)\left|\frac{1 + (1 - x)\cot g\, \gamma}{1 + (1 - x_N)\cot g\gamma}\right|^2$. Between 0 and x_1, using (3) and (4), it is a simple exercise to show :

$$F(x) = A(1 + x \cot g\, \alpha)^2 + B(1 + x \cot g\, \beta)^2 \tag{16}$$

where A and B are determined by $F(x_1)$ and $F'(x_1^-)$, and vanish if $F(x_1) = F'(x_1^-) = 0$.

This achieves the proof of our first point

2° It follows from our first point that no non-trivial linear combination of $2J$ forms like (9) (with all $m_i \neq 0$), can be identically zero. Indeed, if, for $i = 1,\dots J$,

$$\sum_1^{2J} c_n[w_n(x_i)]^2 = \sum_1^{2J} c_n w_n(x_i)[w_n'(x_i^-) + w_n'(x_i^+)] = 0 \tag{17}$$

the function $c(x) = \sum_1^{2J} c_n[w_n(x)]^2$ is identically zero, and so is $c'(x)$ and $c''(x)$. It follows from (7) that the equality $c'(x_i^-) = c'(x_i^+)$ implies that $\sum_1^{2J} \nu_n c_n w_n^2(x_i) = 0$, whereas the equality $c''(x_i^-) = c''(x_i^+)$ implies that $\sum_1^{2J} \nu_n c_n w_n(x_i)[w_n'(x_i^+) + w_n'(x_i^-)] = 0$. Hence, the function $\sum_1^{2J} \nu_n c_n[w_n(x)]^2$ is itself zero. Using this argument several times proves that $\sum_1^{2J} (\nu_n)^p\, c_n[w_n(x)]^2$ is zero for any p. Since the ν_n's are all different, one constructs out of them a non vanishing Van der Monde determinant (for $p = 0,1,2,\dots 2J-1$), which readily yiel's $c_n = 0$ $(n = 1,2,\dots 2J)$.

Consequently, the system of equations :

$$\begin{cases} \sum_{1}^{2J} F_n [w_n(x_i)]^2 = f(x_i) & i = 1,2,\dots J \\ \sum_{1}^{2J} F_n\, w_n(x_i)[w_n'(x_i^-) + w_n'(x_i^+)] = f'(x_i) & i = 1,2,\dots J \end{cases} \tag{18}$$

has one solution $\{F_n\}$, one only. This completes our proof.

3° There is an alternative way to construct the F_n' s. Let $\lambda_1(x)$, $\lambda_2(x)$, $\varphi(x)$, $\psi(x)$ be continuous fonctions, y_1, z_1, y_2, z_2, be solutions of $y_1'' + (\lambda_1 + \varphi)y_1 = 0$, $z_1'' + (\lambda_1 + \psi)\, z_1 = 0$, $y_2'' + (\lambda_2 + \varphi)y_2 = 0$, $z_2'' + (\lambda_2 + \psi)\, z_2 = 0$.

An easy-to-prove generalization of the Borg's identity is

$$2\int_0^X W(y_1 z_1, y_2 z_2)\,[\lambda_1(x) - \lambda_2(x)]dx \tag{19}$$

$$= [W(y_1,z_1)W(y_2,z_2) + W(y_1,y_2)W(z_1,z_2) + W(y_1,z_2)W(z_1,y_2)]_0^X$$

where W stands for the wronskian $(W(u_1,u_2) = u_1u_2' - u_2u_1')$. Let us now define a kind of "Jost solution " ω for the system 2 - 4 by imposing to ω_n to be solutions of (2) satisfying in addition $\omega_n(0) = 1$ and

$$\left.\begin{array}{ll} \omega_n(0)\cos\alpha - \omega_n'(0)\sin\alpha = 0 & \text{for odd } n \\ \omega_n(0)\cos\beta - \omega_n'(0)\sin\beta = 0 & \text{for even } n \end{array}\right\} \tag{20}$$

The identity (19) and the boundary conditions yield

$$(\nu_n - \nu_{n'})\int_0^1 [w_n^2 \frac{d}{dx} w_{n'}\omega_{n'} - w_{n'}\omega_{n'} \frac{d}{dx} w_n^2]\rho(x)dx = 0 \tag{21}$$

In the braces, one should use the average derivatives $\left(\frac{f'(x_i^+) + f'(x_i^-)}{2}\right)$ of the functions at the point masses. Thus the integral vanishes for $n \neq n'$ (but not for $n = n'$), so that it can be used for constructing the coefficients F_n of the expansion $F(x) = \sum_1^{2J} F_n\, w_n^2(x)$:

$$F_n = (w_n^2,\ w_n\omega_n)^{-1}(F,\ w_n\omega_n) \tag{22}$$

where (u,v) stands for $\int_0^1 W(u,v)\rho(x)dx$.

<u>Extremals</u>. We now come back to the condition $M[\rho] \geq M[\rho^o]$ in $C_o \cap B(\rho^o,R)$. So as to simplify the formulas, we introduce new notations, where $x_1,\dots,x_{J'}$, are the abscissas of the non-vanishing point masses of $\rho^o (J' \geq N)$:

$m_i f'(x_i) = f_i$ $(i = 1,\dots J')$ $f(x_i) = f_{J'+i}$ $(i = 1,\dots,J)$

$$m_i w_n(x_i)[w'_n(x_i^+) + w'_n(x_i^-)] = \alpha_i(n) \quad (i = 1,\ldots J') \quad [w_n(x_i)]^2 = \alpha_{J'+i}(n) \quad (i = 1,..J)$$

$$\delta x_i = \delta_i \quad (i = 1,\ldots J') \qquad \delta m_i = \delta_{J'+i} \quad (i = 1,\ldots J)$$

$$r(n,\{\delta_i\}) = \int_0^1 \delta\rho \; w_n^o \, w_n \, dx - \sum_{i=1}^{J+J'} \alpha_i(n)\delta_i \tag{23}$$

First, let us notice that the formula (8), with $\nu = \nu_o$, can be expanded in powers of $\delta\rho$. It suffices to write down the integral equations which gives w by means of the Green's function corresponding to ρ_o, and to solve it by iteration. Let the norm of $\delta\rho$ be defined by $\mathrm{Sup}|\delta_i|$ (in agreement with the distance that has been defined above). Then it is easy to show that the successive iterated terms are of order $\|\delta\rho\|$, $\|\delta\rho\|^2$, etc. In particular, $|r|$ in (23) is of order $\|\delta\rho\|^2$ and is itself a differentiable function (i.e. it can be put into the sum of a continuous quadratic form of $\delta\rho$ and a remainder of order $\|\delta\rho\|^3$ as $\|\delta\rho\| \to 0$).

Now, it follows from our lemma that the linear forms $\{\alpha_i(n)\}$ which appear in $\sum_{i=1}^{2J'} \alpha_i(n)\delta_i$, with $n = 1,2,\ldots 2J'$, are linearly independent. Hence, if n takes only the values $1,2,\ldots 2N$, with $N \leq J'$, we obtain a system of rank $2N$. There exists a set S_1 of values of i, with card $S_1 = 2N$, such that the matrix $\alpha_i(n)$, with $i \in \delta_1$, and $n = 1,2,\ldots 2N$, is invertible. We shall call S_2 the set $(1,\ldots 2J')-S_1$, and T the set $(2J'+1,\ J'+J)$. The set T corresponds to the values of δm_i that are not added to an initial point mass, and which therefore must be positive. The extremal condition and the measurement constraints read :

$$\sum_{i \in S_1} f_i\delta_i + \sum_{i \in S_2+T} f_i\delta_i \geq 0 \tag{24a}$$

$$\sum_{i \in S_1} \alpha_i(n)\delta_i + \sum_{i \in S_2+T} \alpha_i(n)\delta_i + r(n,\{\delta_i\}) = 0 \qquad (n = 1,2,\ldots 2N) \tag{24b}$$

We can determine a set of $2N$ numbers F_n such that

$$i \in S_1 \qquad f_i = \sum_{n=1}^{2N} F_n\, \alpha_i(n) \tag{25}$$

and it follows from (23) and (24) that

$$\sum_{i \in S_2} [f_i - \sum_{n=1}^{2N} F_n\, \alpha_i(n)]\delta_i + \sum_{i \in T} [f_i - \sum_{n=1}^{2N} F_n\, \alpha_i(n)]\delta_i - \sum_{n=1}^{2N} F_n r(n,\{\delta_i\}) \geq 0 \tag{26}$$

is a necessary condition for the existence of a local minimum. Now, suppose we fix the value of δ_i for $i \in S_2+T$, with a common multiplicative coefficient η which can be made as small as we want. The values of δ_i for $i \in S_1$ can be obtained from (24) by inverting the matrix $\alpha_i(n)$ $(i \in S_1)$ and iterating. The process of iteration converges for sufficiently small η because r is both second order in

$\|\delta\rho\|$ and lipschitzian. Now, in our choice of δ_i for $i \in S_2+T$, we can set all δ_i's $= 0$ except one, which can be made successively positive and negative if $i \in S_2$, positive only if $i \in T$. Since the third term in the lhs of (26) is second order in $\|\delta\rho\|$, it follows that the following conditions are <u>necessary</u> for a local minimum :

$$f_i - \sum_{n=1}^{2N} F_n \alpha_i(n) = 0 \qquad (i \in S_2) \tag{27}$$

$$f_i - \sum_{n=1}^{2N} F_n \alpha_i(n) > 0 \qquad (i \in T) \tag{28}$$

Now, since the values of x_i $(i \in T)$ can be chosen anywhere there is no point mass in ρ^o, we obtain the following <u>necessary conditions</u> for ρ^o to be a local minimum of $M[\rho]$:

$$\begin{aligned} f(x_i) &= \sum_1^{2N} F_n w_n^2(x_i) \qquad (i = 1,2\ldots J') \\ f'(x_i) &= \sum_1^{2N} F_n w_n(x_i)[w_n'(x_i^+) + w_n'(x_i^-)] \\ f(x) &> \sum_1^{2N} F_n w_n^2(x) \qquad \text{for any } x \notin \{x_i\} \end{aligned} \tag{29}$$

So as to study the sufficiency of these conditions, we would need to write down the 2nd order terms in (24b) and to study them carefully. We shall not do it here, but instead notice the following point. Suppose that a function ρ^o actually achieves a minimum of $M_f(\rho)$ for a certain function $f(x)$; then it follows from (26) that $M_f(\rho)$ has also a minimum at ρ^o for any function that satisfies (29) and (30).

Thus, if we know ρ^o and f for which the minimum problem is solved, we know an infinity of functions for which it is solved. But if we seek a twice continuously differentiable function f that can be consistent with the existence of a density of the form (2.5) and that makes possible a minimum of $M_f(\rho)$ for all problems (1.4), the class of admissible choices is drastically reduced. Indeed, a glance at (10) and at the conditions (30), which must hold in particular when $x \to x_i$, shows that f should be such that, for any choice of $\{x_i\}$, the function $F(x)$ which is associated to f by (10) and our fundamental lemma should verify $F'(x_i^+)-F'(x_i^-)>0$. This is an easy-to understand condition : it implies that the graph of $F(x)$, xhich generally is not continuously differentiable at x_1, has its two branches which go apart from $f(x_i)$ <u>below $f(x)$</u>. It follows from (13), (14), and (15) that for any x in the interval (0,1), and any couple of points (x,y) inside this interval, and any $\cot\gamma$ $(0 \le \gamma \le \pi/2)$, one must have

$$f'(x) + \frac{2\cot\gamma}{1 + (1-x)\cot\gamma} f(x) > 0 \tag{31}$$

$$\tfrac{1}{2}(f'(x) + f'(y)) \ge \frac{f(y) - f(x)}{y - x} \tag{32}$$

Besides*, $f-F$ must not have any odd order zero in $]x_i, x_{i+1}[$. If we stick at problems for which this interval is so reduced that f'' has a definite sign on it, this last condition simply reduces to $f''-F'' \leq 0$ on (x_i, x_{i+1}) (remember that F'' is constant on this interval). Using formulas (11) to (15) we obtain the condition

$$f''(x) - \frac{f'(x_{i+1})-f'(x_i)}{x_{i+1}-x_i} - \frac{1}{2}\,\frac{\delta_i+\delta_{i+1}}{x_{i+1}-x_i} \leq 0 \tag{33}$$

which should be verified for any $\{x_i\}$ and any spectral problem. Thus it is necessary that for any interval (x,y) where f'' has a definite sign, any number t inside this interval,

$$f''(t \in (x,y)) - \frac{f'(y) - f'(x)}{y - x} \leq 0 \tag{34}$$

The 2nd term is equal to the value of f'' at a certain point ξ of the interval, and $f''(t) \leq f''(\xi)$ for any t. Since $f''(x)$ is continuous this implies that either it is constant or there is a non vanishing subinterval and a positive number ϵ where $f''(t) - f''(\xi) < -\epsilon$, so that $f'(y) - f'(x)$ is smaller than $(y - x)\ f''(\xi)$ which is contradictory. Hence f'' is constant on each interval where it has a given sign Since f'' is continuous, it is constant and $f(x)$ is a 2nd degree polynomial. It is not difficult to extend this necessary condition to the class of continuously differentiable functions by using the density of C_2 in C_1 for the norm $\mathrm{Sup}[|f|, |f'|]$. If $f(x)$ is a 2nd degree polynomial, (32) is trivially verified as an equality. As for (31), it implies that $f(x)$ and $f'(x)$ are positive on $]0, 1[$.

Conclusion The result can be stated : unless $f(x)$ is a 2nd degree polynomial with f and $f' > 0$, there exists minimum problems of the type we described that have no solution in the class (5). Of course, the string problem is academic. But since it is the simplest of many spectral problems connected either with modes or with bound states, one should not hope to find in an arbitrary extremal property a key to class the solutions of these problems when the number of data is not sufficient to determine the solutions. On the other hand this work led us to deal with the products of eigenfunctions in a way which completes the works [(3)] of Borg on these questions in a somewhat unexpected way, and could have interesting connections with the methods generally called "inverse methods for non linear partial differential equations".

---

* In the following it is assumed that the indices have been ordered or reordered in such a way that x_{i+1} is the point mass following x_i, $i = 1,2,\ldots J'$, $(J' \geq N)$ exactly like in the proof of the Lemma (except that J was used in this proof instead of J', but with the same meaning that all the point masses correspond to non vanishing masses).

Acknowledgements

This work has been sponsored by the C.N.R.S. through the RCP 264 : Etudes interdisciplinaires des problèmes inverses.

Bibliography

(1) Barcilon V. Ideal solution of an inverse normal mode problem with finite spectral data. To be publ . in Geoph. J. R. Astr. Soc.

(2) Borg G. Eine Umkehrung der Sturm-Liouvilles-chen Eigenwertaufgabe. Acta math. 78 (1945)

(3) Borg G. On the completeness of some sets of functions. Acta math. 81 264-283 (1949).

ILL-POSED PROBLEMS ARISING IN BOUNDARY CONTROL AND OBSERVATION FOR DIFFUSION EQUATIONS

by

THOMAS I. SEIDMAN*

ABSTRACT For diffusion equations of the form: $\dot{u} = \nabla\cdot D\nabla u - f$ in $Q := (0,T)\times\Omega$ ($\Omega \subset\subset \mathbb{R}^m$) one may pose a variety of questions regarding observation (both state determination and system identification) and control in which the only interaction with this process is through the boundary conditions: $Bu := \alpha u + \beta u_\nu = \varphi$. Many of the standard system theoretic questions involve ill-posed problems when raised in this context. The paper discusses a variety of these ill-posed problems (both linear and nonlinear) and some approaches to approximate computation.

INTRODUCTION The problems considered in this paper all concern processes whose evolution is governed by diffusion equations of the general form

(1) $$\dot{u} = \nabla\cdot D\nabla u - f \quad \text{in} \quad Q := (0,T)\times\Omega .$$

Here $T > 0$ is specified and Ω is a bounded region in $\mathbb{R}^m$ with 'sufficiently smooth' boundary $\partial\Omega$. We take, generally, $D = D(t,x,u) \geq \underline{\delta} > 0$ and $f = f(t,x,u)$ but will primarily be concerned with the linear case: $D = D(t,x)$, $f = q(t,x)u$ so

(1') $$\dot{u} = \nabla\cdot D(t,x)\nabla u - q(t,x)u =: Lu \quad \text{in} \quad Q.$$

Associated with this are initial conditions

(2) $$u(0,\cdot) = u_0 \quad \text{in} \quad \Omega$$

and boundary conditions

(3) $$\alpha u + \beta u_\nu =: Bu = \varphi \quad \text{on} \quad S := (0,T)\times\partial\Omega$$

in which $u_\nu = D\bar{n}\cdot\nabla u$ is the outward D-normal derivative on $\partial\Omega$ and we take $\alpha^2 + \beta^2 \equiv 1$, assuming that (locally on $\partial\Omega$) either $\beta \equiv 0$

*Department of Mathematics, University of Maryland Baltimore County, Baltimore, Md. 21228, U.S.A.

(Dirichlet condition) or $\beta \neq 0$. The <u>complementary boundary data</u> for u is defined to be

$$(4) \qquad \psi := \tilde{B}u := -\beta u + \alpha u_\nu \quad \text{on } S.$$

If Γ is a specified 'boundary patch' ($\Gamma \subset \partial\Omega$) we will let $\tilde{B}_\Gamma u$ be the restriction of $\tilde{B}u$ to $S_\Gamma := (0,T) \times \Gamma \subset S$.

We will consider the following types of problems: CONTROL (given u_0, find φ on S_Γ [i.e., $\varphi \equiv 0$ on $S \setminus S_\Gamma$] such that the resulting solution of (1), (2), (3) has $u(T,\cdot) = u_T$; we say that 'φ steers u_0 to u_T'), STATE DETERMINATION (given φ and $\psi := \tilde{B}_\Gamma u$ — but not u_0 — for u satisfying (1), (3), determine $u(\tau,\cdot)$; a variant is SIGNAL RECOVERY: e.g., given φ on <u>part</u> of S and ψ, determine the 'rest of' φ) and SYSTEM IDENTIFICATION (e.g., given D but not q in (1'), determine the coefficient q from observation of $\psi := \tilde{B}_\Gamma u$ for a solution u of (1') with $\tilde{B}u = 0$ and suitable u_0). In some cases these are ill-posed as they stand, in others well-posed but more conveniently approached through an ill-posed reformulation, in still other cases this is unknown so one wishes a computational approach which does not depend on well-posedness.

We take as 'known' the basic theory of the initial-boundary value problem (1), (2), (3) — see, e.g., [6], [9], [11], etc. A characteristic feature of such equations is that 'solutions are as smooth as possible' (globally or locally). Thus, specification of a 'smoothness class' for the set of data u_0, φ and $\partial\Omega$, D, f, determines (e.g., via trace/extension theorems) a possible smoothness class for the solution and the problem (1'), (2), (3) is, in general, well-posed in this setting. One further has interior regularity results that, say for D, f analytic, all solutions of (1) will be analytic with respect to spatial variables and C^∞ with respect to t in the interior of Q. Thus, the map: $u_0, \varphi \mapsto u_T := u(T,\cdot)$ may be expected to be compact.

1. CONTROL Consider, first, the linear problem (1'), (2), (3). It is convenient to introduce the <u>adjoint</u> problem

$$(5) \qquad -\dot{v} = \nabla\cdot D\nabla v - qv \quad \text{in } Q,$$

$$(6) \qquad Bv = 0 \quad \text{on } S$$

and to let

$$M_\Gamma := \{\tilde{B}_\Gamma v \in L^2(S_\Gamma) : (5), (6) \text{ with } v(T,\cdot) \in L^2(\Omega)\}$$

and let $\overline{M}_\Gamma$ be the closure of M_Γ in $L^2(S)$ (we take $L^2(S_\Gamma)$ embedded in $L^2(S)$ by extension as 0 on $S \setminus S_\Gamma$). We have, then, the identity

$$\int_\Omega uv\Big|_0^T = -\int_S \varphi\psi \tag{7}$$

where u satisfies (1'), (2), (3) and v satisfies (5), (6) with $\psi := \tilde{B}v$. If u_0, u_T are specified and the control φ steering u_0 to u_T has support in S_Γ, then (7) may be taken as defining a functional on $\overline{M}_\Gamma$. This shows that, when u_T is at all reachable from u_0 by such control, there is a unique optimal (= minimum $L^2(S_\Gamma)$-norm) control φ (which will be in $\overline{M}_\Gamma$). Further, for $u_T = 0$, (7) shows that the adjoint of the control map $C: u_0 \mapsto \varphi$ is the map: $\psi \mapsto -v(0,\cdot)$ (i.e., null-control for (1'), (2) is dual to 'terminal' state determination for (5), (6) from observation of $\tilde{B}v$ on S_Γ).

It is known [13], [18], [22] that, under suitable restrictions on the geometry and the coefficients (e.g., assuming $\partial\Omega \setminus \Gamma$ 'visible' from a point in $\mathbb{R}^m \setminus \Omega$ and (1') autonomous and 'almost the heat equation'), the null-control problem is well-posed, i.e., the map: $u_0 \mapsto \varphi$ (selecting the unique optimal control) is defined and continuous on $L^2(\Omega)$ to $\overline{M}_\Gamma$. On the other hand, by linearity we may write

$$u(T,\cdot) = E_T u_0 + F\varphi \tag{8}$$

for solutions of (1'), (2), (3) so that to obtain $u_T = 0$ one must solve

$$F\varphi = -E_T u_0 \tag{9}$$

for φ. Although the results show that $-F^{-1}E_T: L^2(\Omega) \to \overline{M}_\Gamma$ is continuous, it is easy to see that $F: \overline{M}_\Gamma \to L^2(\Omega)$ is compact so (9) — or the more general problem

$$F\varphi = u_T - E_T u_0, \qquad \varphi \text{ in } \overline{M}_\Gamma \tag{9'}$$

when u_T is not zero (but is assumed to be reachable; we do note [21] that the reachable set here is independent of u_0 and of T) — is an ill-posed problem.

We sketch the discussion in [3] of (one approach to) the computation of the optimal control*. Let $\{Y_n\}$ be an expanding sequence of

*We remark, parenthetically, that it was this joint work which introduced the author to consideration of ill-posed problems. At the time, W. Chewning proposed both the scheme discussed here (see also [17]) and a 'least squares' method: taking $\varphi = \varphi_n$ in an expanding sequence of subspaces $X_n \subset \overline{M}_\Gamma$ so as to minimize $\|F\varphi_n + E_T u_0 - u_T\|^2$. For this latter method it has now been shown [23] that poor choice of $\{X_n\}$ will lead to nonconvergence of the computed approximants $\{\varphi_n\}$ even if the right hand

subspaces of $L^2(\Omega)$ and let $X_n = F^* Y_n$ determine a corresponding sequence of subspaces of $\overline{M}_\Gamma$. From (7) we see that, for y in $L^2(\Omega)$, $F^* y = \tilde{B}_\Gamma v$ where v is the solution of (5), (6) with $v(T,\cdot) = -y$. We choose φ_n in X_n so the residual $[F\varphi_n + E_T u_0 - u_T]$ is orthogonal to Y_n; then φ_n will be the orthogonal projection on X_n of the solution φ_* of (9') and $\varphi_n \to \varphi_*$ provided $\{Y_n\}$ is total ($y \perp Y_n$ for all n implies $y = 0$). The actual computation of φ_n proceeds as follows: let $(y_1,\ldots,y_n)$ be a basis for Y_n, assumed n-dimensional, and let $z_k = F^* y_k$ $(k=1,\ldots,n)$; the coefficients $(\xi_1,\ldots,\xi_n)$ of the expansion $\varphi_n = \xi_1 z_1 + \ldots + \xi_n z_n$ can then be obtained by solving the n-dimensional linear system:

$$(10) \qquad \sum_{j=1}^{n} \left[\int_{S_\Gamma} z_j z_k\right]\xi_j = \int_\Omega (E_T u_0 - u_T) y_k \qquad (k=1,\ldots,n).$$

If one sets

$$H_n : L^2(\Omega) \to \mathbb{R}^n : \mapsto \left(\int_\Omega y y_1, \ldots, \int_\Omega y y_n\right),$$

then the map*: $y \mapsto \varphi_n$ will just be the pseudo-inverse of $(H_n F)$.

In general, (10) will be more and more poorly conditioned as n increases but will always be solvable (as F, on $\overline{M}_\Gamma$, has dense range — containing the range of E_T — so F^* is injective, making $\{z_1,\ldots,z_n\}$ a basis of X_n so the Gramian matrix $G_n := ((\int z_j z_k))$ is invertible); the invertibility of the coefficient matrix G_n in (10) implies a limited stability of φ_n under small perturbations including imprecision in the specification of u_T and including the effects of approximate computation of z_j via numerical solution of (5), (6) and of $E_T u_0$ via numerical solution of (1'), (2) with $Bu_0 = 0$ as well as of the integrals in (10)). Nothing can be said <u>in general</u> about the convergence rate, but

---

side of (9') and all computations are exact.

*The problem (9') has an interesting alternative interpretation. Suppose u were known to satisfy (1'), (2) and also (3) with some φ. If observations of u at time T could be made, one might wish to estimate the unknown input φ. For linear observations: $u_T \mapsto \langle u_T, y_k\rangle$ $(k=1,\ldots,n)$ this construction gives φ_n as the estimate of φ (the formulation to use <u>a priori</u> knowledge that φ has support in S_Γ is built into the construction and other <u>a priori</u> information limiting φ to a closed subspace of $L^2(S)$ is handled similarly). In this context, in particular, it is worth noting that y_k need not actually be in $L^2(\Omega)$ but only that the construction of $F^* y_k = z_k$ via (5), (6) and (9) be well-defined — e.g., since the range of F consists of smooth functions there is no difficulty using point evaluations as observations (y_k a generalized function).

if $u_T = 0$ (null-control) and if Y_n can be chosen optimally (so $z_k = Cu_k$ where u_k is the k-th eigenfunction — in descending order of eigenvalues — of C^*C) then one has

$$\|\varphi_n - \varphi\| = \mathcal{O}(\exp[-\tau n^{2/m}]) \qquad \text{as} \quad n \to \infty \tag{11}$$

for every $\tau < T$.

It is clear that the same approach works (with minor caveats) even if the null-control problem were not known to be well-posed provided one somehow knows that (9') has a solution — e.g., under the interpretation posed in the footnote following (10). For nonlinear problems the approach requires modification. We distinguish two forms of nonlinearity: (i) linear equation as above but with nonlinear (but convex) constraints on the control and (ii) a nonlinear equation (1).

In the first case the convex constraints can be formulated as a (finite or) countable set of linear inequalities

$$\langle \varphi, \psi_j \rangle \le \gamma_j \qquad (j = 1,2,\ldots) \tag{12}$$

to be adjoined to the equalities

$$\langle \varphi, z_k \rangle = \eta_k [:= \int_\Omega (E_T u_0 - u_T) y_k] \qquad (k = 1,2,\ldots) \tag{10'}$$

expressed earlier by (10). Letting $\{X_n\}$ be an expanding sequence of subspaces defined as spans of finite subsets of $\{\psi_1,\ldots\} \cup \{z_1,\ldots\}$, one can compute φ_n as the unique minimum norm element of X_n satisfying the appropriate subset of (12) $\cup$ (10') — a standard programming problem. Although φ_n is no longer the orthogonal projection of φ_* on X_n, as in the linear case, one can still show that $\varphi_n \to \varphi_*$.

THEOREM 1.1: Let φ_* be the unique minimum norm solution of (12), (10') (i.e., φ_* is the optimal admissible control steering u_0 to u_T). Let $\{X_n\}$ and $\{\varphi_n\}$ be as above. Then $\varphi_n \to \varphi_*$.

This is a special case of the following more abstract result.

THEOREM 1.2: Let X, Y be Hilbert spaces, $A: X \to Y$ closed and linear, C a closed, convex subset of Y given by a system of inequalities (or equalities):

$$y \in C \iff \langle y, w_k \rangle \begin{Bmatrix} \le \\ = \end{Bmatrix} \eta_k \qquad (k = 1,2,\ldots) \tag{13}$$

with each w_k in $\mathcal{D}(A^*)$ so $x_k := A^*w_k$ is defined. Let φ_* be the unique minimum norm solution of

$$(14)\qquad A\varphi \in C$$

(assume this exists) and let φ_n be the minimum norm solution in $X_n := \mathrm{sp}\{x_1,\ldots,x_n\}$ of the finite system

$$(15)\qquad \langle\varphi, x_k\rangle \begin{Bmatrix}\le\\=\end{Bmatrix} \eta_k \qquad (k = 1,\ldots,n).$$

Then $\varphi_n \to \varphi_*$.

REMARK: Clearly $C_* := \{\varphi\colon (15) \text{ for all } n\}$ is closed and convex; $A^{-1}C = C_* \cap \mathcal{D}(A)$ is convex but not necessarily closed (its closure is C_*). Existence of φ_* as assumed in the theorem requires that the minimum norm element of C_* lies in $\mathcal{D}(A)$ — otherwise one might take membership in C_* to mean (14) is satisfied in a weak sense.

Proof: Since (15) involves the action of $\varphi \in X$ only as a linear functional on X_n, the minimum norm solution will be in X_n — i.e., will be φ_n. Since φ_* satisfies (15) for every n, one has $\|\varphi_n\| \le \|\varphi_*\|$. Thus, $\varphi_{n(k)} \rightharpoonup \hat{\varphi}$ for some subsequence $\{\varphi_{n(k)}\}$ and some $\hat{\varphi}$ with $\|\hat{\varphi}\| \le \underline{\lim}\|\varphi_{n(k)}\| \le \|\varphi_*\|$. Since $\hat{\varphi}$ satisfies (15) for every n (as φ_N does for $N \ge n$), one has $A\hat{\varphi} \in C$ by (13), whence, by the minimality of φ_*, one has $\hat{\varphi} = \varphi_*$. Further, $\|\varphi_{n(k)}\| \to \|\varphi_*\|$ so $\varphi_{n(k)} \to \hat{\varphi} = \varphi_*$. As this holds for a subsequence of any subsequence of $\{\varphi_n\}$, one has $\varphi_n \to \varphi_*$. □

Theorem 1.1 is obtained by taking $X := L^2(S_\Gamma)$, $Y := X \times L^2(\Omega)$, $A := \binom{I}{F}$, $C := \{\varphi\colon (12)\} \times \{E_T u_0 - u_T\}$; clearly convex state constraints could be handled in the same fashion.

In the case of a nonlinear equation (1) one writes $\hat{F}(\varphi) := u(T,\cdot)$ where u is determined by (1), (2) with the specified u_0, (3). One then seeks the optimal control steering u_0 to u_T, i.e., the minimum norm solution of

$$(16)\qquad \hat{F}(\varphi) = u_T.$$

Since $S_* := \{\varphi\colon (16)\}$ is generally not convex, it is no longer clear that such a optimal control exists or is unique but we now *assume**

*Unfortunately, for (1) nonlinear one cannot, at present, verify (17) — there are not even any positive reachability results now available to assert S_* nonempty.

(17) (a) there is a unique minimun norm element φ_* of $S_* \cap \mathcal{D}$

(b) the graph of $\hat{F}$ is closed in $X_w \times Y_w$ (these are $X := L^2(S)$ and $Y := L^2(\Omega)$ with weak topologies)

(note that (17) holds for linear F as above).

Given (17), one possible approximation scheme for (16) is the following: Let $\{X_n\}$ be an expanding sequence of finite-dimensional subspaces of X capable of approximating $\mathcal{D}$ and take $\mathcal{D}_n \subset X_n$ such that $\hat{\varphi}_n \to \varphi_*$ for some $\{\hat{\varphi}_n \in \mathcal{D}_n\}$ while $d_{n(k)} \in \mathcal{D}_{n(k)}$ with $d_{n(k)} \rightharpoonup d$ implies $d \in \mathcal{D}$. Let $\{B_n\}$ in Y be such that $S_n := \{\varphi \in \mathcal{D}_n: \hat{F}(\varphi) \in B_n\}$ is non-empty (indeed, $\hat{\varphi}_n \in S_n$) while $b_n \in B_n$ implies $b_n \rightharpoonup u_T$. Then, taking φ_n to be an element of (approximately) minimum norm in S_n, Theorem 2.3 of [24] gives $\varphi_n \to \varphi_*$. For this to be at all useful, of course, one must assume (17) and construct $\mathcal{D}_n$, B_n so that the minimization problem over S_n which defines φ_n is computationally feasible. Note that one can also take $S_n := \{\varphi \in \mathcal{D}_n: \hat{F}_n(\varphi) \in B_n\}$ where $\{\hat{F}_n\}$ is a suitable sequence of computational approximations to $\hat{F}$.

The introduction of B_n above constitutes a form of regularization of the problem (16): if one takes B_n in the form $B_n := \{y: \|y-u_T\| \leq \alpha_n\}$, then minimization of $\|\varphi\|$ over S_n just corresponds to

$$\text{(18)} \qquad \underline{\text{minimize}} \quad J(\varphi) := \|\varphi\|^2 + \lambda_n\|\hat{F}(\varphi)-u_T\|^2 \qquad \underline{\text{over}} \quad \mathcal{D}_n$$

for λ_n suitably related to α_n (so $\lambda_n \to \infty$ as $\alpha_n \to 0$). This is, of course, the 'standard' quadratic cost criterion and, at least for linear problems, the computational aspects of (18) have been extensively studied (see [10]), although not so much from the present point of view as an approximate approach to exact control $(\lambda \to \infty)$.

2. STATE DETERMINATION For the linear problem, (7) implies [4], [19] the equivalence of nullcontrollability for (1'), (2), (3) as considered above and well-posedness of the state identification problem: $\tilde{B}_\Gamma v \mapsto v(0,\cdot)$ for (5), (6). With time-reversal, this corresponds to (1) and (3) with* $\varphi = 0$. As noted above, this well-posedness is known [13], [18], [22] for the heat equation (and nearby equations) provided Γ is a 'large enough' portion of $\partial\Omega$. It is conjectured that this holds for <u>any</u> linear diffusion equation and <u>any</u> (relatively open) boundary patch Γ; in any case, the 'classical' state identification problem of inferring the <u>initial</u> state u_0 from observations on $[0,T]$ is certainly ill-posed. For C^∞

*If $\beta \neq 0$, there is no essential difficulty [18] in permitting (known) $\varphi \neq 0$ in (3).

linear problems and any Γ of positive measure in $\partial\Omega$, a recent result [15] shows uniqueness ($\tilde{B}_\Gamma u = 0$ implies $u = 0$) and Theorem 1 of [20] then gives well-posedness in the presence of an a priori bound on $\|u(-\varepsilon,\cdot)\|$.

Observe that the state identification problem involves a Cauchy problem from S_Γ (specification, there, of both $B_\Gamma u$ and $\tilde{B}_\Gamma u$ is the same as specifying u and u_ν) and such problems have been studied (e.g., [12], [2]) under a variety of hypotheses. With this point of view in mind we may consider a number of interesting problems, which we now pose in a one-dimensional context for simplicity of exposition.

Signal Recovery: Consider thermal diffusion in a rod $0 \le x \le 1$ with an unknown thermal transient (= signal) $s(t)$ at $x = 1$. Assume the rod is insulated* ($u' = 0$) at the accessible end $x = 0$ and that the temperature u is observable there. Thus $u = \psi(t)$ and $u' = 0$ are known at $x = 0$ and we wish to recover $s(\cdot)$. Assume u satisfies the diffusion equation

(18) $$\dot{u} = (Du')' \quad \text{for} \quad 0 < x < 1, \quad 0 < t < T$$

with the diffusion coefficient $D = D(x,u)$ analytic in x, u and an even function of x (also assume D strictly positive in the relevant domain).

The assumption that D is even in x means that D, u can be reflected across $x = 0$ to give a solution for $-1 < x < 1$. Then $x = 0$ becomes an interior** point so, by 'interior regularity,' u is analytic in x and C^∞ in t on $(0,T] \times [0,1)$. Repeatedly differentiating (18) and the data ψ, one can obtain*** the coefficients in the power series expansion of $u(t,\cdot)$ around $x = 0$ which may be continued to x in $[0,1)$. Thus, even without specification of initial data, the (exact) observation of ψ uniquely determines u in $(0,T] \times [0,1)$ and so, taking

---

*For the linear case, $u'(\cdot,0) = \varphi$ (not necessarily 0) is equally easily handled.

**Actually, with analyticity up to the boundary one could proceed without making 0 an interior point so one need not assume D even (or that $u'(\cdot,0) =: \varphi = 0$ if suitable assumptions are made on φ). An interesting variant is to assume (18) with $u(\cdot,a) =: \varphi_1$ and $u(\cdot,b) = \varphi_2$ for $0 \le a < b < 1$ — corresponding to readings from embedded thermocouples [8] — and seek to 'extrapolate' the signal $s = u(\cdot,1)$. Solving the direct problem for $a < x \le b$, one obtains $u'(\cdot,b) =: \psi$ and, as $x = b$ is interior, could then proceed as described with the Cauchy problem from $x = b$. Note further that the argument as given would apply to more general equations (1) and the m-dimensional case.

***A more detailed exposition of the material on signal recovery described here, with associated numerical experimentation, is now in preparation [25].

traces, <u>uniquely</u> <u>determines</u> <u>the</u> <u>signal</u> s (and also u_0).

In the linear case, if u_0 is known then one can use a computational procedure based on (7) much as for the computation of the optimal control — indeed, one can view the problem of recovering the <u>signal</u> $s(\cdot)$ from the observation ψ as, alternatively, the problem of finding a <u>control</u> s which makes the <u>output</u> ($u(\cdot,0)$ for u satisfying (18) with $u(0,\cdot) = u_0$, $u'(\cdot,0) = 0$ and $u(\cdot,1) = s$) equal ψ. Let y_k be a total sequence in $L^2(0,T)$ and define ω_k and z_k in $L^2(0,T)$ by setting

$$(19) \qquad \omega_k := \int_0^T D(0)\psi y_k + \int_0^1 u_0 v(0,\cdot), \qquad z_k := D(1)v'(\cdot,1)$$

where v is the solution of

$$(20) \qquad -\dot{v} = (Dv')', \quad v(T,\cdot) = 0, \quad v(\cdot,1) = 0, \quad v'(\cdot,0) = y_k.$$

We now define s_n as the solution in $X_n = sp\{z_1,\dots,z_n\}$ of the finite system (for the coefficients σ_k in: $s = \sum \sigma_k z_k$):

$$(21) \qquad \sum_k \left(\int_0^T z_j z_k\right)\sigma_k = \int_0^T s z_j = \omega_j \qquad (j = 1,\dots,n)$$

and, as earlier [17], s_n is the orthogonal projection of s on X_n so $s_n \to s$.

The 'control interpretation' above suggests the following approach: discretize (18) spatially (including the boundary conditions $u'(\cdot,0) = 0$, $u(\cdot,1) = s$ and the observation $y = u(\cdot,0) =: y_*(s)$) to obtain a system of ordinary differential equations

$$(22) \qquad \dot{z} = A_n z + b_n, \qquad y = C_n z, \qquad z(0) \text{ given by } u_0$$

defining the map $y_n\colon \sigma \mapsto y = y_n(\sigma)$ (where the discretization is refined as $n \to \infty$) for which it is a standard problem of lumped parameter (i.e., finite dimensional) system theory to determine the optimal control $\sigma = s_n$, minimizing the quadratic cost criterion

$$(23) \qquad J_n(\sigma) := \|\sigma\|^2 + \lambda_n \|y_n(\sigma) - \psi\|^2$$

for sutiably large λ_n. For application of Theorem 2.3 of [24] to ensure that $s_n \to s$, one must know that the maps $y_n\colon \sigma \mapsto y_n(\sigma)$ defined by (22) converge properly to the 'true' map y_* defined by the partial differential equation (18): one needs that $\hat{s}_n \to \hat{s}$ together with $y_n(\hat{s}_n) \to y_*$ should imply $y_* = y_*(\hat{s})$. It is observed in [24] that this is the case if s_n, s can be taken in $X := H^2(0,T)$ and y_n, y_* in $L^2(0,T)$, thus determining the norms to be used in (23), and if the discretization used is, for example [5], Galerkin's method employing piecewise cubic splines* Note that, if u_0 were not known a priori, the uniqueness result noted earlier permits simultaneous determination of u_0, s — add a term $\|u_0\|^2$ to the cost functional J_n and minimize over the pair (u_0,σ).

The discretization approach as described above could also be used for the nonlinear case with $D = D(x,u)$ (with (22) becoming a nonlinear ODE) but for that it is more difficult to verify the appropriate convergence of y_n to y_* — or even that the map $y_*: s \mapsto u(\cdot,0)$ is continuous for the norms to be employed. It is thus convenient to reformulate the problem to consider the function u as the unknown (obviously s, or the pair (u_0,s) if u_0 is unknown, can be recovered as traces of u) and the pair (y,f) as observed where $f = f[u]$ is obtained from u by the differential operator:

$$(24) \qquad y := u(\cdot,0), \qquad f := \dot{u} - (Du')'.$$

We then must determine u in a suitable space (including also the linear condition: $u'(\cdot,0) = 0$) so that the output pair (y,f) is just $(\psi,0)$. With regularization, this leads to minimization of the quadratic cost criterion

$$(25) \qquad J_n(u) = \|u\|^2 + \lambda_n\|y_n[u] - \psi\|^2 + \mu_n\|f_n[u]\|^2$$

where $y_n[u]$, $f_n[u]$ may represent suitable approximations to (24) — e.g., by restriction to a finite-dimensional domain $\mathcal{D}_n$ (say, given by splines).

We will assume observations in $y := L^2(0,T) \times L^2(Q)$ (the topology for f is not determined by potential accuracy of 'observations' since $f = 0$ is specified exactly by (18) but we take this in $L^2(Q)$ for convenience) and take u in $X := \{u \in H^2(Q) : u(\cdot,0) = 0\}$ — corresponding to an assumption that the initial data u_0 is in $H^{3/2}(0,1)$ and that the signal s is in $H^{3/2}(0,T)$; then the map: $u \mapsto (y,f)$ is continuous from X_w (X with its weak topology) to Y. Taking $\mathcal{D}_n$ to be given**

*This actually gives much stronger convergence of y_n to y_* than is needed here; one could presumably weaken the assumption on X (to $H^1(0,T)$ — or even, perhaps, to $L^2(0,T)$) but it is then not clear how to establish the desired convergence.

**Equivalently one might take y_n, f_n defined on all of X, setting $y_n[u] = y[P_n u]$, $f_n[u] = f[P_n u]$ where P_n is the orthogonal projection on this $\mathcal{D}_n$. Clearly, minimization of J_n over X can then be restricted to as functions in X which are, spatially, piecewise cubic splines, we know that $\{\mathcal{D}_n\}$ becomes dense in X and can easily verify that if u_n minimizes (25) (this need not determine u_n uniquely) and $\lambda_n, \mu_n \to \infty$ at suitable rates (related to the rate of best approximation to the unique 'true' solution u_* by $\mathcal{D}_n$), then $u_n \to u_*$ in X.

Note that this formulation does not depend on any assumption that the nonlinear 'direct' problem ((18) with $u(0,\cdot) = u_0$, $u(\cdot,1) = s$, $u'(\cdot,0) = 0$) is well-posed or even has a unique solution for 'arbitrary' u_0, s — we only need to know that it has some solution u_* in $H^2(Q)$ corresponding to the particular observed $\psi := u_*(\cdot,0)$ and so, as above, necessarily unique.

An interesting variant of this 'signal recovery' problem is consideration of the Stefan problem as a control problem. Writing the free boundary as $\hat{x} = \hat{x}(t)$ so $Q := \{(t,x) : 0 < t < T$ and $0 < x < \hat{x}(t)\}$ we seek boundary data $\varphi = u(\cdot,0)$ (assume $u_0 = u(0,\cdot)$ is given — alternatively we might have taken the boundary condition $u'(\cdot,0) = 0$ and sought an appropriate u_0) to produce, if possible, the specified boundary motion $\hat{x}(\cdot)$. This is equivalent, of course, to observing $\hat{x}(\cdot)$ and then attempting to infer φ. Analyticity considerations similar to those used earlier now give uniqueness of φ and permit approximation procedures along the lines of those just discussed.

3. SYSTEM IDENTIFICATION In many real applications mathematical modeling of the physical situation provides knowledge of the form of the governing equation (1) but leaves as contingent the determination of various parameters and functions (coefficients or nonlinearities) involved in complete specification.

For example, in 'porous flow' problems (arising, e.g., in geophysics and petroleum engineering) one has a nonlinear equation (1) with $D := D(u)$. Here, on physical grounds, we know that (normalized) $0 \le u \le 1$ and, on $(0,1)$, $D > 0$ with $D(0+) = 0$. However, the function $D(\cdot)$ may not be known exactly and we may seek to determine it* from various observations; note, in particular, the work of G. Chavent. Alternatively, one might know $D(\cdot)$ and wish to determine the particular boundary condition in (3).

For the linear problem (1') one might know D (say, $D :=$ const.)

---

minimization over $\mathcal{D}_n$.

*If the material is homogeneous (D not also dependent explicitly on x), then it may be more convenient to set $D =: dG(u)/du$, so $D\nabla u = \nabla G(u)$, and determine $G(\cdot)$.

and seek to determine the coefficient function $q := q(x)$. The 'classical' (finite dimensional = 'lumped parameter') system identification problem assumes trivial initial state ($u_0 = 0$) with observed output for specifiable input or an observed input/output pair. Alternatively, one might observe the output associated with a particular (known or unknown) initial state with trivial input. In the present setting the input would presumably be the boundary data* φ in (3) and the observable output would be complementary data $\psi := \tilde{B}_\Gamma u$ as in (4).

If one assumes that a direct problem gives a map F from the unknowns (coefficients, nonlinearities, parameters and, perhaps, such auxiliary variables as u_0) to the observations, then — for exact observation and computation — the identification problem corresponds to 'inversion' of F. The interpretation implies that the observed function is, automatically, in the range of F so existence of a solution is not at issue; the most difficult aspect is demonstrating uniqueness.

Consider the autonomous one-dimensional linear equation

$$(26) \qquad \dot{u} = u'' - qu, \qquad u(\cdot,0) = u(\cdot,1) = 0$$

with $u(0,\cdot) =: u_0$ unknown and $u'(\cdot,0) =: \psi$ observed. Setting

$$(27) \quad -e_j'' + qe_j = \lambda_j e_j, \qquad e_j(0) = e_j(1) = 0, \qquad \int_0^1 e_j e_k = \delta_{j,k},$$

one has the eigenfunction expansion

$$(28) \qquad u(t,x) = \sum_j \left(\int_0^1 u_0 e_j\right) \exp[-\lambda_j t]\, e_j(x)$$

so

$$(29) \qquad \psi(t) = \sum_j c_j \exp[-\lambda_j t], \qquad c_j := \left(\int_0^1 u_0 e_j\right) e_j'(0).$$

Since $\lambda_j = \mathcal{O}(j^2)$ and $c_j = \mathcal{O}(\lambda_j)$, the Dirichlet series in (29) converges uniformly on $\{\mathrm{Re}\ t \geq t_0\}$ for $t_0 > 0$ so the exponents actually appearing in the series (i.e., those for which $c_j \neq 0$ — meaning $\int u_0 e_j \neq 0$ since $e_j'(0) \neq 0$) are uniquely determined [14]. If we assume** that the (unknown) eigenfunction expansion of the (unknown)

*The input appears in the lumped parameter case as an inhomogeneous 'forcing term' in the equation itself.

**For the 'classical' problem (identification from observation of an input/output pair $\varphi := u(\cdot,1)$ with $u(\cdot,0) = 0$ and $u(0,\cdot) = 0 / \psi := u'(\cdot,0)$) we may more plausibly view φ as at our initial state u_0 is 'full' (i.e., $\langle u_0, e_j\rangle \neq 0$ for each j — which, in some sense, might be expected to hold 'almost surely' for an 'arbitrary' initial state), then the complete eigenvalue sequence $\Lambda(q) := (\lambda_1, \lambda_2, \ldots)$ is available from (exact) observation of ψ on $(0,T)$. If it would be known a priori that $q(\cdot)$ is symmetric about $x = 1/2$, then the 'inverse Sturm-Liouville problem' of determining q from $\Lambda(q)$ is uniquely solvable [1]. We summarize the results of this discussion.

THEOREM 3.1: Let $\psi(t) := u'(t,0)$ for $0 < t < T$ where u is known to satisfy an equation of the form $\dot{u} = u'' - qu$ on $Q := (0,T) \times (0,1)$ with $u(\cdot,0) = 0$ and either $u(\cdot,1) = 0 =: \varphi$ and $u(0,\cdot) = u_0$ such that the eigenfunction expansion of u_0 is full or else $u(0,\cdot) = 0$ and $u(\cdot,1) = \varphi \neq 0$. Then the eigenvalue sequence $\Lambda(q) := (\lambda_1, \lambda_2, \ldots)$ given by (27) is uniquely determined and, assuming $q := q(x)$ is symmetric about $x = 1/2$, the coefficient q is also uniquely determined by knowledge of (φ, ψ).

One could, then, approach the problem of obtaining q from knowledge of ψ (assuming $\varphi = 0$ and u_0 'full') by finding $\Lambda(q)$ from ψ and then applying available computational methods (see [7] and further references there) to obtain q from $\Lambda(q)$. Instead, we follow Example 7.3 of [24] in considering a 'history matching' approach. Assume that it is known that q lies in a specified bounded subset $\mathcal{D}$ in $X := \{\hat{q} \in H^1[0,1] : \hat{q} \text{ even}\}$; take $\mathcal{D}$ closed and convex — hence compact in X_w and so in $L^2(0,1)$. Choose τ in $(0,T)$ and note [16], [22] that there is an 'estimator' $E(\tau,q)\colon \psi|_{[0,\tau]} \mapsto u(\tau,\cdot)$ with $E(\tau,\hat{q})$, viewed an operator: $L^2(0,\tau) \to L^2(0,1)$, continuous in $\hat{q} \in \mathcal{D}$; thus, $\omega_* := u(\tau,\cdot)$ lies in a specifiable compact set U_*, which depends only on τ, ψ, $\mathcal{D}$. Given a sequence of discretizations of the direct problem (26),

---

disposal and argue as follows. From the equation, provided q is reasonably smooth, one has a map: $\varphi \mapsto \psi$ given by

$$(30) \qquad \psi(t) = \int_0^t k(t-s)\varphi(s)\,ds$$

continuously mapping, say, $\varphi \mapsto u \mapsto \psi$: $\{H^{3/4}[0,T] : \varphi(0) = 0\} \to H^{2,1}(Q) \to H^{1/4}[0,T] \hookrightarrow L^2(0,T)$; cf. [11]. One may view (30) as an integral equation (of the first kind) for the 'impulse response' $k = k_q$ and, for any suitable φ, this uniquely determines k on $(0,T)$. Using an eigenfunction expansion for u, one obtains k as a Dirichlet series

$$(31) \qquad k_q(\tau) = \sum_j e_j'(0)e_j'(1)\exp[-\lambda_j\tau] \qquad \tau > 0.$$

Thus, as with (29), specification of k_q uniquely determines $\Lambda(q) := (\lambda_1, \lambda_2, \ldots)$. Note that there is no question of vanishing coefficients in (31); the restriction is on the choice of input φ to determine k via (30).

using some $\hat{q}$ in $\mathcal{D}$ for q, on $(\tau,T)\times(0,1)$ with initial data ω in U_* at $t=\tau$, let $\Psi_k(\hat{q},\omega)$ be the k-th computed approximation to $u'(\cdot,0)$. Now determine the approximation q_k from the variational problem

$$(32)\qquad \underline{\text{minimize}}\ J_k(\hat{q},\omega) \ := \ \frac{1}{2}\int_\tau^T |\psi-\Psi_k(\hat{q},\omega)|^2\,dt \qquad \underline{\text{over}}\ \mathcal{D}\times U_*.$$

THEOREM 3.2: *In the context above,* $q_k\to q$ *as the discretizations are refined.*

Proof: As $\mathcal{D}\times U_*$ is compact, one has a subsequence $q_{k(j)}\to q_0$, $\omega_{k(j)}\to\omega_0$. Refinement of the discretizations means that $\Psi_k(q,\omega_*)\to\Psi_\infty(q,\omega_*)=\psi$ in $L^2(\tau,T)$ so $\inf J_k\to 0$. Further, by continuity, $\Psi_\infty(q_{k(j)},\omega_{k(j)})\to\Psi_\infty(q_0,\omega_0)$ and also $\|\Psi_k(q_k,\omega_k)-\Psi_\infty(q_k,\omega_k)\|\to 0$ whence $\Psi_\infty(q_0,\omega_0)=\psi$. Since the pair (q,ω_*) is uniquely determined by ψ, this means $q_0=q$ (and $\omega_0=\omega_*$) so $q_{k(j)}\to q$. As this holds for a subsubsequence of every subsequence, one has* $q_k\to q$. □

ACKNOWLEDGEMENTS This research was supported by USARO under grant no. DAAG-29-77-G-0061.

REFERENCES

[1] BORG, G., Eine Umkehrung der Sturm-Liouvilleschen Eigenwertaufgabe, Acta Math. 78(1946), pp. 1-96.

[2] CANNON, J.R. and R.E. EWING, A direct numerical procedure for the Cauchy problem for the heat equation, J. Math. Anal. Appl. 56(1976), pp. 7-17.

[3] CHEWNING, W.C., and T.I. SEIDMAN, A convergent scheme for boundary control of the heat equation, SIAM J. Control Opt., 15(1977), pp. 64-72.

[4] DOLECKI, S., and D.L. RUSSELL, A general theory of observation and control, SIAM J. Control Opt. 15(1977), pp. 185-220.

[5] DOUGLAS, jr., J., and T. DUPONT, Galerkin methods for parabolic equations, SIAM J. Numer. Anal. 7(1970), pp. 575-626.

[6] FRIEDMAN, A., Partial Differential Equations of Parabolic Type, Prentice-Hall, New York, 1964.

[7] HALD, O.H., The inverse Sturm-Liouville problem and the Rayleigh-Ritz method, Math. Comp. 32(1978), pp. 687-705.

*This convergence is in $\mathcal{D}\subset L^2(0,1)$ but, as this is homeomorphic to $\mathcal{D}\subset X_w$, we actually have weak convergence in $H^1(0,1)$.

[8] IMBER, M., and J. KHAN, Prediction of transient temperature distributions with embedded thermocouples, AIAA J. 10(1972), pp. 784-789.

[9] LADYŽENSKAJA, O.A., V.A. SOLONNIKOV and N.N. URAL'CEVA, Linear and Quasilinear Equations of Parabolic Type, Nauka, Moscow, 1967.

[10] LIONS, J.L., Contrôle Optimal de Systèmes Gouvernés par des Équations aux Dérivées Partielles, Dunod, Paris, 1968.

[11] LIONS, J.L., and E. MAGENES, Problemes aux Limites Non-Homogènes et Applications (Tome II), Dunod, Paris, 1968.

[12] PUCCI, C., Alcune limitazioni per la soluzioni di equazioni paraboliche, Ann. Mat. Pura Appl. (Ser. IV) 58(1959), pp. 161-172.

[13] RUSSELL, D.L., Exact boundary value controllability theorems for wave and heat processes in star-complemented regions, Differential Games and Control Theory, Marcel Dekker, New York, 1974.

[14] SAKS, S., and A. ZYGMUND, Analytic Functions, Mono. Mat., Warsaw, 1952.

[15] SCHMIDT, G., and N. WECK, On the boundary behavior of solutions to elliptic and parabolic equations — with applications to boundary control for parabolic equations, SIAM J. Control Opt. 16(1978), pp. 593-598.

[16] SEIDMAN, T.I., Observation and prediction for one-dimensional diffusion equations, J. Math. Anal. Appl. 51(1975), pp. 165-175.

[17] ____________, The solution of singular problems, I: linear problems in Hilbert space, Pacific J. Math. 61(1975), pp. 513-520.

[18] ____________, Observation and prediction for the heat equation, III, J. Differential Eqns. 20(1976), pp. 18-27.

[19] ____________, Boundary observation and control for the heat equation, Calculus of Variations and Control Theory, Academic Press, New York, 1976.

[20] ____________, Observation and prediction for the heat equation, IV: patch observability and controllability, SIAM J. Control Opt. 15(1977), pp. 412-427.

[21] ____________, Time invariance of the reachable set for linear control problems, Proc. 2-nd IFAC Symp. on Control of Distributed Parameter Systems (U. Warwick, 1977), Pergamon Press, 1978.

[22] ____________, Exact boundary control for some evolution equations, SIAM J. Control Opt. 16(1978), pp. 979-999.

[23] ____________, Nonconvergence results for the application of least squares estimation to ill-posed problems, J. Opt. Th. Appl., to appear.

[24] ____________, Convergent approximation schemes for ill-posed problems, UMBC Math. Res. Report 78-10, University of Maryland Baltimore Co., 1978.

[25] ____________, Recovery of a diffused signal, (in preparation).

IDENTIFICATION OF A SPATIALLY VARYING COEFFICIENT IN A PARABOLIC EQUATION . A REPORT ON NUMERICAL EXPERIMENTS

Gisbert Stoyan [1)]

We report here from experiments on the identification of a spatially varying heat conduction coefficient a(x) from m discrete measurement numbers (with normally distributed measurement errors of standard deviation σ) of the temperature profile at a fixed time. For representation of a(x) we use piecewise linear functions with n knots and show that there exists a knot number $n=n_{opt}(m,\sigma,a)$ which is optimal with respect to accuracy of approximation and with respect to stability against random errors. We give also an indication on the choice of n for small σ. Moreover, a formula is found relating the error of the estimated a(x) to the standard deviation σ.

In a variety of problems from quite different areas so as aquifer [1], [2] and petroleum problems [3], [4], heat conduction, flow in rivers, there arises the task to determine a spatially varying coefficient a(x) from measurements of u(x,t) at t=T, $x=x_i$, i=1,...,m, where u(x,t) is the solution of the equation

$$(1) \qquad \frac{\partial u}{\partial t} = \frac{\partial}{\partial x}\left(a(x)\frac{\partial u}{\partial x}\right) \quad , \quad 0 < x < 1 \quad , \quad 0 < t \leq T \quad ,$$

along with appropriate initial and boundary conditions.
We report here from a series of numerical experiments aimed to clear up the relation between the input information of the identification problem ($\{u(x_i,T)+\eta_i,\ i=1,\dots,m\}$, standard deviation σ of the assumed normally distributed measurement errors η) and the output information ($\{b^o(x_j),\ j=1,\dots,n\}$, l_2 and maximum norm error of $b^o(x)$), where $b^o(x)$ is the estimate of a(x).

1. For a mathematical background to identification problems of the type described above, see, e.g., [3]-[13]. For example, in [7]-[9] the case of a time dependent coefficient is considered, and integral representations

1) ZIMM der AdW , DDR 108 Berlin , Mohrenstraße 39

of the solution u are employed to prove theorems on solvability (at least for small t) , on uniqueness and stability for several identification problems connected with the heat equation.

In [10] analogous results are obtained by an iterative procedure. In [4], [5], [2], [11], [12], [6] gradient methods are used for identification problems including that of estimation of a spatially varying coefficient. In [12] existence is proved by showing with the help of the generalized Weierstrass theorem that a certain functional takes on its minimum. Considered are more general problems for nonlinear partial differential equations covering the problem investigated here. In [13] a uniqueness result was reported under the assumptions (belong others as smoothness) that a(0) is known and $\frac{\partial u}{\partial x}(0,0) \neq 0$.

As to the practical aspects of solving identification problems for parabolic equations, some numerical experience can be found, for example, in [5], [4], [2].

For surveys on identification problems in partial differential equations, see [14], [15].

2. The aim of the numerical experiments described below was the following: For computation the unknown function a(x) has to be replaced by some finite set of numbers such as function values at selected points, spline or Fourier coefficients. Here, for simplicity, function values at n equidistant points were taken and a(x) was interpolated between them piecewise linearly. The practically important question now is how to choose n if some minimal information on the measurement errors is available. We assume the errors to disturbe $u(x_i,T)$ additively, to be normally distributed with mean zero and standard deviation σ, and to be independent of each other at different points x_i. Then, a reasonable choice of n is such as to reach a compromise between stability against the random errors and accuracy of approximation.

Another question of interest in practical identification is the dependence of the deviation between the true and the estimated a(x) on the standard deviation σ of the measurement errors.

To both questions certain quantitive answers have been found. The experimental results are (being obtained from one special boundary value problem),clearly, of specific nature and may not generalize. They depend also on the numerical methods used.

Nevertheless, it is hoped that guidelines are provided for practical problems where every single measurement may be rather expensive and where there is the stringent requirement to extract the best possible output information.

3. Computations were performed in ALGOL on the computer BESM-6. For the numerical solution of (1), a second order accurate difference scheme of Crank-Nicholson type (see, e.g., [16]) has been employed. This assures a stable and efficient solution of the direct problem to find u(x,t) from (1) along with initial and boundary conditions for given a(x). A central unit time of about one second per 300 gridpoints was required.
In the inverse problem, the function a(x) is estimated via minimization of a functional of the type

$$(2)\qquad f(b_1,\dots,b_n) = \sum_{i=1}^{m} w_i(y_i - f_i(b_1,\dots,b_n))^2 \ , \quad m \geq n \ .$$

The minimizing procedure used was developed from the algorithm described in [17]. The procedure was tested on a great number of problems (including minimization of functionals with one- and two-dimensional sets of minimum points, and solution of systems of nonlinear algebraic equations up to 80th order). The idea of [17] is related to that of [18], [19], and quite similar to the idea presented in [20] for solution of systems of nonlinear equations: First, the gradient and the Newton-Raphson vectors are calculated. Then, several linear combinations of these vectors are tried until a combination is found which, if used as correction vector, diminishes the current functional value. The Newton vector helps to proceed in great steps along deep valleys which are characteristic for parameter estimation problems, and assures fast convergence in the near of the minimum. The gradient stabilizes the procedure, and allows to handle degenerate or nearly degenerate situations.
First order derivatives of the functions f_i (see (2)) needed in the minimizing procedure are approximated by finite differences. The steplengthes for these differences were chosen similar as in [21] and turned out to give a reliable picture of the relative sensivity of the different parameters b_j.

4. For the numerical experiments, equation (1) was considered together with the following initial and boundary conditions:

$$(3)\qquad \begin{aligned} u(x,0) &= (2+\cos 2\pi x)^2 \ , \quad 0 \leq x \leq 1 \ , \\ -a(x)\frac{\partial u}{\partial x} &= 12 \ , \quad x=0 \ , \ x=1 \ , \quad 0 < t \leq T = 0.02 \end{aligned}$$

The function a(x) was chosen to be

(4) $a(x) = a_0(x) = 2 + \cos 2\pi x$

Now, by the difference scheme, an approximation $y = y(x,t,a_0)$ to the solution $u(x,t)$ of (1),(3),(4) was computed at $t=T$, $x=x_i$, $i=1,\dots,m$, $m=20$, where

$$x_i = (i-1)/16\ ,\ i=1,\dots,17,\quad x_{18}= h,\ x_{19}= \tfrac{1}{2} - h,\ x_{20}= 1 - h,\ h = 1/48\ .$$

(The spatial discretization step h of the difference scheme has been chosen here for m=20 - and similarly for other values of m considered - so as to have a relation of three discretization points to one measurement point in the greater part of the x-interval.)
Next, the values $y_i = y(x_i,T,a_0)$ were disturbed additively by pseudonormal random numbers η of mean 0 and standard deviation σ (with σ varying in the experiments from 0 to 2), to give the simulated measurement data $\tilde{y}_i$,

$$\tilde{y}_i = y_i + \eta_i\ .$$

For comparison we mention that the y_i turned out to satisfy

$$3.3 < y_i < 6.5\ .$$

Now, to regain $a_0(x)$ from $\{\tilde{y}_i\}$, piecewise linear functions b(x) with values $\max(b_j,\varepsilon)$ at n knots x_j in $[0,1]$ were taken and a minimum problem was formulated which allows to select one of all possible sets $(b_1,\dots,b_n)$:

$$f(b_1,\dots,b_n) = \min !\ ,$$

where

(5) $$f = \frac{1}{m}\sum_{i=1}^{m}(\tilde{y}_i-y(x_i,T,b))^2 + M\sum_{j=1}^{n}(\max(0,\varepsilon-b_j))^2$$

Here $\varepsilon = 0.1$, $m = 20$, $M = 100$, and $y(x,t,b)$ is the finite difference solution to (1),(3), where b(x) has been inserted for a(x).
The piecewise linear function b(x) may be represented as

(6) $$b(x) = \sum_{j=1}^{n}\max(b_j,\varepsilon)\,\varphi(1-j+x(n-1))\ ,\quad \varphi(z) = \begin{cases} 0 & |z| > 1 \\ 1-|z| & |z| \le 1 \end{cases}\ .$$

From (6) we have $b(x) \ge \varepsilon > 0$ which reflects an important property of heat conduction (or filtration) coefficients and excludes numerical difficulties. For n=1, (6) gives $b(x)=\max(b_1,\varepsilon)=\text{const}$.
The minimizer $(b_1^0,\dots,b_n^0)$ of (5) has the property to define, by (6), a function $b^0(x)$ which produces at $t=T$, $x=x_i$, $i=1,\dots,m$, a finite difference solution $y(x,t,b^0)$ being in the mean (see the first term in f) near to $y(x,t,a_0)$. The second term in f, the penalty term, helps to

prevent the b_j^o from being less ε.
One single experiment may now be described as follows:
A pair (n, σ) is selected, the set $\{y_i\}$ is computed by the difference scheme and with the help of a random number generator disturbed into the set $\{\tilde{y}_i\}$ of simulated measurement data. Next, the functional (5) is minimized by the combination of Newton and gradient steps to give the minimizer $(b_1^o, \ldots, b_n^o)$ which defines $b^o(x)$. Then, $b^o(x)$ is compared with the original function $a_o(x)$ in the l_2 and the maximum norms,

$$\|b^o - a_o\|_l = \left(\frac{1}{m}\sum_{i=0}^{N}(b^o(i/N) - a_o(i/N))^2\right)^{1/2} ,$$

$$\|b^o - a_o\|_c = \max_{0 \le i \le N} |b^o(i/N) - a_o(i/N)| , \quad N = 1/h = 48 ,$$

and, finally, the numbers

$$e_l = \|b^o - a_o\|_l^2 \quad \text{and} \quad e_c = \|b^o - a_o\|_c$$

are obtained.

5. For a fixed pair (n, σ), from five up to twenty experiments were performed. Then, mean values $l(n, \sigma)$ and $c(n, \sigma)$ of the found numbers e_l and e_c were calculated. In the following Table 1 these numbers l and c are collected for $\sigma = 10^{-s}$, s = 3, 2, 1 (and r.s-t is written for $r.s \cdot 10^{-t}$).

Table 1

σ								
σ	n	3	5	7	9	11	13	16
10^{-3}	l	1.21-2	1.30-2	1.93-3	7.43-4	8.04-4	5.21-4	1.99-3
	c	0.3116	0.3799	0.1502	0.0939	0.0863	0.0819	0.1443
σ	n	3	5	6	7	8	9	11
10^{-2}	l	1.25-2	1.62-2	2.20-2	3.74-3	7.16-3	1.70-2	4.14-2
	c	0.3552	0.3756	0.2950	0.1955	0.2629	0.4081	0.6238
σ	n	2	3	4	5	6		
10^{-1}	l	0.5994	0.0391	0.2336	3.6752	4.4101		
	c	1.5168	0.4696	1.1307	4.8713	6.1783		

From this table we see that optimal values n_{opt} of the number of knots in (6) exist, indeed. (Heuristically, this has been clear: Small values of n provide rather poor approximations of a(x) but are quite sure against stochastic disturbances. The converse holds for great values of n, that is, for n in the near of the number of measurement points m.) 13, 7, 3 are the optimal values of n for $\sigma = 10^{-s}$, s = 3, 2, 1.
The next table containes n_{opt} for all values of σ considered. Here n.e.

denotes the number of experiments which led to the mean values l and c shown in the table. We mention that for $\sigma \geq 0.6$, $n \geq 3$ and for $\sigma = 2$, $n = 1$, there were observed some cases of estimation results $b^o(x)$ failing to approach $a_o(x)$ in magnitude even by several orders. These cases have not been excluded.

Table 2

σ	0	1.0-3	2.5-3	5.0-3	1.0-2	1.5-2	2.0-2	5.0-2
$l(n_{opt}, \sigma)$	4.22-5	5.21-4	1.33-3	1.75-3	3.74-3	9.43-3	1.35-2	1.55-2
$c(n_{opt}, \sigma)$	0.0191	0.0819	0.1257	0.1306	0.1955	0.3075	0.3283	0.3445
n_{opt}	16	13	8	8	7	7	3	3
n.e.	1	5	5	5	10	10	10	10
σ	0.1	0.2	0.4	0.6	0.8	1.0	1.4	2.0
$l(n_{opt}, \sigma)$	3.91-2	0.1222	0.4842	0.5947	0.5980	0.8485	0.7780	1.8755
$c(n_{opt}, \sigma)$	0.4696	0.7505	1.1246	1.2391	1.2241	1.3851	1.3652	1.7650
n_{opt}	3	3	3 (l) 1 (c)	1	1	1	1	1
n.e.	15	15	20	15	20	15	20	20

Looking at the table, a further statistical evaluation is called for. A dependence of n_{opt} on σ is obvious. Consequently, $l(n_{opt}, \sigma) = l(\sigma)$, $c(n_{opt}, \sigma) = c(\sigma)$. For $l(\sigma)-l(0)$ and $c(\sigma)-c(0)$ a dependence on σ of the type $a\sigma^b$ seems to be adequate. As usual in statistics, taking logarithms of σ, $l(\sigma)-l(0)$ and $c(\sigma)-c(0)$, a least squares fit is found (for all values of σ considered, and for subsets) resulting in the values of a and b displayed in the following table together with the estimate of the variance.

Table 3

σ	a	b	var.	
0 - 1.5-2	0.4774	1.0091	0.0606	coefficients a and b in $l(n_{opt}(\sigma), \sigma) - l(n_{opt}(0), 0) = a\sigma^b$
0 - 0.2	0.4756	1.0069	0.0813	
0 - 2.0	0.7700	1.1029	0.1052	
0 - 1.5-2	2.0463	0.5095	0.0246	coefficients a and b in $c(n_{opt}(\sigma), \sigma) - c(n_{opt}(0), 0) = a\sigma^b$
0 - 0.2	1.4339	0.4449	0.0254	
0 - 2.0	1.3742	0.4345	0.0201	

If trying directly to fit $l(\sigma)-l(0)$ and $c(\sigma)-c(0)$ by $a\sigma$ and $a\sigma^{1/2}$, respectively, the relations

$$(7) \qquad l(\sigma) = l(0) + 0.64\,\sigma \quad \text{and} \quad c(\sigma) = c(0) + 1.66\,\sigma^{1/2}$$

are obtained.

The optimal value of n, clearly, does not only depend on σ but also on

the used n-parameter function class (6), on the unknown coefficient a(x) and on m. If trying to establish a formula for n_{opt} in the much simpler problem of estimating a function from its given erroneous values, a dependence on higher order derivatives comes in, analogously as in the closure problem of stochastic differential equations.
We remember that the simulated measurement data $\tilde{y}$ and the function $a_o(x)$ attained values of about 4 and 2, respectively, and that m = 20 measurement points were used.
To get an impression on the dependence of n_{opt} on m, further experiments with m = 12 and m = 28 have been carried out leading to the following results:

Table 4 (values of n_{opt} for varying m and σ)

σ	0	1.-3	5.-3	0.01	0.02	0.05	0.1	N
m=12	12	12	8	6	3	3	3	32
m=20	16	13	8	7	3	3	3	48
m=28	17	17	10	9	3	3	3	72

(N denotes the discretization number used in the difference scheme.)

From this table it can be seen that it is advisable to take the number of knots n in (6) according to

$$\frac{m}{3} \leq n < \frac{m}{2} \tag{8}$$

for stochastic errors at the level $0 < \sigma \leq 0.01$. As to the number $\frac{m}{2}$ it is interesting to compare with the result of [22] on linear least squares problems.
For given m and small values of σ, (8) could play the role of a rule of thumb for selection of n.
The table shows also that for σ greater 0.01 the optimal n, seemingly, does not depend on m. Obviously, $n_{opt}=3$ corresponds to the fact that the special function (4) is better approximated by (6) if n is odd. Moreover, n=3 is the minimal number of knots for which functions of the type (6) can display a profile similar to that of (4).
Yet another dependence on the unknown function a(x) became visible in the experiments. The difference between the true and the estimated values of a(x) was not obviously increasing from the inner of the x-interval to its boundaries (which might be plausible, heuristically).
Rather, this difference seemed to reflect the deviation of the current true function value from the mean value.

6. For the investigated identification problem, the following conclusions may be drawn from the experiments.
For given standard deviation σ of the measurement errors and for given number m of measurement points, there exists an optimal number n of knots in (6), depending also on the unknown function a(x).
Further, small measurement errors can be handled rather good:
If one does not try to extract too much output information but confines oneself in a way indicated by(8), then the estimation error may behave in the mean as in (7), where l stands for square of l_2-error and c for maximum norm error.
Moreover, one single parameter (which defines here a constant coefficient) can be identified even in the presence of rather great measurement errors.
In a practical situation, a much more serious problem may be the model error. In the experiments, the model error has been bounded to the replacement of the analytic function (4) by the piecewise linear function (6), whereas the differential equation is the true one.
In a practical identification problem where equation (1) is accepted as a general model, the choice of a suitable n-parameter function class for representation of a(x) is important and should be done on the basis of theoretical considerations concerning modelling of the structure of the real material. If no such basis is available, take the functions (6) or cubic splines.

Earlier evaluations of smaller parts of the experiments have been given in [23], [24]. The author is indepted to Dr. Thiele from the same institute for hints and discussions.

[1] Carotenuto L., di Pillo G., Raiconi G. The doubly cubic spline in the identification of a class of DPS, with application to an underground aquifer. Lecture Notes in Control and Information Sciences 1 (ed. A.V. Balakrishnan, M. Thoma), Springer 1978, 174-195.

[2] Sames D., Häfner F., Heeg W. Modellierung der räumlichen Strömung mischbarer Flüssigkeiten in porösen Stoffen. Preprints des Kolloquiums 2 des Berg- und Hüttenmänn. Tages. Bergakademie Freiberg 1975.

[3] Chavent G., Dupuy M., Lemonnier P. History matching by use of optimal theory. J. Soc. Petroleum Eng., 15(1),(1975),74-86.

[4] Chen W.H., Seinfeld J.H. Estimation of spatially varying parameters in partial differential equations. Int. J. Control 15,3(1972), 487-496.

[5] Chavent G. Estimation de paramétres distribués dans les equations aux dérivées partielles. Lecture Notes in Computer Science 11, Springer 1974, 361-390.

[6] Häfner F. Mathematische Modellierung und Simulation von Strömungsvorgängen in porösen Stoffen unter besonderer Berücksichtung der Umkehraufgabe. Dissertation B, Freiberg, DDR, 1977.

[7] Jones B.F. Determination of a coefficient in a parabolic equation. Part I : Existence and uniqueness. J.Math.Mech. 11(1962),919-926.

[8] Cannon J.R. Determination of an unknown coefficient in a parabolic differential equation. Duke Math.J., 30(1963),313-323.

[9] Безнощенко Н.Я. Об определении коэффициента в параболическом уравнении. Дифф. Уравн. X,I(I974),24-35.

[10] Будак Б.М., Искендеров А.Д. Об одном классе краевых задач с неизвестными коэффициентами. ДАН СССР I75,I(I967),I3-I6.

[11] Kluge R. Optimal control with minimum problems and variational inequalities. Lecture Notes in Computer Science 27, Springer 1975, 377-382.

[12] Kluge R. Über eine Klasse von Minimumproblemen mit Nebenbedingungen. Math. Nachr. 78 (1977),37-64.

[13] Dümmel S. Lecture on the conference EQUADIFF IV, Prague 1977.

[14] Polis M.P., Goodson R.E. Parameter identification in distributed systems : a synthesizing overview. Proc. IEEE 64(1976),45-61.

[15] Kubrusly C.S. Distributed parameter system identification : a survey. Int. J. Control 26,4(1977),509-535.

[16] Самарский А.А. Теория разностных схем. Наука, I977.

[17] Steen N.M., Byrne G.D. The problem of minimizing nonlinear functionals. I : Least squares numerical solution of nonlinear algebraic equations (ed. G.D.Byrne, C.A.Hall) Acad. Press 1973,185-240.

[18] Levenberg K. A method for the solution of certain nonlinear problems in least squares. Quart.Appl.Math. 2(1944),164-168.

[19] Marquardt D.W. An algorithm for least squares estimation of nonlinear parameters. SIAM J. 11(1963),431-441.

[20] Powell M.J.D. A hybrid method for nonlinear equations. Numerical methods for nonlinear algebraic equations (ed. Ph. Rabinowitz). Gordon and Breach, 1970,87-114.

[21] Stewart G.W. A modification of Davidons minimization method to accept difference approximations of derivatives. JACM 14(1967), 72-83.

[22] Einbu J.M. Determination of a basis for linear least squares approximation. BIT 18(1978),67-71.

[23] Stoyan G. Numerical experiments on the identification of heat conduction coefficients. Summerschool on Nonlinear Operators (ed. R. Kluge). Akademieverlag 1978.

[24] Stoyan G. Some results of numerical experiments on identification of a spatially varying heat conduction coefficient. Summerschool of the KAPG 5.2 in Freiberg, GDR, 1978 (ed. J. Webers).

SOME CONSTRUCTIONS OF RIGHT INVERSES AND GENERALIZED INVERSES

Manfred Tasche [1])

In the present paper we sketch some results concerning the construction of right inverses for linear initial value problems and of orthogonal generalized inverses for selfadjoint boundary value problems.

Introduction

It is the aim of our investigations to treat linear initial value problems and boundary value problems from the general point of view of the theory of generalized inverse operators.

The first section is devoted to initial value problems in Banach spaces. We consider a linear closed operator D with a quasinilpotent right inverse. Then we form polynomial operators in D with stationary coefficients. Using the Dunford - Taylor operational calculus right inverses of such polynomial operators are constructed. If the initial value problem is degenerate, then the corresponding right inverse is unbounded.

The section 2 deals with a class of selfadjoint boundary value problems in Hilbert spaces. Let D be a selfadjoint operator with pure point spectrum. Using the Taylor formula a general method of accelerating convergence of eigenvector expansions is described. Then we form polynomial operators in D with scalar coefficients. Using convolution and exponential elements a simple representation of orthogonal generalized inverses of such polynomial operators is given.

---

[1]) Wilhelm-Pieck-Universität Rostock, Sektion Mathematik
25 Rostock, Universitätsplatz 1, DDR.

1. Construction of right inverses

1.1. Suppose that X is a complex Banach space. By L(X) we denote the set of all linear bounded operators on X into X. We consider a linear closed right invertible operator $D: \mathcal{D}(D) \to X$ with dense domain $\mathcal{D}(D) \subset X$ and non-trivial kernel ker D. Let $T \in L(X)$ be a right inverse of D. Assume that T is quasinilpotent. Let $I \in L(X)$ be the identity. By definition, the corresponding initial projector $F: \mathcal{D}(D) \to \ker D$ ([3], [7]) is given by $F = I - TD$.

Examples. a) Let $X = C(0,t_0)$, $(0 < t_0 < \infty)$ and let $\mathcal{D}(D)$ be the set of all continuously differentiable functions $x(t) \in X$. Then $D = \frac{d}{dt}: \mathcal{D}(D) \to X$ has a quasinilpotent right inverse T

$$(Tx)(t) = \int_0^t x(\tau)\, d\tau.$$

The initial projector is given by $(Fx)(t) = x(0)$.

b) Let $X = L_2(-\infty, \infty)$ and let $\mathcal{D}(D)$ be the set of all absolutely continuous functions $x(t) \in X$ with $x'(t) + \frac{t}{2}x(t) \in X$. Then the tempered derivative $D = \frac{d}{dt} + \frac{t}{2}: \mathcal{D}(D) \to X$ has a quasinilpotent right inverse T

$$(Tx)(t) = \frac{1}{\sqrt{2\pi}} e^{-t^2/4} \int_{-\infty}^{\infty} e^{-\sigma^2/2} \int_{\sigma}^{t} e^{\tau^2/4} x(\tau)\, d\tau\, d\sigma.$$

The corresponding initial projector is given by

$$(Fx)(t) = \frac{1}{\sqrt{2\pi}} e^{-t^2/4} \int_{-\infty}^{\infty} e^{-\tau^2/4} x(\tau)\, d\tau.$$

Many other examples can be found in [7].

1.2. An operator $A \in L(X)$ is said to be stationary, if $DA = AD$ on $\mathcal{D}(D)$ and $TA = AT$ on X. It is the aim of this section to construct right inverses of operators of the type $AD + B$ or $D^2 + AD + B$ with stationary operators A and B. Using such right inverses we can describe the solution of the initial value problem

$$(AD + B)x = f, \quad Fx = z_0 \tag{1}$$

respectively

$$(D^2 + AD + B)x = f, \quad Fx = z_o, \quad FDx = z_1 \tag{2}$$

with given $f \in X$ and z_o, $z_1 \in \ker D$.

<u>Theorem 1.</u> If $A^{-1} \in L(X)$, then

$$(A + BT)^{-1}T = \frac{1}{2\pi i} \int_\Gamma (\lambda A + B)^{-1}(I - \lambda T)^{-1}T \, d\lambda$$

is a quasinilpotent right inverse of $AD + B$. The initial value problem (1) has the unique solution

$$x = \frac{1}{2\pi i} \int_\Gamma (\lambda A + B)^{-1}(I - \lambda T)^{-1}(Tf + Az_o) \, d\lambda.$$

By Γ we denote here a simple closed contour, oriented counterclockwise, enclosing the spectrum $\sigma(-A^{-1}B)$.

<u>Theorem 2.</u> Let $L(\lambda) = \lambda^2 I + \lambda A + B$ and let $\sigma(L)$ be the set of all complex λ such that $L(\lambda)$ isn't invertible in $L(X)$. Then

$$(I + AT + BT^2)^{-1}T^2 = \frac{1}{2\pi i} \int_\Gamma L^{-1}(\lambda) \, (I - \lambda T)^{-1}T \, d\lambda$$

is a quasinilpotent right inverse of $D^2 + AD + B$. The initial value problem (2) has the unique solution

$$x = \frac{1}{2\pi i} \int_\Gamma L^{-1}(\lambda)(I - \lambda T)^{-1}(Tf + z_1 + \lambda z_o + Az_o) \, d\lambda.$$

Here Γ is a simple closed contour, oriented counterclockwise, enclosing $\sigma(L)$.

<u>Theorem 3.</u> Let X be a complex Hilbert space. Suppose that A and B are selfadjoint and that stationary operators Z_1, Z_2 exist with following properties:

$$Z_k^2 + AZ_k + B = 0, \quad (k = 1, 2)$$

$$\sigma(L) = \sigma(Z_1) \cup \sigma(Z_2), \quad \sigma(Z_1) \cap \sigma(Z_2) = \emptyset, \quad \sigma(Z_2) = \sigma(Z_2^*).$$

Then

$$(I + AT + BT^2)^{-1}T^2 = ((I - Z_1T)^{-1} - (I - Z_2T)^{-1})(Z_1 - Z_2)^{-1}T.$$

Further, the initial value problem (2) has the unique solution

$$x = ((I - Z_1T)^{-1} - (I - Z_2T)^{-1})(Z_1 - Z_2)^{-1}Tf$$
$$+ (I - Z_1T)^{-1}(Z_1 - Z_2)^{-1}(z_1 - Z_2z_o)$$
$$- (I - Z_2T)^{-1}(Z_1 - Z_2)^{-1}(z_1 - Z_1z_o).$$

The proofs and generalizations of some of these results can be found in [7]. Observe that the initial value problem (2) is a generalization of the equation of vibration.

1.3. If A isn't invertible in L(X), then (1) is called degenerate. For sake of simplicity, we consider the degenerate initial value problem

$$(AD + I)x = f, \quad Fx = z_o \tag{3}$$

with given $f \in X$ and $z_o \in \ker D$. We suppose that the ascent and descent of A are both finite. Then they are equal $d \geqq 1$. In this case A has a unique Drazin inverse $A^- \in L(X)$ of index d ([2]), i.e.

$$A^- = A^-A\,A^-, \quad A^-A = A\,A^-,$$
$$A^d(I - A\,A^-) = 0, \quad A^{d-1}(I - A\,A^-) \neq 0.$$

Let A^- be stationary. Observe that the ascent and descent of A are both finite, if $I - A$ is compact.

Theorem 4. Let $f \in \mathfrak{D}(D^d)$ and

$$x_1 = \sum_{k=1}^{d} (-1)^{k-1} A^{k-1} (I - A\,A^-) D^{k-1}f.$$

Then

$$\sum_{k=1}^{d} (-1)^{k-1} A^{k-1} (I - A A^-) D^{k-1} + (I + A^-T)^{-1} A^-T$$

is a unbounded right inverse of $AD + I$. The degenerate initial value problem (3) is solvable if and only if

$$(I - A A^-)z_0 = Fx_1.$$

Under this condition, the unique solution of (3) is given by

$$x = x_1 + (I + A^-T)^{-1} A^-Tf + (I + A^-T)^{-1} A A^- z_0.$$

We see that degenerate initial value problems are ill-posed. More general results can be found in [7].

2. Constructions of orthogonal generalized inverses

2.1. Let X be a complex separable Hilbert space and $D: \mathfrak{D}(D) \to X$ a linear closed operator with dense domain $\mathfrak{D}(D) \subset X$. Assume that D is selfadjoint with pure point spectrum. Let λ_k, $(k=0,1,\ldots)$ be its eigenvalues (inclusively their multiplicities) with $|\lambda_k| \leq |\lambda_{k+1}|$ and $|\lambda_k| \to \infty$ as $k \to \infty$, and let $\{x_k;\ k=0,1,\ldots\}$ be the corresponding complete system of orthonormal eigenvectors. Further, it holds

$$\mathfrak{D}(D) = \{x \in X;\ \sum_{k=0}^{\infty} |\lambda_k|^2 |(x,x_k)|^2 < \infty\}.$$

Examples. a) Let $X = L_2(0,1)$, and let $\mathfrak{D}(D)$ be the set of all absolutely continuous functions $x(t) \in X$ with $x'(t) \in X$ and $x(0) = x(1)$. Then $D = i\,\frac{d}{dt}: \mathfrak{D}(D) \to X$ is selfadjoint with pure point spectrum $\{\lambda_k = 2k\pi;\ k \text{ integer}\}$. Corresponding orthonormal eigenfunctions are $x_k(t) = e^{-2k\pi i t}$.

b) Chebyshev operational calculus ([5])

c) Dyadic Walsh calculus ([1]) and dyadic Haar calculus ([4])

d) Boundary value problems for second-order differential operators (of Sturm - Liouville-, Hermite-, Legendre-, Laguerre- or Bessel-type) ([6], pp. 287 - 369)

e) Boundary value problems for second-order elliptic differential operators ([6], pp. 387 - 400)

Under our assumptions, $D - \mu: \mathfrak{D}(D) \to X$ can be represented for any complex μ in the form

$$(D - \mu)x = \sum_{k=0}^{\infty} (\lambda_k - \mu)(x,x_k)\, x_k$$

for all $x \in \mathfrak{D}(D)$. By

$$(D - \mu)^+ x = \sum_{k=0}^{\infty}{}' (\lambda_k - \mu)^{-1}(x,x_k)\, x_k \tag{4}$$

for all $x \in X$ we introduce a completely continuous operator $(D - \mu)^+$: $X \to X$. The series $\sum'$ is defined as the series of all elements with non-vanishing denominator. It is easy to check that $(D - \mu)^+$ is the inverse of $D - \mu$ in the case $\mu \neq \lambda_k$ for all k. Otherwise, $(D - \mu)^+$ is the orthogonal generalized inverse of $D - \mu$ ([2]). Therefore, $(D - \mu)^+$ is called <u>generalized resolvent</u> of D. Further,

$$\|(D - \mu)^+\| = \sup\{|\lambda_k - \mu|^{-1};\ \mu \neq \lambda_k,\ k=0,1,\dots\}.$$

Let D_1 be an extension of D. We denote by $F_1 = I - D^+D_1$. By the help of the well-known Taylor formula of operational calculus ([3], [7]) one obtains

$$x = \sum_{k=0}^{p-1} (D^+)^k F_1 D_1^k x + (D^+)^p D_1^p x$$

for all $x \in \mathfrak{D}(D_1^p)$, (p=1,2,...) and therefore following general method of accelerating convergence of eigenvector expansions ([8]):

<u>Theorem 5.</u> Under these assumptions, every $x \in \mathfrak{D}(D_1^p)$, (p=1,2,...) can be represented in the form

$$x = \sum_{k=0}^{p-1} (D^+)^k F_1 D_1^k x + \sum_{k=0}^{\infty}{}' \lambda_k^{-p}(D_1^p x, x_k)\, x_k$$

and it holds the error estimation ($\lambda_{n+1} \neq 0$)

$$\Big\|x - \sum_{k=0}^{p-1} (D^+)^k F_1 D_1^k x - \sum_{k=0}^{n}{}' \lambda_k^{-p}(D_1^p x, x_k)\, x_k\Big\| \leq |\lambda_{n+1}|^{-p}\, \|D_1^p x\|.$$

More details and applications of this method in numerical analysis can be found in [8], [9].

2.2. Throughout this section, we assume now that

$$\sum_{k=0}^{\infty}{}' |\lambda_k|^{-2} < \infty . \qquad (5)$$

Then D^+ is a Hilbert - Schmidt operator. It is useful to introduce a multiplication in X, defined by

$$x * y = \sum_{k=0}^{\infty} (x,x_k)(y,x_k)\, x_k \qquad (6)$$

for any $x,\ y \in X$. Evidently, this multiplication is continuous. We call (6) a convolution. For complex μ we consider the element

$$e_\mu = \sum_{k=0}^{\infty}{}' (\lambda_k - \mu)^{-1} x_k$$

which is called the exponential element of D. According to (6), we get

$$(D - \mu)^+ x = e_\mu * x$$

for all $x \in X$. In particular,

$$D^+ x = e_0 * x.$$

In the case $0 < |\mu| < \|D^+\|^{-1}$, one obtains the expansion

$$e_\mu + \sum_{\substack{k=0 \\ (\lambda_k=0)}}^{\infty} \mu^{-1} x_k = \sum_{j=0}^{\infty} \mu^j (D^+)^j e_0 .$$

This means that the left side of the last equation is the "generating function" for the set $(D^+)^j e_0$, $(j=0,1,\dots)$.
If e_μ^n is defined as

$$e_\mu^1 = e_\mu, \quad e_\mu^{n+1} = e_\mu^n * e_\mu, \ (n=1,2,\dots),$$

it follows that

$$e_\mu^n = \sum_{k=0}^{\infty}{}' (\lambda_k - \mu)^{-n} x_k .$$

If $\mu \neq \lambda_k$ for all k, then e_μ has derivatives of all orders with

$$e_\mu^n = \frac{1}{(n-1)!} \frac{d^{n-1}}{d\mu^{n-1}} e_\mu$$

for n=1,2,... . As immediate consequence of (4) and (6), we obtain:

Theorem 6. Under our assumptions, it holds

$$((D-\mu)^+)^n x = e_\mu^n * x$$

for all $x \in X$ and n=1,2,... .

2.3. Let p(s) be the complex polynomial of degree $n \geqq 2$

$$p(s) = s^n + \alpha_{n-1}s^{n-1} + \dots + \alpha_0 = \prod_{j=1}^{r} (s-\mu_j)^{n_j}$$

with pairwise different roots μ_j. We decompose $p(s)^{-1}$ into vulgar fractions

$$p(s)^{-1} = \sum_{j=1}^{r} \sum_{k=1}^{n_j} \alpha_{jk}(s-\mu_j)^{-k}. \tag{7}$$

Let $\{k_1, k_2, \dots, k_m\}$ be the set of all k with $\lambda_k = \mu_j$ for some j= 1,2,...,r.

It is the aim to give simple representations of the orthogonal generalized inverse of p(D). It is easy to prove that the operator $p(D)^+: X \to X$

$$p(D)^+x = \sum_{k=0}^{\infty}{}' \frac{(x,x_k)}{p(\lambda_k)} x_k \tag{8}$$

has the following properties:

$$p(D)^+p(D)x = x - \sum_{i=1}^{m} (x,x_{k_i}) x_{k_i},$$

$$p(D)p(D)^+p(D)x = p(D)x$$

for all $x \in \mathfrak{D}(D^n)$ and

$$p(D)p(D)^+x = x - \sum_{i=1}^{m} (x,x_{k_i}) x_{k_i}, \tag{9}$$

$$p(D)^+p(D)p(D)^+x = p(D)^+x$$

for all $x \in X$. Hence, $p(D)^+$ is the orthogonal generalized inverse of p(D) ([2]). By definition of $p(D)^+$ and by (6), we get the equation

$$p(D)^+x = p * x$$

with

$$p = {\sum_{k=0}^{\infty}}' \frac{1}{p(\lambda_k)} x_k. \tag{10}$$

Now we extend the notion of exponential element and introduce the elements

$$e_{\mu_j}(\mu_1,\dots,\mu_r) = \sum_{\substack{k=0 \\ k \neq k_1,\dots,k_m}}^{\infty} (\lambda_k - \mu_j)^{-1} x_k, \quad (j=1,2,\dots,r).$$

The connection with the corresponding exponential element can be described by

$$e_{\mu_j}(\mu_1,\dots,\mu_r) = e_{\mu_j} - {\sum_{i=1}^{m}}' (\lambda_{k_i} - \mu_j)^{-1} x_{k_i}.$$

From (7), (10) and (11), we obtain the result

$$p = \sum_{j=1}^{r} \sum_{k=1}^{n_j} \alpha_{jk}\, e^k_{\mu_j}(\mu_1,\dots,\mu_r)$$

with

$$e^1_{\mu_j}(\mu_1,\dots,\mu_r) = \sum_{\substack{k=0 \\ k \neq k_1,\dots,k_m}}^{\infty} (\lambda_k - \mu_j)^{-1} x_k.$$

<u>2.4.</u> Now we consider the equation

$$p(D)x = f \tag{12}$$

for given $f \in X$. This equation is solvable if and only if $f = p(D)p(D)^+f$, this means by (9)

$$(f, x_{k_i}) = 0, \quad (i=1,2,\dots,m). \tag{13}$$

Now we obtain following theorems from the properties of $p(D)^+$ ([2]):

<u>Theorem 7.</u> Under the assumptions (5) and (13), any solution of (12)

possesses the form

$$x = p(D)^{+}f + \sum_{i=1}^{m} \gamma_i x_{k_i} \tag{14}$$

with arbitrary complex γ_i. The element

$$\tilde{x} = p(D)^{+}f = \sum_{j=1}^{r} \sum_{k=1}^{n_j} \alpha_{jk}(e^{k}_{\mu_j}(\mu_1,\dots,\mu_r) * f) \tag{15}$$

is the unique solution of (12) with minimal norm, i.e., $\|x\| > \|\tilde{x}\|$ for any solution $x \neq \tilde{x}$.

An element $\bar{x} \in \mathcal{D}(D^n)$ is called a <u>least squares solution</u> of (12), if

$$\|p(D)\bar{x} - f\| = \inf\{\|p(D)x - f\|;\ x \in \mathcal{D}(D^n)\}.$$

Evidently, every solution is also a least squares solution.

<u>Theorem 8.</u> Under the assumption (5), any least squares solution of (12) can be represented in the form (14). The element (15) is the unique least sqares solution of minimal norm, i.e., $\|x\| > \|\tilde{x}\|$ for any least sqares solution $x \neq \tilde{x}$.

If condition (5) isn't fulfilled, then exponential elements don't exist such that $(D - \mu)^{+}x$ and $p(D)^{+}x$ are only given by (4), (8) respectively.

References

[1] Butzer, P.L. and H.J. Wagner, Walsh - Fourier series and the concept of a derivative, Applicable Anal. 3 (1973), 29 - 46.

[2] Nashed, M.Z. and G.F. Votruba, A unified operator theory of generalized inverses, in: Generalized Inverses and Applications, edited by M.Z. Nashed, New York - San Francisco - London 1976, 1 - 109.

[3] Przeworska-Rolewicz, D., Algebraic theory of right invertible operators, Studia math. 48 (1973), 129 - 144.

[4] Splettstösser, W. and H.J. Wagner, Eine dyadische Infinitesimalrechnung für Haar-Funktionen, Z. Angew. Math. Mech. 57 (1977), 527 - 541.

[5] Stens, R., Beste algebraische Approximation, fraktionierte Ableitungen und Lipschitzklassen im Rahmen der Chebyshevtransformation, Diss. TH Aachen 1976.

[6] Triebel, H., Höhere Analysis, Berlin 1972.

[7] Tasche, M., Funktionalanalytische Methoden in der Operatorenrechnung, Nova Acta Leopoldina (N.F.) Nr. 231, Bd. 49 (1978).

[8] Tasche, M., Zur Konvergenzbeschleunigung von Fourierreihen, Math. Nachr. (in print).

[9] Tasche, M., Eine einheitliche Herleitung verschiedener Interpolationsformeln mittels der Taylorschen Formel der Operatorenrechnung, Z. Angew. Math. Mech. (in print).

NEW APPROACH TO THE SOLUTION OF THE INTEGRAL EQUATION OF FIRST KIND

D. ZIDAROV [+)]

1. Introduction

We shall treat the numerical solution of the integral equation of 1st kind

$$(1) \qquad U(t) = \int_{Q_0} A(x,y,t)\; z(x,y)\; dxdy$$

or

$$U = A Z$$

where Q_0 is a rectangle on the plane x,y defined by the inequalities $a < x < b$, $c < y < d$, the function $A(x,y,t)$ is known, $U(t)$ is the function of the data, while $z(x,y)$ is the unknown function. We shall treat the case when there are infinite number of solutions of (1).

Lavrentiev [1-3] eliminates the ambiguity of the solution of (1) applying the equation

$$U = A Z + E Z,$$

where E is the unit operator. Tichonov [4-6] replaces (1) by the "regulated" equation

$$U(t) = \int_{Q_0} A(x,y,t)\; z(x,y)\; dxdy + \alpha\, \Omega(z, \frac{\partial z}{\partial x}, \frac{\partial z}{\partial y}),$$

where Ω is an additional "smoothing" expression, $\alpha > 0$.

The solution of (1) which is going to be suggested represents a generalization of the corresponding solution of the inverse gravimetric problem. That is why we shall use the gravimetric terminology [7-18]: the function $z(x,y)$ will be called "density" of the masses, the expression $\int_{\Delta S} z(x,y)\, dxdy$ means the "quantity of the masses" on the square ΔS; each region (or set of regions) S, in the points of which the density function is defined, will be called "body" and be denoted with τ; the given function $U(t)$ will be called "field". If

---

+) Geophysical Institute
Sofia 1000 / Bulgaria
6 Moskovska Str.

the region S and the density z(x,y) satisfy the equation

$$U(t) = \int_S A(x,y,t)\, z(x,y)\, dxdy ,$$

we shall say that τ represents a body-solution of the formulated inverse problem. Instead of S we shall use the equivalent denotation $\check{\tau} \equiv S$.

Theorem 1 [7,8]. If the kernel A(x,y,t) of equation (1) is an analytical function of t, then the necessary and sufficient condition for the bodies τ_1 and τ_2 with densities $z_1(x,y)$ and $z_2(x,y)$ to be solution of the equation (1) is the equality of their moments:

$$(2) \qquad \int_{\check{\tau}_1} \left(\frac{\partial^n A}{\partial t^n} \right)_{t=0} z_1(x,y)\, dxdy = \int_{\check{\tau}_2} \left(\frac{\partial^n A}{\partial t^n} \right)_{t=0} z_2(x,y)\, dxdy,$$

$$n = 0,\ 1, \ldots .$$

The expressions (2) follow from the uniqueness of the development of A(x,y,t) in Mc Loren's series with respect to t. On the contrary, on the base of (2) multiplying with $\frac{1}{n!} t^n$ and summing with respect to n we shall get

$$\int_{\check{\tau}_1} A(x,y,t)\, z_1(x,y)\, dxdy = \int_{\check{\tau}_2} A(x,y,t)\, z_2(x,y)\, dxdy.$$

Thus, theorem 1 is proved.

We shall note that at n = 0 we get from (2)

$$(3) \qquad \int_{\check{\tau}_1} z_1(x,y)\, dxdy = \int_{\check{\tau}_2} z_2(x,y)\, dxdy ,$$

i.e. (in gravimetric terminology) the masses of the two body-solutions are equal. Equation (2) (especially (3)) can be applied as conditional equation after Lagrange, when the function z(x,y) is determined by minimization of the expression

$$\int_{t_1}^{t_2} \left[U(t) - \int_{Q_0} A(x,y,t)\, z(x,y)\, dxdy \right]^2 dt .$$

2. Direct Solutions [12-16,18].

These are solutions of (1) in case one searches a body-solution at given density $z_0(x,y)$ in advance (in the sense that $z_0(x,y)$ is defined in all points belonging to Q_0). We suppose that $z_0(x,y)$ can be developed in Mc Loren's series when the origin of the coordinate system is in the centre of Q_0. Besides, we suppose that there is a solution of (1) which is equal to $z_0(x,y)$ in the neighbourhood of the point O(0,0) [14,16,18]. We replace the unknown function z(x,y) by

$$\tilde{z}(x,y) = \sum_{\alpha=0}^{1} \sum_{\beta=0}^{1} \sum_{l=0}^{l_0} \sum_{m=0}^{m_0} a_{\alpha\beta lm} f_{\alpha\beta lm}(x,y) ;$$

$$f_{\alpha\beta lm}(x,y) = \cos\left(\alpha\frac{\pi}{2} + \frac{2\pi lx}{L_0}\right) \cos\left(\beta\frac{\pi}{2} + \frac{2\pi my}{M_0}\right) ,$$

$$L_0 = b - a,\ M_0 = c - d$$

and form the functional

$$\begin{aligned} W = & \int_{t_1}^{t_2} \left[U(t) - \int_{Q_0} A(x,y,t)\, \tilde{z}(x,y)\, dxdy\right]^2 dt \ + \\ & + \lambda \int_{Q_0} R^4\, \tilde{z}(x,y)\, dxdy + \mu \sum_{p=0}^{l_0+1} \sum_{q=0}^{m_0+1} \left(\frac{\partial^{p+q}(\tilde{z}-z_0)}{\partial x^p\, \partial y^q} \right)_0 \\ & + \nu \sum_{p=0}^{2(l_0+1)} \sum_{q=0}^{2(m_0+1)} \lambda_{pq}^2 \\ = & W_0 + \lambda W_1 + \mu W_2 + \nu W_3 , \end{aligned} \tag{4}$$

where λ, μ, ν are positive numbers, R is the distance between the points (x,y) and O(0,0); we suppose that the point O(0,0) belongs to the unknown body τ, see Fig. 1.

The expression W_1 is introduced into (4) in order to "force" the masses of the unknown body τ to concentrate maximum upon point O(0,0), the expression W_2 in order to "force" the density $\tilde{z}(x,y)$ to tend to $z_0(x,y)$ in the neighbourhood of O(0,0) at $l_0 \to \infty$ and $m_0 \to \infty$; the role of the expression W_3 will become clear further on. The points of the transition of $\tilde{z}(x,y)$ from zero to $z_0(x,y)$ will represent the boundary points of the unknown body τ.

We introduce suitable relations between the indeces $\alpha\beta lm$ and pq and denote $a_{\alpha\beta lm}$, λ_{pq}, $f_{\alpha\beta lm}$, and $\int_{Q_0} R^4 f_{\alpha\beta lm} f_{\alpha'\beta' l'm'}\, dxdy$ with X_s, Y_s, f_s, and $E_{ss'}$, $s,s' = 1,\dots,4(l_0+1)(m_0+1)$. Differentiating W with respect to X_s and Y_s and equalizing to zero the obtained derivatives, we get the following linear system

$$\left\| \begin{array}{ccc|ccc} H_{11}+\lambda E_{11} & \cdots & H_{1\varepsilon}+\lambda E_{1\varepsilon} & B_{11} & \cdots & B_{1\varepsilon} \\ \vdots & & \vdots & \vdots & & \vdots \\ H_{\varepsilon 1}+\lambda E_{\varepsilon 1} & \cdots & H_{\varepsilon\varepsilon}+\lambda E_{\varepsilon\varepsilon} & B_{\varepsilon 1} & \cdots & B_{\varepsilon\varepsilon} \\ \hline B_{11} & \cdots & B_{1\varepsilon} & \nu\ 0 \cdots & & 0 \\ & & & 0\ \nu \cdots & & 0 \\ \vdots & & \vdots & \vdots & \ddots & \vdots \\ B_{1\varepsilon} & \cdots & B_{\varepsilon\varepsilon} & 0\ 0 \cdots & & \nu \end{array} \right\| \cdot \left\| \begin{array}{c} X_1 \\ \vdots \\ X_\varepsilon \\ Y_1 \\ \vdots \\ Y_\varepsilon \end{array} \right\| = \left\| \begin{array}{c} D_1 \\ \vdots \\ \\ \\ \\ D_{2\varepsilon} \end{array} \right\|$$

or

$$\alpha Z = D$$

where

$$Z = \left\| \begin{array}{c} Z_1 \\ \vdots \\ Z_{2\varepsilon} \end{array} \right\|, \quad D = \left\| \begin{array}{c} D_1 \\ \vdots \\ D_{2\varepsilon} \end{array} \right\|, \quad Z_s = X_s,\ Z_{\varepsilon+s} = Y_s,\ s = 1,\dots,\varepsilon .$$

Let us examine the function $W^I = W_0 + \lambda W_1$. The second varial $\delta^2 W_0$ of W_0 is positive or equal to zero. It can be proved that the second varial of W_1 is positive [16,18]. Consequently $\delta^2 W^I > 0$ and the determinant of the matrix

$$\alpha_{11} = \left\| \begin{array}{ccc} H_{11}+\lambda E_{11} & \cdots & H_{1\varepsilon}+\lambda E_{1\varepsilon} \\ \vdots & & \vdots \\ H_{\varepsilon 1}+\lambda E_{\varepsilon 1} & \cdots & H_{\varepsilon\varepsilon}+\lambda E_{\varepsilon\varepsilon} \end{array} \right\|$$

is different from zero. It can be proved furthermore that the determinant of the matrix

$$\alpha_{12} = \left\| \begin{array}{ccc} B_{11} & \cdots & B_{1\varepsilon} \\ \vdots & & \vdots \\ B_{\varepsilon 1} & \cdots & B_{\varepsilon\varepsilon} \end{array} \right\|$$

is equal to zero. From this it follows that $\det \alpha = \nu^\varepsilon \det \alpha_{11} \neq 0$, i.e. the formulated problem has a unique solution under the introduced constraining conditions. If at least one solution of (1) exists, then at $\lambda, \mu, \nu \to 0$ and $\varepsilon \to \infty$ the function $\tilde{z}(x,y)$ will tend to a solution of the equation (1). This solution coincides with the given function $z_0(x,y)$ in the neighbourhood of the point $O(0,0)$; the masses of the corresponding body τ are concentrated maximum upon the point $O(0,0)$. We obtain an analogous "point" variant of the solution of (1) in the following way. We divide the rectangle Q_0 into equal squares ΔS_k using two families of lines parallel to the axis x and y; the distance between two neighbouring lines in the corresponding family is Δl. We denote the coordinates of the centre of the square ΔS_k with (x_k, y_k) and the masses on the same square with $\int_{\Delta S_k} z(x,y)\,dxdy$, see Fig. 2.

The set of all these points (x_k, y_k) belonging to Q_0 will be called ground netting point set and be denoted with $\breve{T}_0$. We shall call the set of point masses which we shall obtain when we put point masses μ_k in all points of $\breve{T}_0$ ground netting point set of masses; we shall denote this set of point masses with T_0. Each set $\breve{T}$ which represents a subset of $\breve{T}_0$ will be called netting point set. When point masses m_k are put in the points (x_k, y_k) of $\breve{T}$, we shall say that a netting set of point masses or a netting point body is defined and denote it with T. The sets $\breve{T}$ and $\breve{T}_0$ (or T and T_0) belong to one and the same netting and they are called co-netting sets. The four points $(x_k+\alpha\Delta l, y_k+\beta\Delta l)$, $\alpha = \pm 1$, $\beta = \pm 1$, $\alpha^2+\beta^2 = 1$, are called neighbouring points of the point (x_k, y_k). The points of $\breve{T}$ (or of T) which have at least one neighbouring point not belonging to $\breve{T}$ are boundary points of $\breve{T}$; the rest of the points of $\breve{T}$ are interior points of $\breve{T}$. The set of the interior points of $\breve{T}$ (or of point masses of T) will be denoted by $]\breve{T}[$ (or $]T[$).

The integral U(t) can be replaced by the sum of the "individual contributions"

$$(5) \qquad \Delta U_k(t) = \int_{\Delta S_k} A(x,y,t)\, z(x,y)\, dxdy \approx A(x_k, y_k, t)\, m_k$$

of the squares ΔS_k:

$$U(t) = \sum_k \int_{\Delta S_k} A(x,y,t)\, z(x,y)\, dxdy \approx \sum_k A(x_k, y_k, t)\, m_k \,.$$

Substituting the surface distribution z(x,y) of the masses by point masses distribution m_k we make a general error

$$(6) \qquad \sigma_1 = N_{x0}\, N_{y0}\, \Delta A\, \Delta z\, \Delta l^2 ,$$

where $N_{x0} = \frac{L_0}{\Delta l}$ is the number of squares ΔS_k in a horizontal row of Q_0, $N_{y0} = \frac{M_0}{\Delta l}$ is the number of squares in a vertical column of Q_0, while ΔA and Δz are the maximal values of the differences $|A(x_k, y_k, t) - A(x,y,t)|$ and $|z(x_k, y_k) - z(x,y)|$, $(x,y) \in \Delta S_k$.

We denote the values of $U(t)$ in the points t_l, $l = 1,\dots,l_t$, with U_l and the values of $A(x_k,y_k,t_l)$ with A_{lk}. Then the unknown masses $m_k = \int_{\Delta S_k} z(x,y)\,dxdy$ will be determined by the solution of the linear system

$$(7) \qquad \sum_{k=1}^{n} A_{lk}\, m_k = U_l \;, \quad l = 1,\dots,l_t > n \;,$$

where $n = N_{x0}N_{y0}$.

In case $\det\{A_{lk}\}$ is equal to zero, it is necessary to introduce additional constraining conditions in order to get a unique solution of (7). A suitable (for the further considered transformation of the masses) approximate solution of (7) can be obtained by minimizing the function

$$(8) \qquad \dot{W} = \sum_{l=1}^{l_t} \Big[\sum_{k=1}^{n} A_{lk}\, m_k - U_l\Big]^2 + \lambda \sum_{k=1}^{n} m_k{}^2 + \mu \sum_{k \neq i} R_{ik}{}^2 + \nu \exp\Big(b \sum_{k=1}^{n} \varphi_k\Big) \;,$$

where $\varphi_k = A(-a-m_k)(a-m_k)(m_{k_0}-a-m_k)(m_{k_0}+a-m_k)$, $A > 0$, $b > 0$, while R_{ik} is the distance between the points (x_k,y_k) and a certain point (x_i,y_i), $i \neq k$.

Let us put $\mu = \nu = 0$ in (8) and suppose that at least one of the matrices $\begin{Vmatrix} A_{1i} \\ \vdots \\ A_{l_t i} \end{Vmatrix}$ has rank equal to 1. We substitute the sum $\sum_{k=1}^{n} m_k{}^2$ by $\sum_{k \neq i} m_k{}^2$. In this case it is proved that W has one and only one minimum and the maximal quantity of masses is concentrated in the chosen point (x_i,y_i); in the points (x_k,y_k), $k \neq i$, the masses are the possible least ones. If we put $\lambda = 0$ and $\mu > 0$, $\nu > 0$ in (8), then the expression $\exp(b \sum_{k=1}^{n} \varphi_k)$ will "force" the point masses m_k to get into one of the intervals $(-a,a)$ and $(m_{k_0}-a, m_{k_0}+a)$, where m_{k_0} are numbers given in advance; the functions $\exp(\varphi_k)$ obtain their least values only, when the numbers m_k are in these intervals. At the same time the presence of the sum $\sum_{k \neq i} R_{ik}{}^2$ in (8) will concentrate the masses m_k, $k \neq i$, upon the point (x_i,y_i).

If we only put $\nu=0$ in (8) (λ and μ are positive) we get a solution at which the sum $\sum_{k=1}^{n} m_k^2$ will be the least possible and the masses m_k will be situated upon the point (x_i,y_i) so that the sum $\sum_{k\neq i} R_{ik}^2$ will become the least possible.

3. Derivative Solutions [7,8,11,16,18].

For a lot of problems not only one solution of (1) has to be found but a family of solutions satisfying certain additional conditions. This makes us invent a method for constructing new solutions when one of them is known [7,8,11,18]. We suppose that we know one of the mentioned "point" solutions of (1). We shall use the already introduced division of the rectangle Q_0 into little squares ΔS_k, $k=1,\ldots,n$. After summing up the equations

$$
(9)\quad
\begin{aligned}
A(x_k+\Delta l,y_k,t)\ m_k &= m_k\Big[A(x_k,y_k,t)+\Delta l\Big(\frac{\partial A}{\partial x}\Big)_{\Delta l=0}+\frac{\Delta l^2}{2!}\Big(\frac{\partial^2 A}{\partial x^2}\Big)_{\Delta l=0}+\ldots\Big]\\
A(x_k-\Delta l,y_k,t)\ m_k &= m_k\Big[A(x_k,y_k,t)-\Delta l\Big(\frac{\partial A}{\partial x}\Big)_{\Delta l=0}+\frac{\Delta l^2}{2!}\Big(\frac{\partial^2 A}{\partial x^2}\Big)_{\Delta l=0}+\ldots\Big]\\
A(x_k,y_k+\Delta l,t)\ m_k &= m_k\Big[A(x_k,y_k,t)+\Delta l\Big(\frac{\partial A}{\partial y}\Big)_{\Delta l=0}+\frac{\Delta l^2}{2!}\Big(\frac{\partial^2 A}{\partial y^2}\Big)_{\Delta l=0}+\ldots\Big]\\
A(x_k,y_k-\Delta l,t)\ m_k &= m_k\Big[A(x_k,y_k,t)-\Delta l\Big(\frac{\partial A}{\partial y}\Big)_{\Delta l=0}+\frac{\Delta l^2}{2!}\Big(\frac{\partial^2 A}{\partial y^2}\Big)_{\Delta l=0}+\ldots\Big]
\end{aligned}
$$

and neglecting the terms of second degree and even higher degrees with respect to Δl, we obtain

$$
\begin{aligned}
\Delta U_k(t) \approx A(x_k,y_k,t)\ m_k \approx \frac{m_k}{4}\Big[&A(x_k,y_k+\Delta l,t)+A(x_k,y_k-\Delta l,t)+\\
&+A(x_k+\Delta l,y_k,t)+A(x_k-\Delta l,y_k,t)\Big].
\end{aligned}
$$

It means that the expression (5) will not be varied if we distribute uniformly the mass put in point (x_k,y_k) in the neighbouring four points $(x_k+\alpha\Delta l,y_k+\beta\Delta l)$, $\alpha=\pm1$, $\beta=\pm1$, $\alpha^2+\beta^2=1$.

Let us suppose that we want to obtain another netting point body $T'=\sum_k m_k'$ from a given netting point body $T=\sum_k m_k^{(0)}$ almost equivalent to T in a sense that the sum $\sum_k A(x_k,y_k,t)\ m_k^{(0)}$ is almost equal to $\sum_k A(x_k,y_k,t)\ m_k'$. We shall denote the initial masses of T with $m_k^{(0)}$, the masses of T during the transformation of its masses with m_k, and the final masses of T (the final values of m_k) with m_k'. We shall suppose that in the points (x_k,y_k) of $\check{T}_0$ the values of the masses μ_k, by which we want to substitute the initial masses $m_k^{(0)}$ of T, are defined in advance, i.e. the ground netting set T_0 of point masses μ_k is given; $\mu_k < m_k^{(0)}$ in the points (x_k,y_k) belonging to the initial body T. We shall construct the substituting body T' in the following way. At the point (x_k,y_k) belonging to T we form the difference $m_k^{(0)}-\mu_k$ (this difference is positive by condition) and distribute it uniformly in the four neighbouring points $(x_k+\alpha\Delta l, y_k+\beta\Delta l)$, $\alpha = \pm 1$, $\beta = \pm 1$, $\alpha^2+\beta^2 = 1$; a point mass equal to μ_k remains in the point (x_k,y_k). This procedure is carried out in all points (x_k,y_k) of $\check{T}_0$, where $m_k^{(0)} - \mu_k > 0$ or $m_k - \mu_k > 0$; if $m_k - \mu_k < 0$ the mass m_k remains the same. It is clear that during the transformation of the masses new masses arrive to the point (x_k,y_k), so that the difference $m_k - \mu_k$ will become positive again. We repeate the procedure of the mass transformation infinitely until we obtain $m_k = m_k' \leqslant \mu_k$ in all points of $\check{T}_0$. The set of point masses m_k' obtained in this way will represent the unknown body T', the substituting body of T. The used manner of construction of T' shows that T' is a least point body possessing the characteristic $m_k = m_k' \leqslant \mu_k$. This netting body T' will be called a derivative body of T at a given average density μ_k (or at a given ground netting point body T_0). Besides, we shall say that T' is obtained by partial transformation of the masses of T to the masses of T_0 (or to the masses with average density μ_k). Further on we shall show that the result of the partial mass transformation does not depend on the order of making the rounds of the points of $\check{T}_0$ (or, shortly: on the order of making the rounds of $\check{T}_0$). We shall consider at first the independence of the perfect transformation of the masses of a given netting point body T of the order of making the rounds of]T[. This transformation is obtained when we substitute the mass $m_k(0)$ (or m_k) of (x_k,y_k) by four equal masses $\frac{m_k}{4}$, put in the neighbouring points $(x_k+\alpha\Delta l, y_k+\beta\Delta l)$, $\alpha = \pm 1$, $\beta = \pm 1$, $\alpha^2 + \beta^2 = 1$, while the mass remains equal to zero in the point (x_k,y_k). During this procedure of perfect mass transformation in the point (x_k,y_k) new masses arrive and we must transform them again. Mass transformation cannot take place in the boundary points of T; only arriving of masses is possible in them. The procedure of perfect transformation of the masses of T to its boundary points is finished when there are no more masses in the interior points of T: all masses of T are accumulated in its boundary points.

Lemma 1. The mass m_{1k}, which will arrive from a point P_k interior of a given netting point body T in the boundary point $\overline{P}_1$ of T at perfect transformation of the interior masses of T and at given order of making the rounds of]T[, does not depend on the existence of other masses in the interior points of T; it is proportional to the initial mass $m_k^{(0)}$ in the point P_k.

In order to prove the lemma and further on we shall use the method of the "colouring" of the masses [17]. We "colour" the mass $m_k^{(0)}$ in blue and carry out each operation of perfect mass transformation of]T[in a way that the blue masses distribute uniformly in the corresponding neighbouring points. After that, the rest of the masses in the same point (x_k, y_k) (if there exist any of them) are also distributed uniformly in the neighbouring points of (x_k, y_k), so that the general quantity of masses in the point (x_k, y_k) is distributed uniformly in the corresponding neighbouring points. Thus, bearing in mind the given order of making the rounds of]T[, we shall obtain the same quantity blue masses in the point $\overline{P}_1$ which arrived from the point P_k as if other masses do not exist. Evidently, if we multiply the initial mass $m_k^{(0)}$ α times, the quantity m_{1k} of the mass which arrives in the point $\overline{P}_1$ from the point P_k will be multiplied α times, too. Thus, the lemma is proved.

Lemma 2. The quantity of the mass which arrives in each of the boundary points of the netting point body T at the perfect transformation of its interior masses does not depend on the order of making the rounds of]T[; this quantity is proportional to the initial quantity $m_k^{(0)}$ put in the points of]T[.

We "colour" the four newly obtained masses at the perfect transformation of the masses $m_k^{(0)}$ of T in four various colours. Thus, from $m_k^{(0)}$ at (x_k, y_k) we obtain four masses $\frac{m_k^{(0)}}{4}$ situated in the neighbouring points $(x_k+\alpha\Delta l, y_k+\beta\Delta l)$, $\alpha = \pm 1$, $\beta = \pm 1$, $\alpha^2 + \beta^2 = 1$, and denote their colours with $K(\alpha_1, \beta_1)$ (in fact, these denotations of the "colours" represent the indeces of the newly obtained masses). After that (supposing that all points $(x_k+\alpha\Delta l, y_k+\beta\Delta l)$ belong to]T[), we shall obtain 16 new masses in the points $(x_k+\alpha_1\Delta l+\alpha_2\Delta l, y_k+\beta_1\Delta l+\beta_2\Delta l)$, $\alpha_1 = \pm 1$, $\alpha_2 = \pm 1$, $\beta_1 = \pm 1$, $\beta_2 = \pm 1$, $\alpha_1^2 + \beta_1^2 = 1$, $\alpha_2^2 + \beta_2^2 = 1$, and denote their colours with $K(\alpha_1,\alpha_2,\beta_1,\beta_2)$. These masses will be obtained independently from the order of making the rounds of the masses coloured with $K(\alpha_1,\beta_1)$. The masses arriving in some of the boundary points $\overline{P}_1$ of T from the point (x_k,y_k) will have colours $K(\alpha_1,\alpha_2,\ldots,\alpha_q,\beta_1,\beta_2,\ldots,\beta_q)$, $\alpha_k = \pm 1$, $\beta_k = \pm 1$, $\alpha_k^2+\beta_k^2 = 1$, $k = 1,\ldots,q$, and their quantities will be equal to $m_k/4^q$. Each of these masses represents only a function of the indeces $K(\alpha_1,\ldots,\alpha_q,\beta_1,\ldots,\beta_q)$ but it is not a function of the order of transformation of the masses (or of the order of making the rounds of]T[). When we multiply m_k a times the arriving in the point $\overline{P}_1$ will be multiplied a times, too. Thus, the lemma is proved.

Lemma 3. In case we have stopped some quantity $\overline{m}_k$ of mass at a given point (x_k,y_k) of]T[from further transformation the result of the perfect transformation of the netting point body T to its boundary points does not depend on the moment at which this quantity is (or its portions are) stopped.

We suppose that we have perfectly transformed the masses of]T[to the boundary points of T in two ways and have stopped the quantity $\overline{m}_k$ of mass in the point (x_k,y_k). The point masses obtained in these two ways in the boundary points $\overline{P}_1$ of T are denoted with $m_{1_0}^{I}$ and $m_{1_0}^{II}$; at supposition at least one of the masses $m_{1_0}^{I}$ differs from the corresponding mass $m_{1_0}^{II}$. Further on additionaly we perfectly transform the point mass m_k and get new point masses m_{1k}^{I} and m_{1k}^{II} in the boundary points of T. According to lemma 2 $m_{1k}^{I} = m_{1k}^{II}$ and the sum $m_{1_0}^{I} + m_{1k}^{I}$ must be equal to $m_{1_0}^{II} + m_{1k}^{II}$. That means that $m_{1_0}^{I} = m_{1_0}^{II}$. Thus, lemma 3 is proved.

Lemma 4. The result of the perfect transformation of the masses $m_k^{(0)}$ of a given netting point body T to the boundary points of another co-netting point body T_1 which includes T ($T \subset T_1$) does not depend on the fact that at first we have perfectly transformed the masses of T to its boundary points and after that we have perfectly transformed the obtained masses to the boundary points of T_1, or we have perfectly transformed the masses of T directly to the boundary points of T_1.

In order to be close to the case which has been treated by lemma 2 we must suppose that our problem is the perfect transformation of the masses of a new body T^*. It is composed of the initial body T and zero point masses situated in those points of T_1 which do not coincide with the points of T. This gives us the right to consider the first perfect transformation of the masses of T to its boundary points as a temprorary ditaining of the corresponding boundary masses from further transformation. After that we continue the perfect transformation of these boundary masses to the boundary points of T_1. According to lemma 3 the result of these two manners (direct and indirect) of perfect transformation of the masses of T is the same. Thus, lemma 4 is proved.

Theorem 2. Let us suppose that we have partially to transform the masses of a given netting point body $T = \sum m_k$ to the masses of a given ground co-netting point body $T_0 = \sum \mu_k$ and that $\mu_k < m_k$ in the points belonging to T. The result of this mass transformation (the newly obtained point body $T' = \sum m_k'$, $m_k' \leqslant \mu_k$) does not depend on the order of making the rounds of $\breve{T}_0$.

As a matter of fact the problem consists in the perfect transformation of the surplus masses $m_k - \mu_k$ in all points at which the mass transformation procedure has been reached if the difference $m_k - \mu_k$ is positive. To this end we form a new point set $\breve{T}_1$ adding its neighbouring points to the body T. Further on we perfectly transform the masses $m_k^{(0)}$ of T (the initial masses of T are denoted with $m_k^{(0)}$) to the boundary points of $\breve{T}_1$. If all newly obtained masses in the boundary points of $\breve{T}_1$ are less than the corresponding masses μ_k of T_0 then according to lemma 3 we can say that the result of this mass transformation is unique, and the theorem is proved. We consider the contrary case: in some boundary points of $\breve{T}_1$ the newly obtained masses m_k are greater than the corresponding masses μ_k of T_0; we denote the set of these points with $\breve{M}_1$. Evidently, m_k is the least quantity of mass which would arrive in the corresponding point of $\breve{M}_1$ when any masses do not remain in the points of T. At all other mass transformations at which the mass transformation procedure has taken place in points not belonging to T and, at the same time, no masses have remained in the points of T a greater quantity of masses would arrive in the points of $\breve{M}_1$ (greater than in case the procedure of perfect mass transformation has only taken place in the points of T). This means that when there is another manner of mass transformation (besides the accepted, to the boundary points of $\breve{T}_1$) a sufficient quantity of masses being greater than the corresponding masses μ_k of T_0 would arrive in the points of $\breve{M}_1$. We detain the further transformation of the necessary quantity μ_k in the points of $\breve{M}_1$. We perfectly transform the surplus masses $m_k - \mu_k$ to the boundary points of the point set $\breve{T}_2$ which is formed by adding those neighbouring points of $\breve{T}_1$ which are at the same time neighbouring points of $\breve{M}_1$ to $\breve{T}_1$. In this way we shall reach the point set $\breve{T}_n$ in the boundary points of which the mass quantity is less than the corresponding quantity μ_k. Particularly, all points of $\breve{T}_n$ or some points of $\breve{T}_n$ can be boundary points of the ground co-netting point set $\breve{T}_0$. According to lemma 3 the result of this mass transformation (at detaining some of the interior point masses from further transformation) is unique. This result does not depend on the order of making the rounds of $\breve{T}_0$. Thus, theorem 2 is proved.

Lemma 5. Let us consider the co-netting point bodies $T_1 = \sum m_{1k}$, $T_2 = \sum m_{2k}$, and the corresponding co-netting ground body $T_0 = \sum \mu_k$ supposing that $\mu_k < m_{1k}$ and $\mu_k < m_{2k}$ in the points belonging to T_1 or T_2, and $m_{1k} > 0$, $m_{2k} > 0$. Let us partially transform the masses of T_1 and T_2 separately to the masses of T_0 (supposing the rectangle Q_0 is rather great so that the mass transformation procedure would not reach the boundary points of T_0) and denote the obtained point bodies with A_1 and A_2 respectively. Let us partially transform the body $T_1 + T_2$ to the masses of T_0 and denote the obtained point body with A. If the initial point sets $\breve{T}_1$ and $\breve{T}_2$ have at least one common point or at least one of the boundary points of $\breve{T}_1$ has a boundary point of $\breve{T}_2$ as a neighbouring point then either the points of $\breve{A}_1 + \breve{A}_2$ are interior points of $\breve{A}$ or if some of the boundary points of $\breve{A}$ represent boundary points of $\breve{A}_1$ or $\breve{A}_2$ in each of these points the mass of A will be greater than the corresponding point mass of A_1 or A_2.

Lemma 5 is proved by colouring the masses of T_1 in blue while the masses of T_2 are coloured in red. The masses of $T_1 + T_2$ are transformed in such a way that the transformation of the blue masses is the same as at the separate transformation of the blue masses.

Lemma 6. Let us consider a given netting point body $T = \sum m_k$, $m_k > 0$. Let us partially transform the masses of T to the masses of the ground co-netting point body $T_0 = \sum \mu_k$ supposing that $\mu_k < m_k$ in the points belonging to T and denote the obtained derivative point body with $T' = \sum m_k'$, $m_k' \leqslant \mu_k$. The maximal differences $|x_k' - x_l'|$, $|y_p' - y_q'|$ between the coordinates of the points of T' and the maximal differences $|x_k - x_l|$, $|y_p - y_q|$ between the coordinates of the points of the initial body T satisfy the inequalities

$$(10_\alpha) \qquad |x_k' - x_l'| < |x_k - x_l| \frac{m}{\mu} + 2\,\Delta l,$$

$$(10_\beta) \qquad |y_p' - y_q'| < |y_p - y_q| \frac{m}{\mu} + 2\,\Delta l$$

where m is the maximal value of the masses m_k of T, μ is the minimal value of the masses μ_k of T' while Δl is the edge length of the squares of the used quadratic netting.

We form an auxiliary infinite point set $\breve{T}_{x,\infty}$ using the given ground point set $\breve{T}_0$ and making $a = -\infty$ and $b = +\infty$. Besides this, we suppose that at least one boundary point of $\breve{T}_{x,\infty}$ from the highest row of it (when y_k has the least value) coincides with some boundary points of T and at least one boundary point of $\breve{T}_{x,\infty}$ from the lowest row of it (when y_k has the greatest value) coincides with some boundary points of T. We suppose that the point set $\breve{T}$ belongs to $\breve{T}_{x,\infty}$. Now we form an auxiliary infinite point body $T_{x,\infty}$ putting point masses equal to m in the points of $T_{x,\infty}$.

Further on, we transform the masses of $T_{x,\infty}$ to the masses of the corresponding co-netting infinite ground point body $T_{x0,\infty}$ possessing equal point masses μ. Bearing in mind that the quantity of mass in one vertical column of the initial infinite body $T_{x,\infty}$ is equal to the quantity of the mass in the corresponding vertical column of the derivative body $T'_{x,\infty}$ (which will be obtained by transformation of $T_{x,\infty}$ to the masses of $T_{x0,\infty}$) we get the inequality

$$|y_p' - y_q'| < |y_p - y_q| \frac{m}{\mu} + 2\,\Delta l$$

where y_p' and y_p are the least ordinates of $T'_{x,\infty}$ and $T_{x,\infty}$, respectively, while y_q' and y_q are the greatest ordinates of the points of $T'_{x,\infty}$ and $T_{x,\infty}$, respectively. According to lemma 5, this inequality can be used for the derivative body T'. In the same way we can prove the other inequality. Thus, lemma 6 is proved.

4. Estimation of the Errors at the Mass Transformation [11].

The error σ made at the mass transformation represents a sum of the error σ_1 made at the transformation $\tau \longrightarrow T$, of the error σ_2 made at the transformation $T \longrightarrow T'$, and of the error σ_3 made at the transformation $T \longrightarrow b$ where b is a body with the density D(x,y) satisfying the equations $\mu_k = \int_{\Delta S_k} D(x,y)\, dxdy$. The error σ_1 is defined by (6). The error σ_3 can be estimated in an analogous way (similar to the error σ_1) keeping in view the inequalities (10_α) and (10_β):

$$\sigma_3 = (N_{x0} \frac{m}{\mu} + 2)\ (N_{y0} \frac{m}{\mu} + 2)\ \Delta A\ \Delta z\ \Delta l^2,$$

where $N_{x0} = \frac{|x_k - x_1|}{\Delta l}$, $N_{y0} = \frac{|y_k - y_1|}{\Delta l}$.

We denote the error which is made at the transformation of the unit point mass in the neighbouring points of it by σ_{20} and the common quantity of a transformed mass during all operations of mass transformation at the transformation $T \longrightarrow T'$ by M. Then, the error σ_2 will be equal to the product $\sigma_{20}M$.

The error σ_{20} is estimated by expressions of the remaining terms (after Lagrange) of the series (9):

$$\sigma_{20} < 4 \frac{\Delta l^2}{3!} \left(\frac{\partial^2 A}{\partial l^2} \right)_{max},$$

where $\left(\frac{\partial^2 A}{\partial l^2} \right)_{max}$ is the maximal value of the derivatives $\frac{\partial^2 A}{\partial x^2}$ and $\frac{\partial^2 A}{\partial l^2}$ in the points $(x_k, y_k) \in \check{T}_0$.

The mass M will be estimated by means of the common quantity of the transformed mass M_q for obtaining the quadratic co-netting point body T^* possessing $N = N_{10} \frac{m}{\mu} + 2$ points in each row and in each column (N_{10} is the greatest of the numbers N_{x0} and N_{y0}) and point masses equal to μ. The mass M_q is equal to $(N-2)(N-1)N\mu$ and therefore,

$$\sigma_2 < (N-2)(N-1)N\ \mu\ \sigma_{20} < \frac{2}{3} (N_{10} \Delta l)^3 \frac{m^3}{\mu^2} \left(\frac{\partial^2 A}{\partial l^2} \right)_{max} \Delta l.$$

Evidently, at $\Delta l \longrightarrow 0$ the common error $\sigma = \sigma_1 + \sigma_2 + \sigma_3$ will tend to zero as well.

5. Some Special Initial and Derivative Solutions [12,16,18].

1. Let us suppose that there exists a solution of the equation (1) in the chosen rectangle. This solution can be transformed into a point body and after that, the obtained point body can be transformed into a "simple point layer" situated on the boundary points of the corresponding co-netting ground body using the perfect mass transformation.

2. Let us change the rectangle Q_0 in such a way that $a \to -\infty$ and $b \to \infty$ and suppose that there exists a solution $z(x,y)$ of the equation (1) in this infinite "rectangle". The same solution can be transformed into two simple "point layers" situated on the lines $y_1 = c$ and $y_2 = d$, respectively, using the perfect mass transformation. The "fields" of these two simple layers will be approximately equal to the given "field" $U(t)$. If $d \to \infty$, the point masses of the simple layer situated on $y_2 = d$ will tend to zero. In this case, if $A(x,y,t)$ tends to zero sufficiently quick, the field $U(t)$ will be expressed by means of the field of one simple layer situated on the line $y_1 = c$.

Note. The properties (1. and 2.) of the solutions of (1) mentioned above are valied only when the unknown function z depends on at least two (or more) parameters.

3. If there exists a solution $z(x,y)$ of the equation (1) in the rectangle Q_0 (or between the lines $y_1 = c$ and $y_2 = d$), on the base of this solution we can obtain a special, "stratified", solution in the following way [13-16,18]. Let $d - c < b - a$. We form rectangles $Q_0, Q_1, \ldots, Q_{n-1}$ with the edge length equal to $l_{xk} = L_0 - k\Delta l$, $l_{yk} = M_0 - k\Delta l$, $L_0 = b - a$, $M_0 = d - c$, $k = 0,\ldots,n-1$, respectively, where $n = \frac{d-c}{2\Delta l}$ is an odd number. Each of the rectangles Q_k includes all rectangles Q_{k+p} possessing greater indeces. We form the sum

$$\overline{W}_l = \int_{t_1}^{t_2} \left[U(t) - \sum_{k=0}^{l-1} \int_{C_k} A(x,y,t)\, \zeta_k^{(\lambda)}(x,y)\, ds \right]^2 dt$$

$$\lambda \int_{C_{l-2}} (\zeta_{l-2}^{(\lambda)})^2 ds\ , \ \lambda > 0,$$

where ds is a linear element of the contour line C_k of Q_k. We put

$l = 2$ and determine the functions $\zeta_0^{(\lambda)}$ and $\zeta_1^{(\lambda)}$ by minimizing the function $\overline{W}_2$. After that, we put $l = 3$ and consider $\overline{W}_3$ as a function of $\zeta_1^{(\lambda)}$ and $\zeta_2^{(\lambda)}$ only. Continuing this way, we shall determine all functions $\zeta_k^{(\lambda)}$ till $\zeta_{n-2}^{(\lambda)}$ and $\zeta_{n-1}^{(\lambda)}$. The obtained body $\overline{\tau} = \sum_{k=0}^{n-1} \int_{C_k} \zeta_k^{(\lambda)} ds$ will be called a stratified body. The manner of construction of the stratified body allows us to assert:

a. at $\lambda \rightarrow 0$, the simple layer on the contour line C_k of Q_k only depends on the position of the following rectangle Q_{k+p},

b. at $\Delta l \rightarrow 0$,

$$\sum_{k=0}^{n-1} \int_{C_k} A(x,y,t) \zeta_k^{(\lambda)}(x,y) \, ds \longrightarrow U(t)$$

will be valied, and

c. at $\lambda \rightarrow 0$, each of the simple layers on the contour lines C_k will have a field $\Delta U_k(t)$ corresponding to the field of those masses of which cannot be transported on the contour line C_{k+1} of Q_{k+1} or into Q_{k+1}.

The stratified body can be used as an initial body-solution from which one can obtain all other solutions of the same equation.

6. Conclusions.

The suggested method for the transformation of mass distribution when one of them is known allows us to consider the improperly posed inverse problem in another aspect. Instead of looking for some solution of (1), we have to look for a family of solutions subordinated to additional conditions. These conditions depend on the physical considerations which we have in advance. For example, on the base of one homogeneous body-solution of the inverse gravimetric problem we can obtain the whole family of homogeneous body-solutions of the same problem using mass transformation method. Further on, we can choose one body of this family which corresponds mostly to other additional geophysical information. The mass transformation method can be applied as well for the solution of some multiparametrical linear equations arising from the numerical solution of the integral equations.

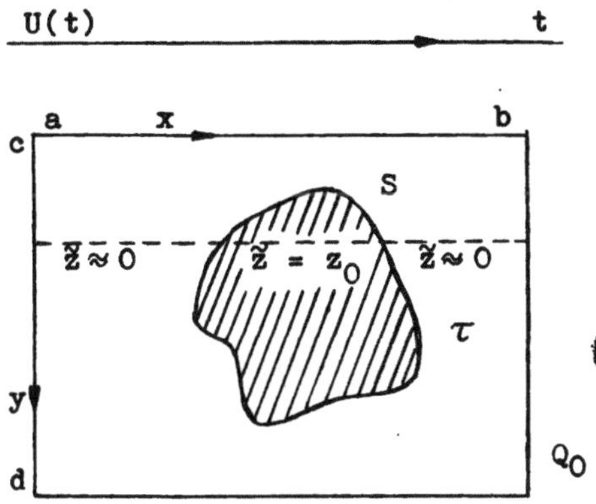

Fig. 1. Direct solution of the inverse problem
z_0 - the given function of the density; $\tilde{z}$ - the function of the density which has to be found; S - the points at which $\tilde{z}$ transfers from zero to $z_0(x,y)$.

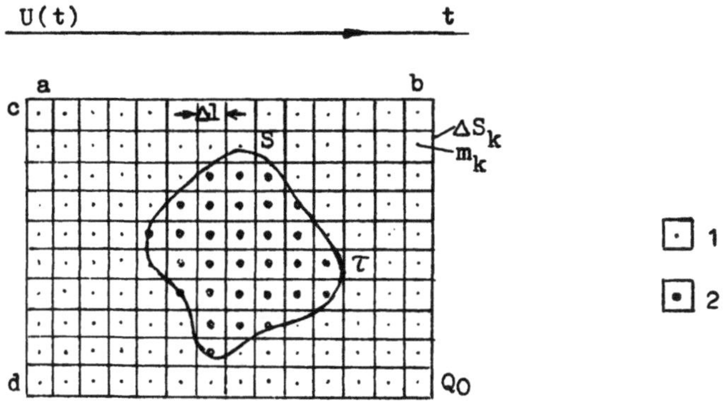

Fig. 2. The "point" variant of the direct solution
1. points at which the masses m_k satisfy the inequalities $-a < m_k < a$; 2. points at which the masses m_k satisfy the inequalities $m_{k0} - a < m_k < m_{k0} + a$, m_{k0} - numbers given in advance.

References

[1] Лаврентьев М.М.,1959,Об интегральных уравнениях первого рода, ДАН СССР,127,1.

[2] Лаврентьев М.М.,1960,Об интегральных уравнениях первого рода, ДАН СССР 133,2.

[3] Лаврентьев М.М.,1962,О некоторых некорректных задачах математической физики,Изд-во СО АН СССР.

[4] Тихонов А.Н.,1963, О решении некорректно поставленных задач, ДАН СССР 153,1.

[5] Тихонов А.Н.,О регуляризации некорректно поставленны задач, ДАН СССР,1 , 1963.

[6] Тихонов А.Н.,В.а.Арсенин, методы решения некорректных задач, Изд-во "Наука", Москва.

[7] Зидаров Д.П.,1968,О решении некоторых обратных задач потенциальных полеи и его применения к вопросам геофизики,Изд-во БАН,София (with an English summary).

[8] Зидаров Д.П.,1965,Решение на обратната гравиметрична,магнитна, и електрична задача,Докторска дисертация,Геофизичен институт на БАН,София.

[9] Zidarov, D. P.: A Solution of the Inverse Gravimetric (and Magnetic) Problem and its Application to Study of Earth Structure, Compt. Rend. Acad. Bulg. Sci., 17,9(1964).

[10] Zidarov, D.: Solution of Some Inverse Problem of Applied Geophysics, Geoph. Prosp., 13,2(1965).

[11] Zidarov, D. and Zhelev, Zh.: On Obtaining a Family of Bodies with Identical External Field-Method of Bubling, Geoph. Prosp., 18,1(1970).

[12] Zidarov, D.: Point (Dipole) Solution of the Inverse Potential Problem-Mass-Pressing Method, Compt. Rend. Acad. Bulg. Sci., 25,10(1972).

[13] Zidarov, D.: Characteristic Solution of the Inverse Gravimetric and Magnetic Problem, Compt. Rend. Acad. Bulg. Sci., 27,5(1974).

[14] Zidarov, D.: Solution of Inverse Gravimetric Problem by Determining the Density of the Unknown Masses, Compt. Rend. Acad. Bulg. Sci., 27,2(1974).

[15] Zidarov, D.: Reduction of Gravimetric Data with According Nonhomogenities of the Intermediary Layer, Compt. Rend. Acad. Bulg. Sci., 28,8(1975).

[16] Zidarov, D. P.: A Direct Solution of the Inverse Gravimetric Problem at a Given Law of the Density-Trap Method, Compt. Rend. Acad. Bulg. Sci., 31,9(1978).

[17] Zidarov, D. P.: Partially Regulated Solution of Overdetermined Systems of Linear Equations Used in Geophysics, Compt. Rend. Acad. Bulg. Sci., 31,10(1978).

[18] Zidarov, D.: Direct, Characteristic, and Derivative Solutions of the Inverse Gravimetric and Magnetic Problem. In 23-rd Geophysical Simposium, Varna, October 4 - 7, 1978, State Geophysical Enterprise, Sofia, 1978.

REGULAR METHOD FOR SOLVING NONLINEAR PROBLEMS

V.A. MOROZOV [x)]

S u m m a r y. Solving nonlinear operator equations of the first kind belongs to not sufficiently studied region of the set of incorrectly posed problems. The concept of regularity suggested and studied by the author in connection with linear problems is presented. This concept is applied to the operator equations of the first kind and proved to be valuable from the theoretical and practical points of view.

I. P r o b l e m

The problem of solving operator equation

$$Au = f, \quad f \in A(\mathcal{D}_A) \tag{I}$$

is considered, where $\mathcal{D}_A \subset U, A: U \to F$, U and F are complete metric spaces, with approximate data $\tilde{\mathcal{A}} = \{\tilde{A}, \tilde{f}\}$. Here $\tilde{f} \in F: \rho_F(f, \tilde{f}) \leq \delta$, an approximation set of operators is the following:

$$\rho_F(\tilde{A}u, Au) \leq h\ \Omega(u), \quad \forall u \in \mathcal{D} \subseteq \mathcal{D}_A, 0 < h \leq h_0 \tag{2}$$

We shall call Ω an estimation functional. We assume, that a set of precise solutions $\bar{\mathcal{D}} = \{u \in \mathcal{D}_A : Au = f\}$ is so, that a set of "smooth" solutions $\hat{\mathcal{D}} = \{\bar{\mathcal{D}} \cap \mathcal{D}\} \neq \emptyset$. Let's note, that

$$|\rho(u) - \tilde{\rho}(u)| \leq h\ \Omega(u) + \delta, \quad \forall u \in \mathcal{D}, \tag{3}$$

where $\rho(u) = \rho_F(Au, f)$, $\tilde{\rho}(u) = \rho_F(\tilde{A}u, \tilde{f})$ are residual functionals on precise and approximate data.

D e f i n i t i o n I. Any mapping $T_\sigma : T_\sigma(\tilde{\mathcal{A}}) = \tilde{\mathcal{D}} \subset \mathcal{D}$, where $\sigma = (h, \delta)$ - precision vector of input data $\tilde{\mathcal{A}}$, we shall call an approximate method for solution (I).

D e f i n i t i o n 2. An approximate method T_σ residualy converges if

$$\lim_{\sigma \to 0} \sup_{u \in \tilde{\mathcal{D}}} \rho(u) = 0. \tag{4}$$

[x)] Prof. Dr. V. A. Morozov, Scientific-Research Computing Centre, Moscow State University "M. V. Lomonosov", Moscow V-234, USSR.

The report is based on [I7].

D e f i n i t i o n 3. An approximate method T_σ we shall call Ω -regular [3], if it residualy converges and

$$\overline{\lim_{\sigma\to 0}} \sup_{u\in\tilde{\mathfrak{D}}} \Omega(u) \le \Omega^* = \inf_{u\in\hat{\mathfrak{D}}} \Omega(u). \qquad (5)$$

We shall call sets $\tilde{\mathfrak{D}} = T_\sigma(\tilde{A})$ which satisfies (4), (5) Ω - regular approximate solutions (I).

D e f i n i t i o n 4. An approximate method T_σ we shall call weakly Ω -regular if it residualy converges [3] and

$$\overline{\lim_{\sigma\to 0}} \sup_{u\in\tilde{\mathfrak{D}}} \Omega(u) \le R^* < +\infty. \qquad (6)$$

Such sets $\tilde{\mathfrak{D}}$ we shall call weakly Ω -regular approximate solutions (I).

Our goal is to study regular and weakly regular methods and to present their characteristics. We note, that Hadamard's correctness of problem (I) is not assumed.

2. Regularity conditions of approximate methods.

ε -optimal residual method [5]. Set $\mathfrak{D}_\sigma = \{u\in\mathfrak{D} : \tilde{\rho}(u) \le h\Omega(u)+\delta\}$, which contains $\hat{\mathfrak{D}}$ we call set of formal solutions (I). $\mathfrak{D}_\sigma$ is not an approximation of $\hat{\mathfrak{D}}$ in any sensible way. Let $\tilde{\Omega}^* = \inf_{u\in\mathfrak{D}_\sigma} \Omega(u)$ and define a set

$$\tilde{\mathfrak{D}}_1 = \{u\in\mathfrak{D}_\sigma : \Omega(u) \le \tilde{\Omega}^* + \varepsilon_1, \ \varepsilon_1 > 0\}.$$

Obviously $\tilde{\Omega}^* \le \Omega^*$, so

$$\rho(u) \le 2h(\Omega^* + \varepsilon_1) + 2\delta, \quad \Omega(u) \le \Omega^* + \varepsilon_1, \ \forall u\in\tilde{\mathfrak{D}}_1. \qquad (7)$$

ε -quasisolutions method [7]. Let $R > \Omega^*$. $\mathfrak{D}_R = \{u\in\mathfrak{D} : \Omega(u) \le R\}$, $\tilde{\rho}_R^* = \inf_{u\in\mathfrak{D}_R} \tilde{\rho}(u)$. ε -quasisolutions set is

$$\tilde{\mathfrak{D}}_2 = \{u\in\mathfrak{D}_R : \tilde{\rho}(u) \le \tilde{\rho}_R^* + \varepsilon_2, \ \varepsilon_2 > 0\}.$$

Since $\tilde{\rho}_R^* \le h\Omega^* + \delta$, then

$$\rho(u) \le h(\Omega^* + R) + 2\delta + \varepsilon_2, \quad \Omega(u) \le R, \ \forall u\in\tilde{\mathfrak{D}}_2. \qquad (8)$$

A set of generalised quasisolutions

$$\tilde{\mathfrak{D}}_3 = \{u\in\mathfrak{D}_R : \tilde{\rho}(u) \le hR + \delta\}.$$

Then

$$\rho(u) \le 2(hR+\delta) \quad , \quad \Omega(u) \le R \quad , \forall u \in \tilde{\mathcal{D}}_3 . \tag{9}$$

ε -minimax residual method [I6]. Let $\tilde{\omega}(u) = \tilde{\rho}(u) + h\,\Omega(u), u \in \mathcal{D}$ and $\tilde{\omega}^* = \inf_{u \in \mathcal{D}} \tilde{\omega}(u)$. Define sets

$$\tilde{\mathcal{D}}_4 = \{ u \in \mathcal{D} : \tilde{\omega}(u) \le \tilde{\omega}^* + \varepsilon_3 \quad , \varepsilon_3 > 0 \} .$$

Since $\tilde{\omega}^* \le h\,\Omega^* + \delta$, then

$$\rho(u) \le 4h\,\Omega^* + 3\delta + 2\varepsilon_2 \quad , \quad \Omega(u) \le 2\,\Omega^* + \frac{\delta + \varepsilon_3}{h} . \tag{IO}$$

ε -regularization method [I]. Define a functional

$$\tilde{\Phi}_\alpha [u] = \tilde{\rho}^2(u) + \alpha\, \Omega^2(u) \,,\; u \in \mathcal{D} ,$$

where $\alpha > 0$ - a regularizations parameter. Let $\tilde{\Phi}^*_\alpha = \inf_{u \in \mathcal{D}} \tilde{\Phi}_\alpha [u]$.

Set of ε-regularized solution

$$\tilde{\mathcal{D}}_\alpha = \{ u \in \mathcal{D} : \tilde{\Phi}_\alpha [u] \le \tilde{\Phi}^*_\alpha + \varepsilon_4^2 \alpha \} .$$

Since $\tilde{\Phi}^*_\alpha \le (h\Omega^* + \delta) + \alpha\Omega^{*2}$, then $\forall u \in \tilde{\mathcal{D}}_\alpha$ we have:

$$\rho(u) \le 2(h \;\; \Omega^* + \delta)^2 + \sqrt{\alpha}\,(\Omega^* + \varepsilon_4) + \frac{h\,(h\Omega^* + \delta)}{\sqrt{\alpha}} , \tag{II}$$

$$\Omega(u) \le \Omega^* + \frac{h\,\Omega^* + \delta}{\sqrt{\alpha}} + \varepsilon_4 .$$

Note, that in derivation of (7)-(II) we use the only conditions of the approximations (2), residual functionals inequality (3) and obvious properties of mathematical programming.

T h e o r e m I. Let $\delta \to 0$, $\alpha \to 0$.

If $\quad h\varepsilon_1 + hR + \varepsilon_2 + \varepsilon_3 + \sqrt{\alpha}\,\varepsilon_4 + \dfrac{h^2 + h\delta}{\sqrt{\alpha}} \to 0,$

than all above defined approximate methods residualy converge. If, in addition,

$$\varepsilon_1 + R + \frac{\delta + \varepsilon_3}{h} + \varepsilon_4 + \frac{h + \delta}{\sqrt{\alpha}} \le const < \infty ,$$

then all these methods are weakly regular. If also

$$\varepsilon_1 + (R - \Omega^*) + \varepsilon_4 + \frac{h + \delta}{\sqrt{\alpha}} \to 0 ,$$

then all these methods exept minimax residual method are regular.

The proof is obvious.

3. C o n v e r g e n s o f a p p r o x i m a t e m e t h o d s .

Convergence is an important characteristic of approximate method.

T h e o r e m 2. If problem (I) is classical correctly posed than for any residualy convergent method:

$$\lim_{\sigma\to 0} \sup_{u\in\tilde{\mathscr{D}}} \rho_V(u,\bar{u}) = 0, \quad \tilde{\mathscr{D}} = T_\sigma(\tilde{A}), \; A\bar{u} = f.$$

Theorem I shows that we do not use the stability of operators $\tilde{A}$ (which is usually apriory assumed) and so our statement is nontrivial.

If problem (I) is incorrectly posed, then (2) is obviously not enough to guarantee convergence of approximations in V.

Further we shall need the following definitions.

D e f i n i t i o n 5. The sequence of elements $u_n \in \mathscr{D}, n\to\infty$ is called A-convergent if a sequence Au_n converges in F, and A-bounded if $\rho_F(Au_n, f) \leq const < \infty$ for some $g \in F$. We call sequence $u_n \in \mathscr{D}, n\to\infty$, Ω-bounded if $\Omega(u_n) \leq const < \infty$.

Let define some topology τ in V (may be not the same as initial one). If V is Banach reflexive space than, usually, is its weak topology.

D e f i n i t i o n 6. We shall call an operator A and functional Ω weakly τ-complementory on $\mathscr{D}_A$ if for any A-convergent and Ω-bounded sequence $u_n \in \mathscr{D}, n\to\infty$, there exists subsequence $u_{n_k}, k\to\infty$, which τ-converges to some element $u_0 \in \mathscr{D}_A : Au_0 = \lim_{k\to\infty} Au_{n_k}$, and τ-complementary on $\mathscr{D}$ if, in addition, $u_0 \in \mathscr{D}$, $\Omega(u_0) \leq \underline{\lim}_{k\to\infty} \Omega(u_{n_k})$.

D e f i n i t i o n 7. An approximate method T_σ is τ-convergent to some subset of $\bar{\mathscr{D}}$ - set of precise solutions of problem (I), if for any σ sets $\tilde{\mathscr{D}} = T_\sigma(\tilde{A})$ are in any open set (in topology τ), which contains the same subset of solutions.

T h e o r e m 3. Let A and Ω be weakly τ-complementary on $\mathscr{D}_A$. Any weakly Ω-regular method τ-converges to the set of precise solutions $\bar{\mathscr{D}}$.

P r o o f. Assuming the opposite statement there exist such sequence $u_n \in \tilde{\mathscr{D}}_n = T_{\sigma_n}(A\sigma_n), n\to\infty$, and an open set O, which contains $\bar{\mathscr{D}}$, that $u_n \notin O$. Sequence $\{u_n\}$ is A-convergent and Ω-bounded because method T_σ is weakly regular. So this sequence τ-converges to some element $u_0 \in \mathscr{D}_A : Au_0 = \lim_{n\to\infty} Au_n = f$, i.e. $u_0 \in \bar{\mathscr{D}}$

which is opposite to our assumption about $\{u_n\}$ and O . This contradiction proves the theorem.

The set $\tilde{\mathcal{D}}$ can happen to be too "wide". Next statement adds some new conditions.

T h e o r e m 4. Let $\hat{\mathcal{D}}_R = \{u \in \hat{\mathcal{D}} : \Omega(u) \le R\}$. If A and Ω are τ -complementary on $\mathcal{D}$, then any weakly Ω -regular method τ -converges to $\hat{\mathcal{D}}_R$ with some R .

P r o o f. Analogous to the previous one.

Let $\hat{\mathcal{D}}^* = \{u \in \hat{\mathcal{D}} : \Omega(u) = \Omega^*\}$. This set is called set of Ω-normal [8] or Ω-minimal [4,p.I4I] solutions (I).

T h e o r e m 5. If A and Ω are τ -complementary on $\mathcal{D}$,then $\hat{\mathcal{D}}^* \neq \emptyset$.

T h e o r e m 6. If A and Ω are τ -complementary on $\mathcal{D}$, then any Ω -regular method τ -converges to the set $\hat{\mathcal{D}}^*$, and

$$\lim_{\sigma \to 0} \sup_{u \in \tilde{\mathcal{D}}} |\Omega(u) - \Omega^*| = 0, \quad \tilde{\mathcal{D}} = T_\tau(A_\sigma).$$

P r o o f. The same as in theorem 3.

R e m a r k. In some cases theorem 6 quarantees convergence of approximate solutions to the set $\hat{\mathcal{D}}^*$ in some stronger topology than τ (or initial topology of U). The concept of regular method was introduced by author in [4, p.I49] and was widely used in [3]. This concept allowes us to formulate the convergence in "natural" terms of characteristics of some specific spaces and to prove convergence of different methods rather likely.We consider this way rather valuable for nonlinear problems, because it makes possible to prove convergence in the most strong topology [4,p.I40]. Also important is the fact that regularity conditions are satisfied under minimal requirements on data (theorem I).

4. R e g u l a r i z a t i o n p a r a m e t e r c h o o s i n g p r o b l e m.

Conditions on α in theorem I are asymptotic and of little use in computing practice. We shall discuss some algorithmic methods, developed by us, for choosing regularization parameter [3]. In particular, we shall show in the following theorems that regularization method can be used for finding approximate solutions, which can be received by optimal residual and quasisolutions methods.

T h e o r e m 7. Let for any $\alpha > 0$ the only element $u_\alpha \in \mathcal{D}$: $\tilde{\Phi}_\alpha[u_\alpha] = \tilde{\Phi}^*_\alpha$ exists. If, in addition there is such α, $0 < \alpha < \infty$, that

$$\tilde{\rho}(u_\alpha) = h\, \Omega(u_\alpha) + \delta , \tag{I2}$$

than $\Omega(u_\alpha) = \tilde{\Omega}^*$, that is optimal residual method is resolved.

If $\alpha, 0 < \alpha < \infty$:

$$\Omega(u_\alpha) = R , \tag{I3}$$

than $\tilde{\rho}(u_\alpha) = \tilde{\rho}^*_R$, that is quasisolution method is resolved.

Choice of regularization parameter according to (I2) is called residual principle (criterion) and was studied in [3], [I0] for linear problems. Choice of parameter according to (I3) can naturaly be called quasisolutions principle (criterion). Algorithms for choosing regularization parameter according to (I2) and (I3) for wide class of linear problems were considered in [3], [I4].

Let us present the solvable conditions regularization and minimax residual methods. We need some definitions.

D e f i n i t i o n 8. Operator A and functional Ω are called strongly τ-complementary on $\mathcal{D}$ if for any A and Ω-bounded sequence $u_n \in \mathcal{D}, n \to \infty$, there exists subsequence u_{n_k} which τ-converges to some element $u_0 \in \mathcal{D}$:

$$\rho_F(Au_0, g) \leq \varliminf_{k\to\infty} \rho_F(Au_{n_k}, g), \quad \Omega(u_0) \leq \varliminf_{k\to\infty} \Omega(u_{n_k})$$

for any element $g \in F$.

T h e o r e m 8. Let $\tilde{A}$ and Ω be strongly τ-complementary on $\mathcal{D}$. Then optimal residual and regularization methods are solvable, that is we can set $\varepsilon_3 = \varepsilon_4 = 0$.

Now we consider another method of choosing regularization parameter offered by the author for linear problems in [II] and developed for nonlinear problems in [I2], [I8]. This method is called smoothing functional principle [I5]. Function $\varphi(\alpha) = \tilde{\Phi}^*_\alpha$ is defined for $\alpha > 0$ under minimal restrictions on input data (I).

L e m m a I. Function $\varphi(\alpha)$ is continuous and increasing. In fact for $\forall \varepsilon, 0<\varepsilon\leq 1$, elements exist $u^{\varepsilon}\in \mathfrak{D}$:

$$\varphi(\alpha') \leq \Phi_{\alpha'}(u^{\varepsilon}) \leq \varphi(\alpha')+\varepsilon \quad ,\alpha'>0.$$

Then

$$\varphi(\alpha)-\varphi(\alpha') \leq (\alpha-\alpha')\,\Omega^{2}(u^{\varepsilon})+\varepsilon .$$

Since $\varphi(\alpha')\leq(h\Omega^{*}+\delta)^{2}+\alpha'\Omega^{*2}$, then $\Omega^{2}(u^{\varepsilon})\leq \frac{(h\Omega^{*}+\delta)^{2}}{\alpha_0}+\Omega^{*2}+\frac{1}{\alpha_0}=C_0^2, \forall\alpha'\geq\alpha_0>0.$
Therefore $\varphi(\alpha)-\varphi(\alpha')\leq|\alpha-\alpha'|C_0^2+\varepsilon$. Analogous $\varphi(\alpha')-\varphi(\alpha)\leq|\alpha-\alpha'|C_0^2+\varepsilon', \varepsilon'\leq 1.$

R e m a r k. The function $\varphi(\alpha)$ actually satisfies Lipshitz's condition and under some definite conditions is stricfly monotonous.

L e m m a 2. Let $R^{*}\geq\Omega^{*}$ and $(hR^{*}+\delta)^{2}\leq\tilde{\varphi}_{\infty}=\lim_{\alpha\to\infty}\varphi(\alpha)$. Then the equation $\varphi(\alpha)=\Delta$, $\Delta\in I=(hR^{*}+\delta,\tilde{\varphi}_{\infty})$ has at least one root α_{Δ}: $0<\alpha_{\Delta}<\infty$.

In accordance with the smoothing functional principle let:

$$\tilde{\mathfrak{D}}_{\delta}=\{u\in\mathfrak{D}:\ \Delta\leq\tilde{\Phi}_{\alpha_{\Delta}}[u]\leq\Delta+\varepsilon_{\delta}\ ,\varepsilon_{\delta}>0\}.$$

T h e o r e m 9. Let δ be small enough and $\inf_{\delta}\tilde{\varphi}_{\infty}=\varphi>0$.
If $\Delta=2(hR^{*}+\delta)^{2}$, $\varepsilon_{\delta}/(h+\delta)^{2}\to 0$, sets $\tilde{\mathfrak{D}}_{\delta}$ are weakly regular.
If $\Delta=(hR^{*}+\delta)^{2(1-s)}, 0<s<1, \frac{\varepsilon_{\delta}}{(h+\delta)^{2(1-s)}}\to 0$, sets $\tilde{\mathfrak{D}}_{\delta}$ are regular.

We further assume set $\mathcal{U}_{\infty}=\{u\in\mathfrak{D}:\Omega(u)=0\}\neq\emptyset$. This is correct for some important cases. Let

$$\mathcal{U}^{*}_{\infty}=\{u^{*}\in\mathcal{U}_{\infty}:\ \rho(u^{*})=\inf_{u}\rho(u)\ ,u\in\mathcal{U}_{\infty}\}.$$

L e m m a 3. If $A(K)$ and Ω are strongly τ-complementary on $\mathfrak{D}$ than $\mathcal{U}^{*}_{\infty}\neq\emptyset$ and $\tilde{\varphi}_{\infty}=\rho_F^2(AV^{*}_{\infty},\tilde{f})\geq(\rho^{*}_{\infty}-\delta)^{2}$.

S e q u e n c e . Under lemma's assumtions from $\rho^{*}_{\infty}\geq\rho>0$ follows that $\varphi>0$ and one of the conditions of theorem 9 is satisfied.

C o n c l u s i o n. Method for solving incorrectly posed nonlinear operator equations were studied earlier in [1], [5-7], [9]. Ideas of Ω-regularity for approximate methods were developed in author's articles not only for solving nonlinear operator equations but also for approximate solving rather general optimization problems.

BIBLIOGRAPHY

I. А.Н.Тихонов. О нелинейных уравнениях первого рода. ДАН СССР, т.I6I, №5, I965, с. I023-I026.

2. В.К.Иванов. Интегральные уравнения первого рода и приближенное решение обратной задачи потенциала. ДАН СССР, т.I42, №5, I962, с. 998-I000.

3. В.А.Морозов. Регулярные методы решения некорректно поставленных задач. М., Изд. МГУ, I974.

4. Труды I-ой конф. мол. ученых факультета вычислит. матем. и кибернетики, под ред. В.А.Морозова, М.С.Никольского. М., Изд. МГУ, I973.

5. В.А.Морозов. О принципе оптимальности невязки при приближенном решении уравнений с нелинейными операторами. Ж. вычисл. матем. и матем. физ., т.I4, №4, I974, с. 8I9-827.

6. В.А.Морозов. О решении методом регуляризации некорректно поставленных задач с нелинейными неограниченными операторами. Дифференц. уравнения. т.УI, №8, I970, с. I453-I458.

7. В.К.Иванов. О приближенном решении операторных уравнений первого рода. Ж. вычисл. матем. и матем. физ. т.6, №6, I966, с. I089-I094.

8. А.Н.Тихонов, Ф.П.Васильев. Методы решения некорректных экстремальных задач. Banach Center Publication, v.3, I978, 297-342.

9. В.А.Морозов. Методы решения неустойчивых задач. М., Изд. МГУ, I967.

I0. А.В.Гончарский, А.С.Леонов, А.Г.Ягола. О регуляризации некорректных задач с приближенно заданным оператором, Ж. вычисл. матем. и матем. физ. т.I4, №4, I974, с. I022-I027.

II. В.А.Морозов. О регуляризации некорректно поставленных задач и выборы параметра регуляризации, Ж. вычисл. матем. и матем. физ., т.6, №I, I966, с. I70-I75.

I2. В.А.Морозов. Метод регуляризации и некоторые его приложения. кандид. диссерт., Москва, I967.

I3. В.А.Морозов. О задаче дифференцирования и некоторых алгоритмах приближения экспериментальной информации, в сб. "Вычисл. методы и программ.", I970, вып. XIУ, Изд. МГУ, с. 46-62.

I4. В.А.Морозов, В.И.Гордонова. Численные алгоритмы выбора параметра в методе регуляризации, Ж. вычисл. матем. и матем. физ. т.I3, №3, I973, с. 539-645.

I5. О.А.Лисковец. Оптимальные свойства принципа сглаживающего функционала, ДАН СССР, т.234, №2, I977, с. 288-30I.

I6. М.К.Лихт. О вычислении функционалов на решениях линейных уравнений I рода, Ж. вычисл. матем. и матем. физ. т.7, №3, I967, с. 667-672.

I7. В.А.Морозов. Регулярные методы решения нелинейных операторных уравнений, Изв. ВУЗов, Матем., №II(I98), I978.

I8. В.А.Морозов. Регулярные методы решения линейных и нелинейных некорректно поставленных задач, Докт. диссерт., М., I978.

Most problems in applications, for instance in physics, chemistry, biology, engineering and so on, are inverse problems. Previously these problems were studied as direct problems. Often the solution of an inverse problem does not depend continuously on the measured data and is therefore improperly posed. Each physical system must be stable, i. e. must depend continuously on the exterieur data accessible to measurements. This is one reason why improperly posed problems have been the subject to such extensive study recently. The practical importance of inverse and improperly posed problems is such that they may be considered among the pressing problems of current mathematical research. Many problems remain unsolved in the study of inverse problems.

Die meisten Probleme in den Anwendungen, z. B. in der Physik, der Chemie, der Biologie, der Ingenieurwissenschaften usw., sind inverse Probleme. Früher wurden diese Probleme als direkte Probleme behandelt. Oft hängt die Lösung eines inversen Problems nicht stetig von den gemessenen Werten ab und ist daher inkorrekt gestellt. Jedes physikalische System muß aber stabil sein, d. h., es muß stetig von den äußeren Daten, die gemessen werden können, abhängen. Dieser Sachverhalt ist ein Grund, warum in der letzten Zeit inkorrekt gestellte Probleme so intensiv studiert wurden. Die praktische Bedeutung von inversen und inkorrekt gestellten Problemen ist so groß, daß man sie zu den unbedingt zu lösenden mathematischen Problemen der Gegenwart zählen muß. Es gibt noch viele offene Fragen bei inversen Problemen.

Bei Fragen zur Produktsicherheit wenden Sie sich bitte an:
If you have any questions regarding product safety,
please contact:

Walter de Gruyter GmbH
Genthiner Straße 13
10785 Berlin
productsafety@degruyterbrill.com